THE CARTOON GUIDE TO BIOLOGY

ALSO BY LARRY GONICK

THE CARTOON HISTORY OF THE UNIVERSE, VOLUMES 1–7
THE CARTOON HISTORY OF THE UNIVERSE, VOLUMES 8–13
THE CARTOON HISTORY OF THE UNIVERSE, VOLUMES 14–19
THE CARTOON HISTORY OF THE MODERN WORLD, PART 1
THE CARTOON HISTORY OF THE MODERN WORLD, PART 2
THE CARTOON HISTORY OF THE UNITED STATES
HYPERCAPITALISM (WITH TIM KASSER)
THE CARTOON GUIDE TO ALGEBRA
THE CARTOON GUIDE TO CALCULUS
THE CARTOON GUIDE TO CHEMISTRY (WITH CRAIG CRIDDLE)
THE CARTOON GUIDE TO THE COMPUTER
THE CARTOON GUIDE TO THE ENVIRONMENT (WITH ALICE OUTWATER)
THE CARTOON GUIDE TO GENETICS (WITH MARK WHEELIS)
THE CARTOON GUIDE TO (NON)COMMUNICATION
THE CARTOON GUIDE TO PHYSICS (WITH ART HUFFMAN)
THE CARTOON GUIDE TO SEX (WITH CHRISTINE DEVAULT)
THE CARTOON GUIDE TO STATISTICS (WITH WOOLLCOTT SMITH)
THE ATTACK OF THE SMART PIES

"GONICK'S *CARTOON GUIDE TO STATISTICS*...[IS] THE ONLY REFERENCE TEXT FOR MY GENERAL EDUCATION COURSE 'REAL-LIFE STATISTICS: YOUR CHANCE FOR HAPPINESS (OR MISERY).'"
—XIAO-LI MENG, CHAIRMAN, STATISTICS DEPARTMENT, HARVARD UNIVERSITY

"SO CONSISTENTLY WITTY AND CLEVER THAT THE READER IS BARELY AWARE OF BEING GIVEN A THOROUGH GROUNDING IN THE SUBJECT." —*OMNI* MAGAZINE

"[*THE CARTOON HISTORY OF THE UNIVERSE, BOOK 3*, IS] A MASTERPIECE!" —STEVE MARTIN

"LARRY GONICK SHOULD GET AN OSCAR FOR HUMOR AND A PULITZER FOR HISTORY."
—RICHARD SAUL WURMAN, CREATOR OF THE TED CONFERENCES

GONICK'S CARTOON HISTORIES AND CARTOON GUIDES HAVE BEEN REQUIRED READING IN COURSES AT BISMARCK HIGH SCHOOL, BISMARCK, NORTH DAKOTA; BLOOMSBURG UNIVERSITY; BOSTON COLLEGE; BUCKINGHAM BROWN & NICHOLS SCHOOL, CAMBRIDGE, MASSACHUSETTS; CALIFORNIA INSTITUTE OF THE ARTS; CALIFORNIA STATE UNIVERSITY AT CHICO; CARNEGIE-MELLON UNIVERSITY; COLUMBIA UNIVERSITY; CORNELL UNIVERSITY; DARTMOUTH COLLEGE; DUKE UNIVERSITY; GIRVAN ACADEMY, SCOTLAND; HARVARD UNIVERSITY; HUMBOLDT STATE UNIVERSITY; HUNTINGDON COLLEGE; ILLINOIS STATE UNIVERSITY; JOHN JAY COLLEGE; THE JOHNS HOPKINS UNIVERSITY; KENT SCHOOL DISTRICT, KENT, WASHINGTON; KENYON COLLEGE; LANCASTER UNIVERSITY, ENGLAND; LICK-WILMERDING HIGH SCHOOL, SAN FRANCISCO, CALIFORNIA; LIVERPOOL UNIVERSITY, ENGLAND; LOGAN HIGH SCHOOL, LOGAN, UTAH; LONDON SCHOOL OF ECONOMICS; LOUISIANA STATE UNIVERSITY; LOWELL HIGH SCHOOL, SAN FRANCISCO, CALIFORNIA; THE MARIN ACADEMY; MARQUETTE HIGH SCHOOL, CHESTERFIELD, MISSOURI; MIT; NEW YORK UNIVERSITY; NORTH CAROLINA STATE UNIVERSITY; NORTHWESTERN UNIVERSITY; NUEVA SCHOOL, HILLSBOROUGH, CALIFORNIA; OHIO STATE UNIVERSITY; PENNSYLVANIA STATE UNIVERSITY; PHILIPPINE HIGH SCHOOL, DILMAN, PHILIPPINES; REDBUD ACADEMY, AMARILLO, TEXAS; ROCHESTER INSTITUTE OF TECHNOLOGY; RUTGERS UNIVERSITY; SAINT IGNATIUS HIGH SCHOOL, SAN FRANCISCO, CALIFORNIA; SAN DIEGO STATE UNIVERSITY; SAN DIEGO SUPERCOMPUTER CENTER; SOUTHEAST MISSOURI STATE UNIVERSITY; SOUTHWOOD HIGH SCHOOL, SHREVEPORT, LOUISIANA; STANFORD UNIVERSITY; SWARTHMORE COLLEGE; TEMPLE UNIVERSITY; UNIVERSITEIT UTRECHT, NETHERLANDS; UNIVERSITY OF ALABAMA; THE UNIVERSITY OF CALIFORNIA AT BERKELEY, LOS ANGELES, SANTA BARBARA, SANTA CRUZ, AND SAN DIEGO; THE UNIVERSITY OF CHICAGO; THE UNIVERSITY OF EDINBURGH, SCOTLAND; THE UNIVERSITY OF FLORIDA; THE UNIVERSITY OF IDAHO; THE UNIVERSITY OF ILLINOIS; THE UNIVERSITY OF LEICESTER, ENGLAND; THE UNIVERSITY OF MARYLAND; THE UNIVERSITY OF MIAMI, FLORIDA; THE UNIVERSITIES OF MICHIGAN, MISSOURI, NEBRASKA, NEW BRUNSWICK, SCRANTON, SOUTH FLORIDA, TEXAS, TORONTO, WASHINGTON, AND WISCONSIN; YALE UNIVERSITY; AND MANY MORE INSTITUTIONS OF HIGHER AND LOWER EDUCATION!

THE CARTOON GUIDE TO BIOLOGY

LARRY GONICK &
DAVE WESSNER

WILLIAM MORROW
An Imprint of HarperCollins*Publishers*

HarperCollins books may be purchased for educational, business, or sales promotional use. For information, please email the Special Markets Department at SPsales@harpercollins.com.

FIRST EDITION

Library of Congress Cataloging-in-Publication Data has been applied for.

ISBN 978-0-06-239865-9

24 25 26 27 28 LBC 11 10 9 8 7

CONTENTS

Chapter 1
ORGANIZING A RIOT

THE RIOT OF LIFE, THAT IS

BIOLOGY ISN'T WHAT IT USED TO BE...

IN TIMES GONE BY, OUR ANCESTORS SAW LITTLE DIFFERENCE BETWEEN LIVING AND NONLIVING THINGS.

THEY SAW VOLCANOES BELCH, RIVERS RUSH, THE SUN RISE IN THE MORNING, THE WIND MOAN AND SIGH, AND CLOUDS DROP WATER.

SO THE ANCIENTS CAN BE FORGIVEN FOR SEEING A WORLD ANIMATED BY SPIRITS, CAN'T THEY?

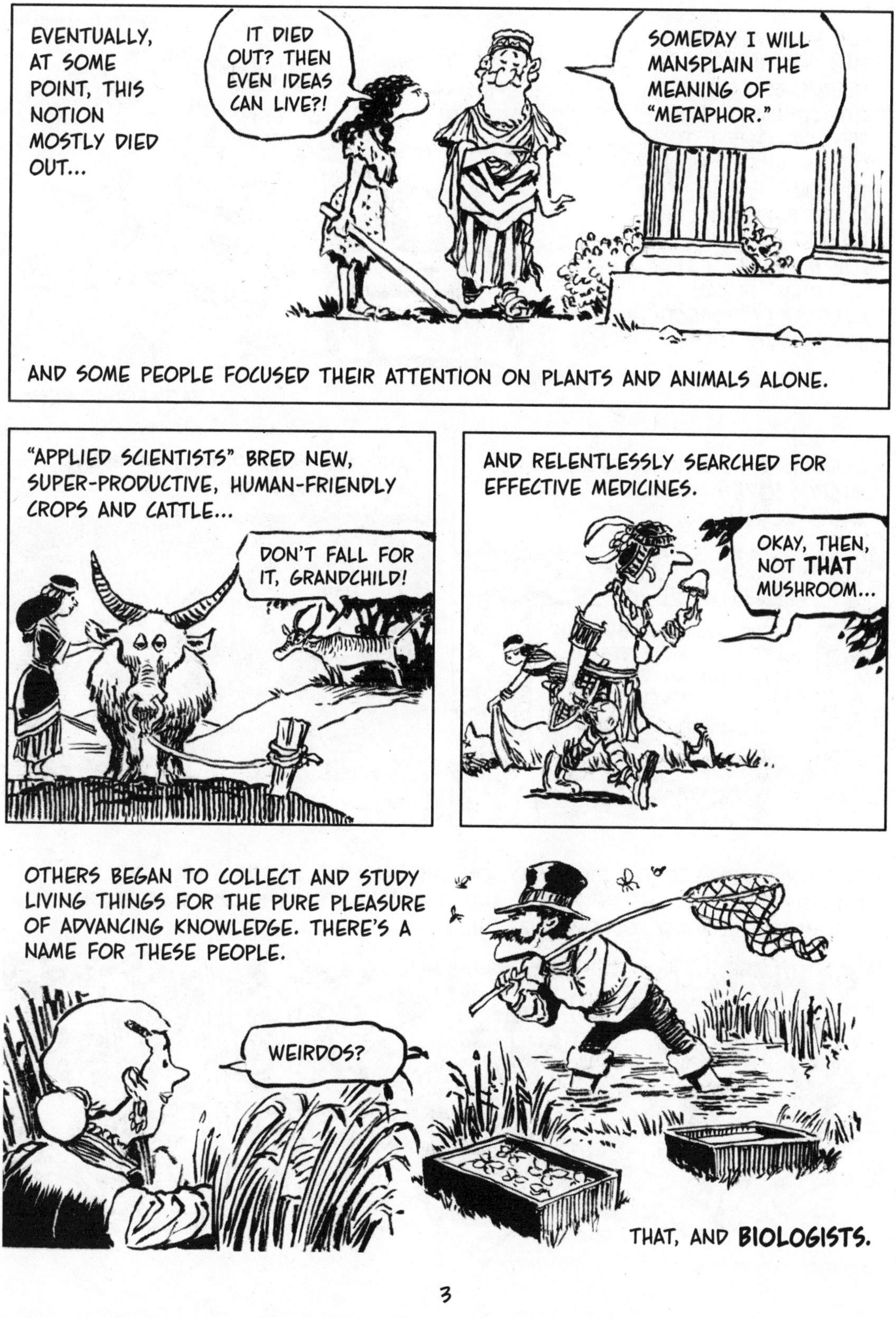
EVENTUALLY, AT SOME POINT, THIS NOTION MOSTLY DIED OUT...
IT DIED OUT? THEN EVEN IDEAS CAN LIVE?!
SOMEDAY I WILL MANSPLAIN THE MEANING OF "METAPHOR."
AND SOME PEOPLE FOCUSED THEIR ATTENTION ON PLANTS AND ANIMALS ALONE.
"APPLIED SCIENTISTS" BRED NEW, SUPER-PRODUCTIVE, HUMAN-FRIENDLY CROPS AND CATTLE...
DON'T FALL FOR IT, GRANDCHILD!
AND RELENTLESSLY SEARCHED FOR EFFECTIVE MEDICINES.
OKAY, THEN, NOT THAT MUSHROOM...
OTHERS BEGAN TO COLLECT AND STUDY LIVING THINGS FOR THE PURE PLEASURE OF ADVANCING KNOWLEDGE. THERE'S A NAME FOR THESE PEOPLE.
WEIRDOS?
THAT, AND BIOLOGISTS.

FOR MANY CENTURIES, THIS WAS BIOLOGY: SEARCH, COLLECT, KILL, CUT, COMPARE, CLASSIFY. BIOLOGISTS TOOK ON THE WORLD FROM THE **OUTSIDE IN.** THE GREEK PHYSICIAN **GALEN** (130–210), FOR INSTANCE, "LEARNED" HUMAN ANATOMY BY DISSECTING BARBARY APES.

SURPRISINGLY FUN, TOO!
BACKGROUND MUSIC

BIOLOGISTS WENT EVEN DEEPER WHEN **MICROSCOPES** REVEALED AN INVISIBLE LIVING WORLD...

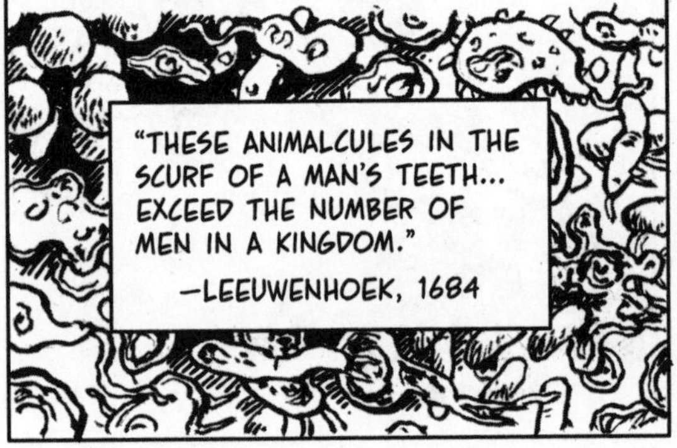

"THESE ANIMALCULES IN THE SCURF OF A MAN'S TEETH... EXCEED THE NUMBER OF MEN IN A KINGDOM."
—LEEUWENHOEK, 1684

AND THE FINER DETAILS OF PLANT AND ANIMAL ANATOMY.

DOGS HAVE EGGS!*

*DISCOVERED BY K. E. VON BAER IN 1827

THIS ONLY DOUBLED THE DIFFICULTY. ON THE OUTSIDE, LIFE IS A WILD RIOT OF MILLIONS AND MILLIONS OF FORMS, AND IT ONLY GETS MORE COMPLICATED ON THE INSIDE. HOW DO YOU ORGANIZE THIS?

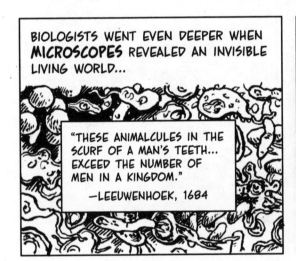

WHERE IS THE UNIFYING PRINCIPLE? WHAT MAKES SOMETHING ALIVE? WHO KNEW?

THERE MUST BE A "VITAL ESSENCE," WHICH I CAN NEITHER SEE, HEAR, SMELL, TASTE, TOUCH, MEASURE, NOR IMAGINE, REALLY.
NO, NO, NO, IT'S ALL IN THE BREATH!
IT'S A MYSTERY NOT BORN OF THIS WORLD.
"ELECTRIC FLUID."

DESPITE MANY IDEAS, NO SCIENTIST COULD FIND "**THE** SECRET OF LIFE," AS "IT" WAS OPTIMISTICALLY CALLED.

AT LEAST, I HOPE THERE'S ONLY ONE...
VOLTA

IN THE MEANTIME, BIOLOGY HAD TO SETTLE FOR A **DESCRIPTION** OF LIFE INSTEAD OF A DEFINITION.

HERE'S WHAT ALL ORGANISMS (I.E., LIVING THINGS) HAVE IN COMMON:

IF SOMETHING HITS ALL FIVE POINTS, YOU CAN BE PRETTY SURE IT'S ALIVE—AND IF NOT, NOT.

THEY ARE **CELLULAR**. EVERY LIVING THING IS MADE UP OF SELF-CONTAINED BLOBS SEPARATED FROM THE OUTSIDE WORLD BY A MEMBRANE. SOME ORGANISMS HAVE BILLIONS OF CELLS; SOME CONSIST OF ONLY ONE. NO ORGANISM IS A PART-CELL.

I'M IN HERE!

AND YOU'RE NOT!

MMM

ORGANISMS **REGULATE** THEMSELVES. THEY WORK TO KEEP THEIR OWN BODIES IN OPTIMAL CONDITION, KNOWN AS **HOMEOSTASIS**.

ORGANISMS **REACT** TO THE OUTSIDE WORLD. THEY SEEK FAVORABLE ENVIRONMENTS AND TRY TO ESCAPE OR NEUTRALIZE THREATS.

ROAR

ORGANISMS **EAT**. THEY REQUIRE—AND GET—NUTRIENTS AND ENERGY FROM THE OUTSIDE WORLD.

AND OF COURSE, ALL ORGANISMS **REPRODUCE**.

NOT BAD, ACTUALLY, BUT MODERN BIOLOGY CAN DO BETTER. THE TWENTIETH CENTURY TURNED THIS OUTSIDE-IN SCIENCE INSIDE OUT! (THE TWENTIETH CENTURY TENDED TO DO THINGS LIKE THAT.)

WHAT A CENTURY!

THANKS TO THE MICROSCOPE'S HIGH-TECH DESCENDANTS LIKE ELECTRON MICROSCOPES AND X-RAY DIFFRACTION CRYSTALLOGRAPHY—NOT TO MENTION RADICAL ADVANCES IN CHEMISTRY—LIFE'S SECRETS BEGAN TO SPILL OUT.

AND YES, THERE TURNED OUT TO BE MORE THAN ONE OF THEM. SOME SECRETS ARE STILL SECRET, BUT WE NOW HAVE A MUCH MORE PRECISE IDEA OF WHAT IT MEANS TO BE ALIVE ON EARTH.

A MORE MODERN, INSIDE-OUT DEFINITION OF EARTHLY LIFE LOOKS LIKE THIS:

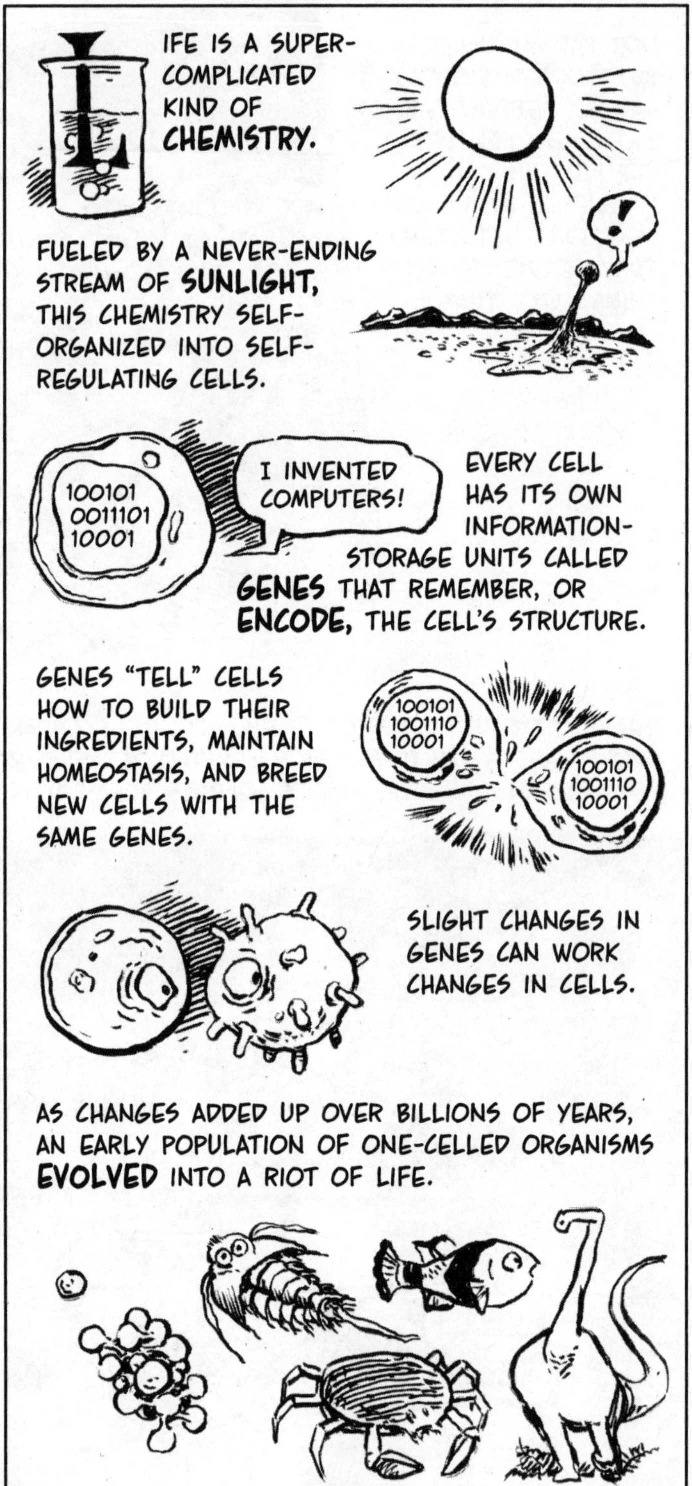
LIFE IS A SUPER-COMPLICATED KIND OF **CHEMISTRY.**
!
FUELED BY A NEVER-ENDING STREAM OF **SUNLIGHT,** THIS CHEMISTRY SELF-ORGANIZED INTO SELF-REGULATING CELLS.
100101
0011101
10001
I INVENTED COMPUTERS!
EVERY CELL HAS ITS OWN INFORMATION-STORAGE UNITS CALLED **GENES** THAT REMEMBER, OR **ENCODE,** THE CELL'S STRUCTURE.
GENES "TELL" CELLS HOW TO BUILD THEIR INGREDIENTS, MAINTAIN HOMEOSTASIS, AND BREED NEW CELLS WITH THE SAME GENES.
100101
1001110
10001
100101
1001110
10001
SLIGHT CHANGES IN GENES CAN WORK CHANGES IN CELLS.
AS CHANGES ADDED UP OVER BILLIONS OF YEARS, AN EARLY POPULATION OF ONE-CELLED ORGANISMS **EVOLVED** INTO A RIOT OF LIFE.

OF COURSE, THIS MODERN APPROACH AFFECTS HOW A BIOLOGY BOOK IS GOING TO LOOK.
HI, I'M A TINY INK VERSION OF DAVE WESSNER, AND I'LL BE YOUR GUIDE THROUGH THIS INSIDE-OUT TOUR OF BIOLOGY.
LIKE MANY BIOLOGISTS, I FIRST GOT HOOKED ON SCIENCE BY SEEING THE WORLD AROUND ME—THINGS LIKE DOGS, AND FROGS, AND BOGS. THE VISIBLE IS ALWAYS A GREAT PLACE TO START!
UNFORTUNATELY, THIS OLD OUTSIDE-IN APPROACH IS A HARD WAY TO FIND THE "BIG PICTURE."
IT'S A FOREST-FOR-THE-TREES THING!
THE NEW WAY, AS YOU'RE ABOUT TO SEE, STARTS WITH THE INVISIBLE.
SORRY!
IN OUR FIRST FEW CHAPTERS, WE DISCUSS THINGS YOU'LL NEVER SEE IN EVERYDAY LIFE.

WE START SMALL, ON THE INSIDE, AND WORK OUR WAY OUT.
WE'LL LOOK AT CELL **CHEMISTRY,** HOW CELLS HARNESS **ENERGY,** HOW THEY **REGULATE** THEMSELVES AND COOPERATE THROUGH **COMMUNICATION,** HOW THEY STORE AND PROCESS **INFORMATION** AND USE IT TO **REPRODUCE,** AND HOW LIFE **EVOLVES.**
IN THE END, I PROMISE, EVERYTHING WILL FIT TOGETHER IN A BEAUTIFUL STRUCTURE—AND YES, IT EVEN INCLUDES PLANTS AND ANIMALS!
SO WHAT DO YOU SAY? LET'S GET ORGANIZED!

Chapter 2
RAW MATERIALS

DESPITE ITS PECULIAR FEATURES, LIFE HAS MUCH IN COMMON WITH THE NON-LIVING WORLD: AT SOME LEVEL, LIFE IS NOTHING BUT **CHEMISTRY.**

ALL CHEMICALS ARE MADE UP OF **ELEMENTS,** SUBSTANCES THAT CAN'T BE BROKEN DOWN BY HEAT, ELECTRICITY, SOLVENTS, OR HAMMERING.

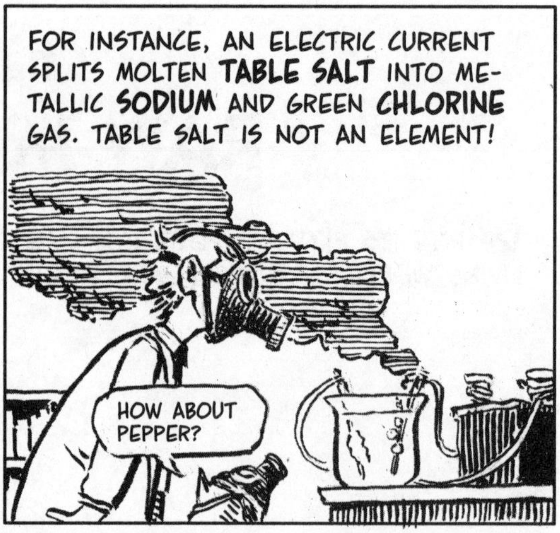
FOR INSTANCE, AN ELECTRIC CURRENT SPLITS MOLTEN **TABLE SALT** INTO METALLIC **SODIUM** AND GREEN **CHLORINE** GAS. TABLE SALT IS NOT AN ELEMENT!
HOW ABOUT PEPPER?

SODIUM AND CHLORINE **ARE** ELEMENTS, BECAUSE THEY DEFY CHEMICAL DECOMPOSITION. SO ARE SILVER, GOLD, IRON, AND MORE THAN 100 OTHER MATERIALS.
EACH ELEMENT IS MADE UP OF TINY CHEMICALLY UNBREAKABLE PARTICLES CALLED **ATOMS.** HOW TINY? ONE GRAM OF PURE IRON CONTAINS SOME **10.78 BILLION TRILLION ATOMS.**
A HUMAN HAIR IS ABOUT A BILLION ATOMS ACROSS.
SODIUM'S CHEMICAL SYMBOL, Na, FOR *NATRIUM,* DERIVES FROM ANCIENT EGYPTIAN *NETJERI,* SODA.

EACH ATOM HAS A CORE, OR **NUCLEUS,** MADE OF SMALLER PARTICLES: ELECTRICALLY CHARGED **PROTONS** (CHARGE +1) AND UNCHARGED **NEUTRONS.** LIGHTER CHARGED PARTICLES, **ELECTRONS** (CHARGE −1), SWARM AROUND THE NUCLEUS. THE NUMBER OF PROTONS AND ELECTRONS IS EQUAL; THEIR CHARGES BALANCE; AND THE ATOM IS ELECTRICALLY NEUTRAL.

ATOMS ARE MOSTLY EMPTY!

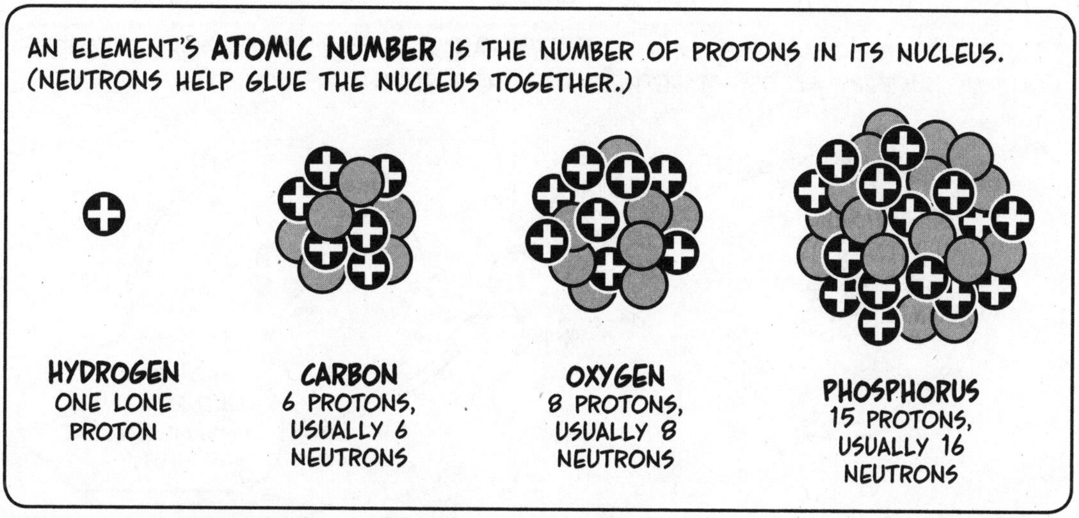
AN ELEMENT'S **ATOMIC NUMBER** IS THE NUMBER OF PROTONS IN ITS NUCLEUS. (NEUTRONS HELP GLUE THE NUCLEUS TOGETHER.)
HYDROGEN ONE LONE PROTON
CARBON 6 PROTONS, USUALLY 6 NEUTRONS
OXYGEN 8 PROTONS, USUALLY 8 NEUTRONS
PHOSPHORUS 15 PROTONS, USUALLY 16 NEUTRONS

OF ALL THE 100-PLUS ELEMENTS, LIFE MAKES MAJOR USE OF JUST **EIGHT** OF THEM.

ATOMIC NO.	NAME	SYMBOL
1	HYDROGEN	H
6	CARBON	C
7	NITROGEN	N
8	OXYGEN	O
11	SODIUM	Na
15	PHOSPHORUS	P
16	SULFUR	S
19	POTASSIUM	K

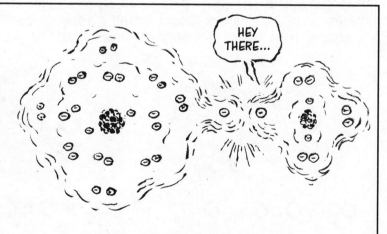

AND THEIR COMBINATIONS, OF COURSE! ATOMS HAVE A HABIT OF GETTING TOGETHER, OF FORMING **BONDS**.

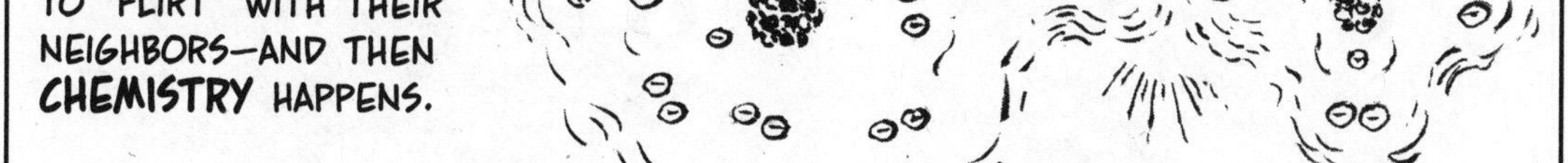
THESE BONDS COME FROM THE ELECTRONS SWARMING AROUND EACH NUCLEUS. THE OUTERMOST ELECTRONS OF MOST ATOMS TEND TO "FLIRT" WITH THEIR NEIGHBORS—AND THEN **CHEMISTRY** HAPPENS.
HEY THERE...

ELECTRONS, IT SO HAPPENS, LIKE TO **FORM PAIRS.** PUT TWO **HYDROGEN ATOMS** (ATOMIC NUMBER 1) CLOSE TOGETHER, AND THEIR ELECTRONS COUPLE UP.

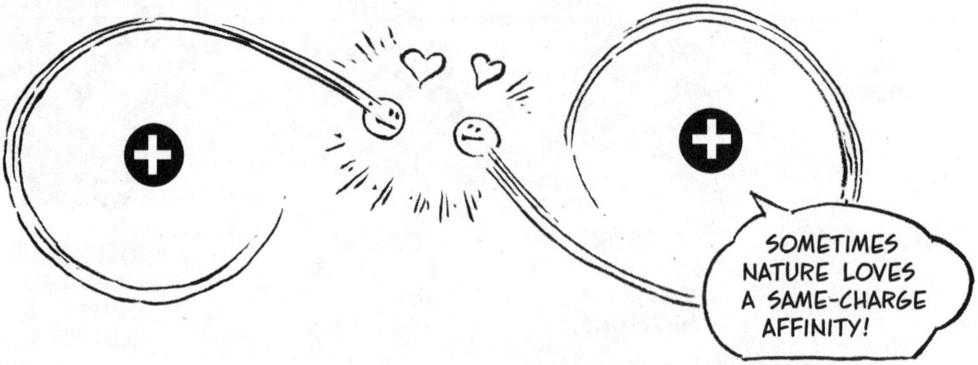

SOMETIMES NATURE LOVES A SAME-CHARGE AFFINITY!

THE SHARED PAIR CEMENTS A **COVALENT BOND** BETWEEN THE TWO ATOMS, WHICH ENTER A STABLE, TWO-ATOM UNIT CALLED A **HYDROGEN MOLECULE.** HYDROGEN GAS CONSISTS ENTIRELY OF THESE H_2 MOLECULES, WITH ALMOST NO SOLO ATOMS.

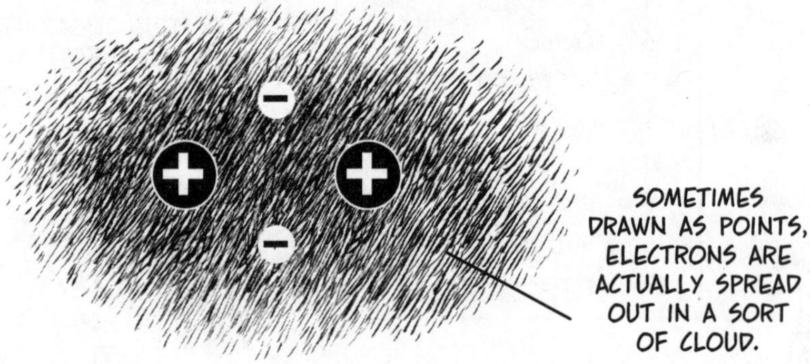

SOMETIMES DRAWN AS POINTS, ELECTRONS ARE ACTUALLY SPREAD OUT IN A SORT OF CLOUD.

WE WRITE:

H_2

OR

H:H

OR

H–H

TWO **OXYGEN** ATOMS JOIN TO FORM AN O_2 MOLECULE, BUT THEY SHARE **TWO PAIRS** OF ELECTRONS, FORMING A **DOUBLE** BOND. IN **CARBON DIOXIDE,** CO_2, A CARBON ATOM FORMS A DOUBLE BOND WITH EACH OF TWO OXYGEN ATOMS.

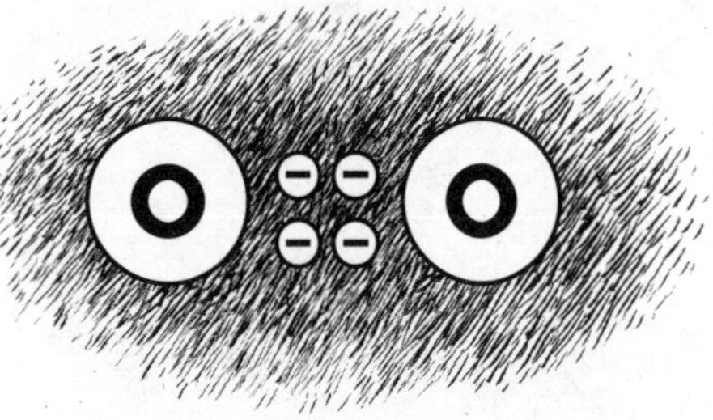

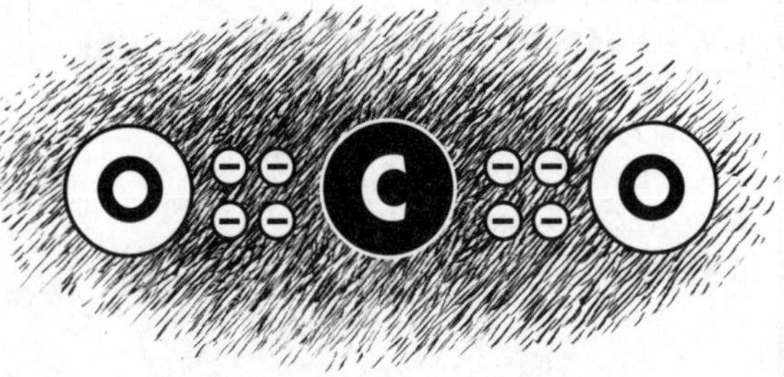

O_2 OR O::O OR O=O

CO_2 OR O::C::O OR O=C=O

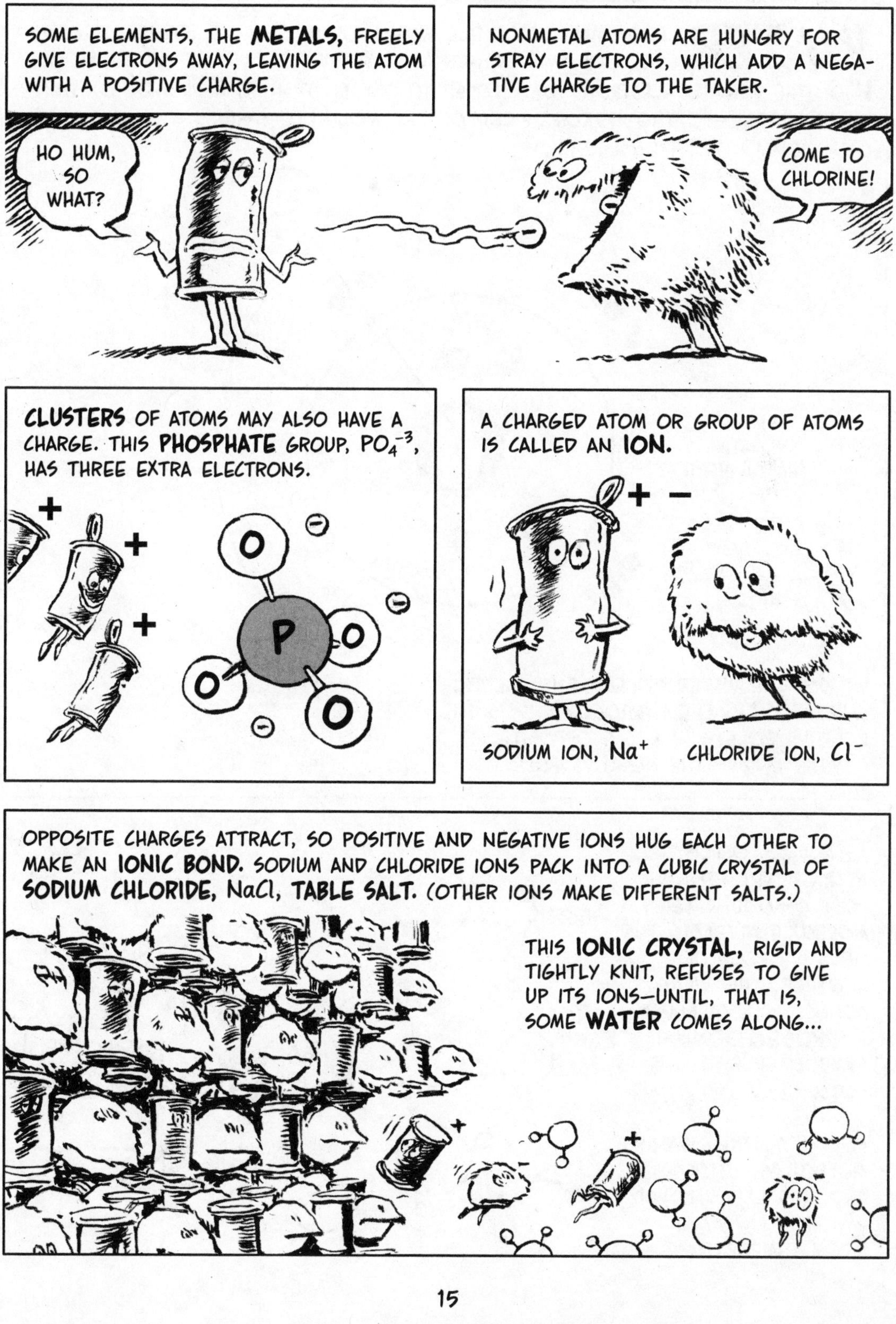
SOME ELEMENTS, THE METALS, FREELY GIVE ELECTRONS AWAY, LEAVING THE ATOM WITH A POSITIVE CHARGE.
HO HUM, SO WHAT?
NONMETAL ATOMS ARE HUNGRY FOR STRAY ELECTRONS, WHICH ADD A NEGATIVE CHARGE TO THE TAKER.
COME TO CHLORINE!
CLUSTERS OF ATOMS MAY ALSO HAVE A CHARGE. THIS PHOSPHATE GROUP, PO_4^{-3}, HAS THREE EXTRA ELECTRONS.
+
+
+
O
P
O
O
O
A CHARGED ATOM OR GROUP OF ATOMS IS CALLED AN ION.
+ −
SODIUM ION, Na^+
CHLORIDE ION, Cl^-
OPPOSITE CHARGES ATTRACT, SO POSITIVE AND NEGATIVE IONS HUG EACH OTHER TO MAKE AN IONIC BOND. SODIUM AND CHLORIDE IONS PACK INTO A CUBIC CRYSTAL OF SODIUM CHLORIDE, NaCl, TABLE SALT. (OTHER IONS MAKE DIFFERENT SALTS.)
THIS IONIC CRYSTAL, RIGID AND TIGHTLY KNIT, REFUSES TO GIVE UP ITS IONS—UNTIL, THAT IS, SOME WATER COMES ALONG...

WATER IS A SMALL MOLECULE WITH HUGE IMPORTANCE TO LIFE. IT CONSISTS OF ONE OXYGEN ATOM BOUND TO TWO HYDROGEN ATOMS, H_2O. THE BONDING ELECTRONS STAY CLOSER TO OXYGEN THAN TO HYDROGEN—OXYGEN PULLS HARDER—SO THE HYDROGEN END OF THE MOLECULE CARRIES A SLIGHT (FRACTIONAL) POSITIVE CHARGE.

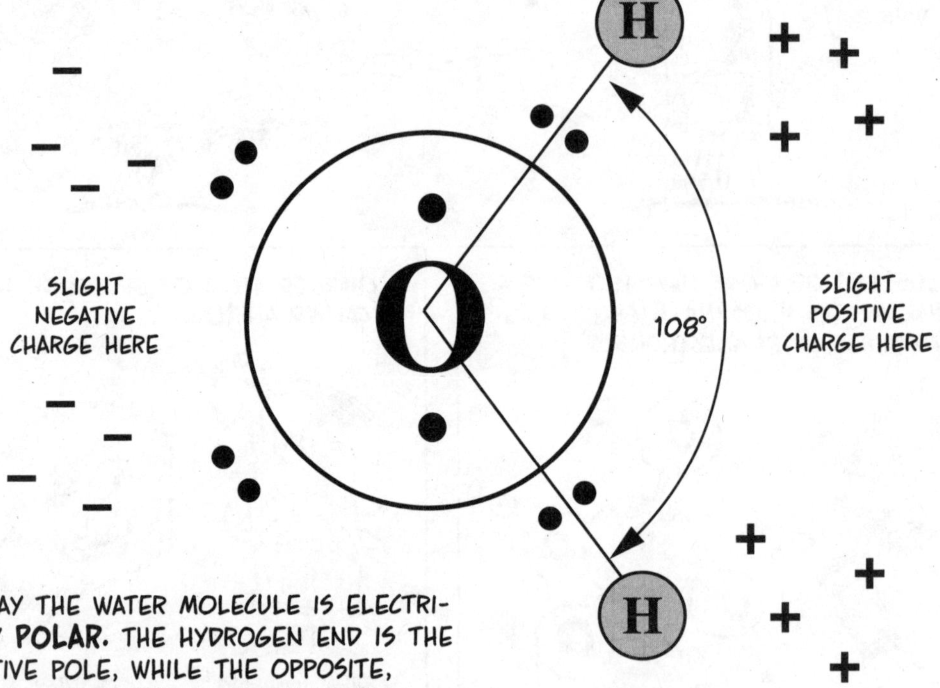

WE SAY THE WATER MOLECULE IS ELECTRICALLY **POLAR.** THE HYDROGEN END IS THE POSITIVE POLE, WHILE THE OPPOSITE, OXYGEN END IS THE NEGATIVE POLE.

AS ALWAYS, OPPOSITES ATTRACT. THE POSITIVE POLE OF ONE WATER MOLECULE ATTRACTS THE NEGATIVE POLE OF ANOTHER. THIS WEAK CONNECTION, CALLED A **HYDROGEN BOND,** IS SYMBOLIZED BY THREE DOTS, • • • .

POLARITY MAKES WATER MOLECULES "STICKY," AND THIS EXPLAINS WHY WATER IS LIQUID AT ROOM TEMPERATURE.

WHEN WATER MEETS TABLE SALT, SOME Na^+ AND Cl^- IONS BREAK FREE OF THE CRYSTAL. WATER MOLECULES FORM HYDROGEN BONDS WITH THE IONS, "CAGING" THEM. EVENTUALLY, THE WHOLE CRYSTAL DISSOLVES.

WE'LL SEE MANY OTHER IONS DISSOLVING IN WATER.

ANY ION OR MOLECULE THAT MIXES WITH WATER IS CALLED WATER-LOVING, OR

HYDROPHILIC.

ON THE OTHER HAND, SUBSTANCES THAT SHUN WATER ARE CALLED

HYDRO-PHOBIC.

METHANE

THE METHANE MOLECULE, CH_4, IS **NONPOLAR:** ELECTRONS ARE SPREAD AROUND SYMMETRICALLY, AND NO POLES FORM. THIS EXPLAINS WHY OIL AND WATER DON'T MIX.

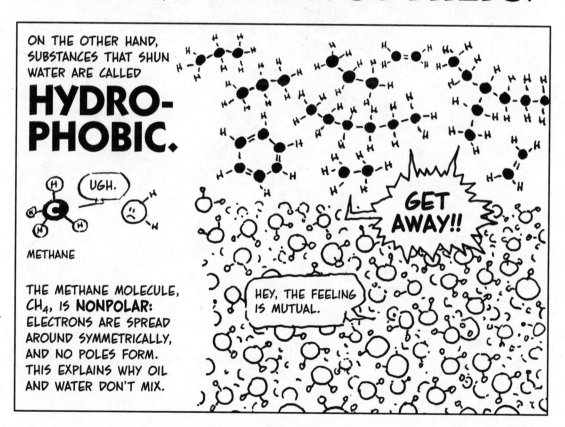

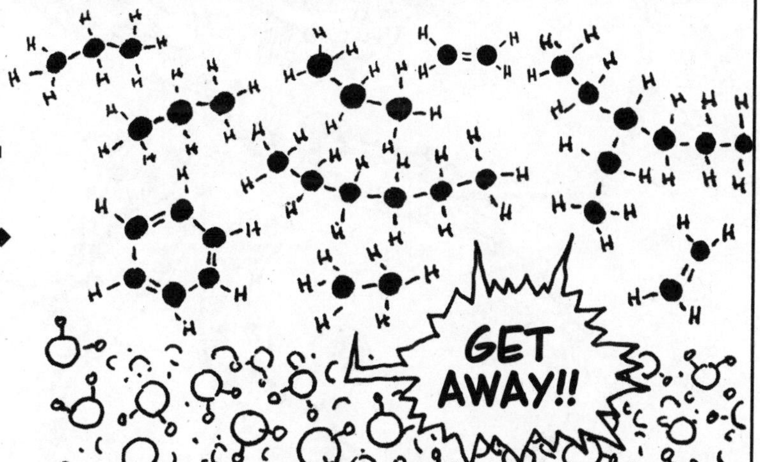

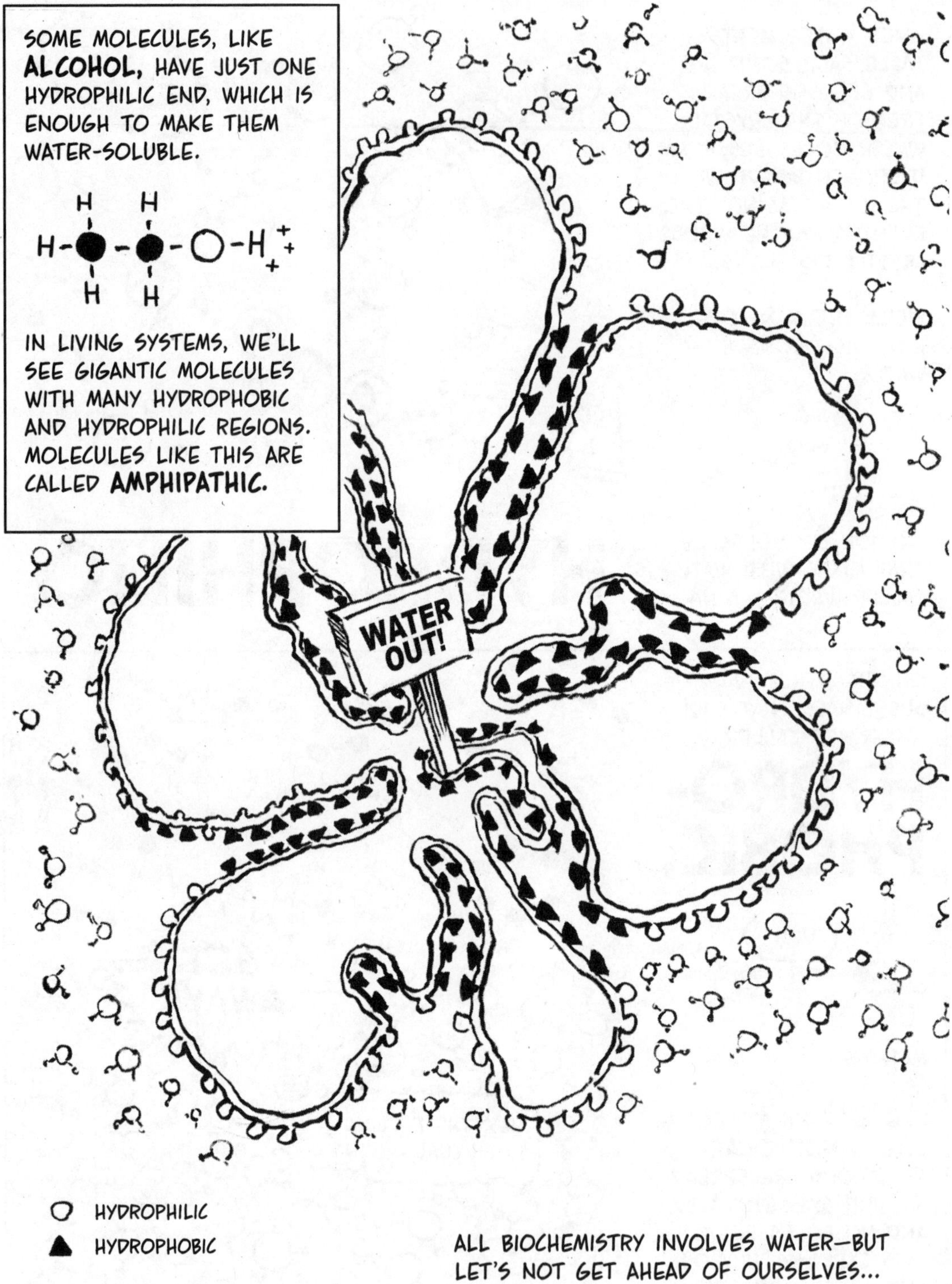
SOME MOLECULES, LIKE **ALCOHOL,** HAVE JUST ONE HYDROPHILIC END, WHICH IS ENOUGH TO MAKE THEM WATER-SOLUBLE.
IN LIVING SYSTEMS, WE'LL SEE GIGANTIC MOLECULES WITH MANY HYDROPHOBIC AND HYDROPHILIC REGIONS. MOLECULES LIKE THIS ARE CALLED **AMPHIPATHIC.**
WATER OUT!
HYDROPHILIC
HYDROPHOBIC

ALL BIOCHEMISTRY INVOLVES WATER—BUT LET'S NOT GET AHEAD OF OURSELVES...

Chapter 3
THE CHEMICALS OF LIFE

CARBON, CARBON, CARBON, AND SOME OTHER STUFF

ONE ELEMENT RULES THE CHEMISTRY OF LIFE. ORGANIC CHEMISTRY IS **CARBON** CHEMISTRY, AND CARBON-BASED SUBSTANCES COME, AS ANY TORMENTED PRE-MED WILL TELL YOU, IN A MADDENING VARIETY OF SHAPES, SIZES, TEXTURES, AND SMELLS.

WITH ATOMIC NUMBER 6, CARBON HAS FOUR OUTER ELECTRONS THAT BECKON TO OTHER ATOMS.

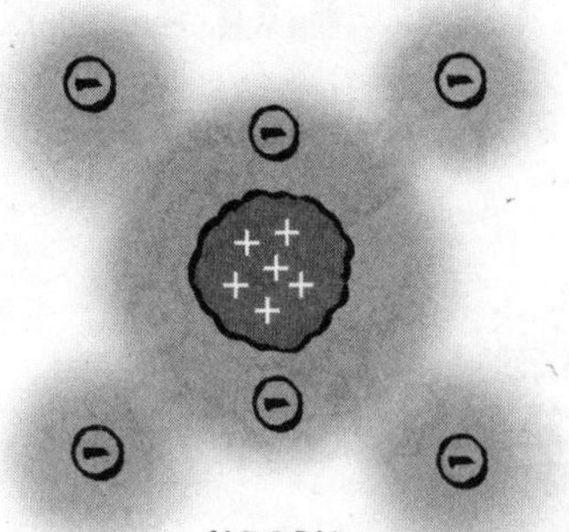

CARBON

HAVING FOUR OUTER ELECTRONS GIVES CARBON GREAT VERSATILITY: IT CAN BOND TO AS MANY AS FOUR OTHER ATOMS AT ONCE.

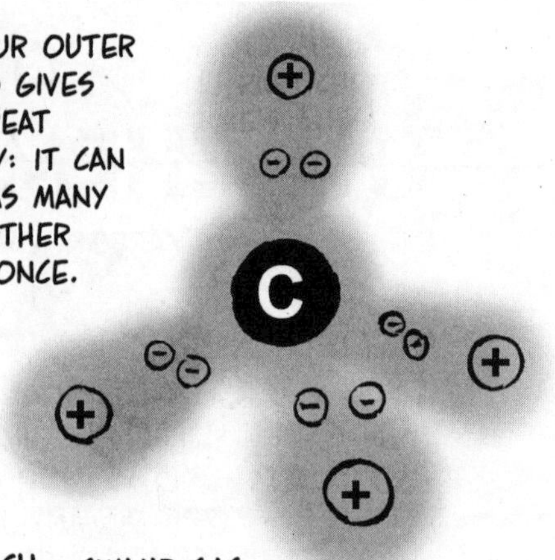

METHANE, CH_4, SWAMP GAS

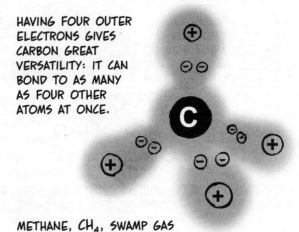

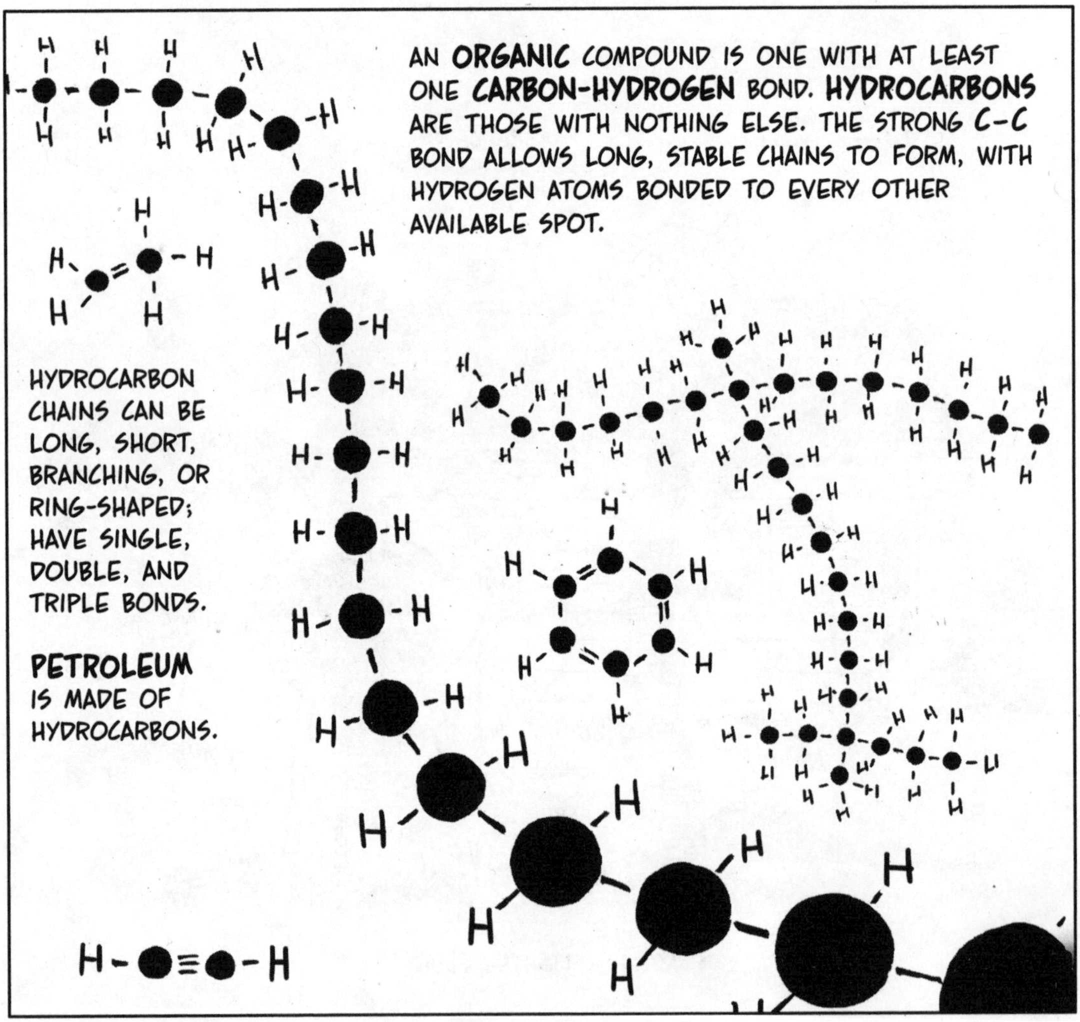
AN **ORGANIC** COMPOUND IS ONE WITH AT LEAST ONE **CARBON-HYDROGEN** BOND. **HYDROCARBONS** ARE THOSE WITH NOTHING ELSE. THE STRONG **C–C** BOND ALLOWS LONG, STABLE CHAINS TO FORM, WITH HYDROGEN ATOMS BONDED TO EVERY OTHER AVAILABLE SPOT.
HYDROCARBON CHAINS CAN BE LONG, SHORT, BRANCHING, OR RING-SHAPED; HAVE SINGLE, DOUBLE, AND TRIPLE BONDS.
PETROLEUM IS MADE OF HYDROCARBONS.

Add Oxygen, Get Fat

PUTTING A LITTLE OXYGEN INTO THE MIX GIVES RISE TO MORE POLAR OR AMPHIPATHIC MOLECULES LIKE **GLYCEROL**, USED IN COSMETICS...

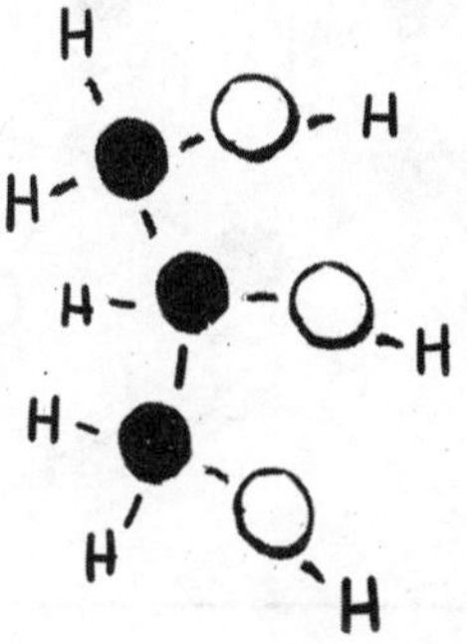

AND **ORGANIC ACIDS.** (COOH IS THE SIGNATURE OF ORGANIC ACIDS.)

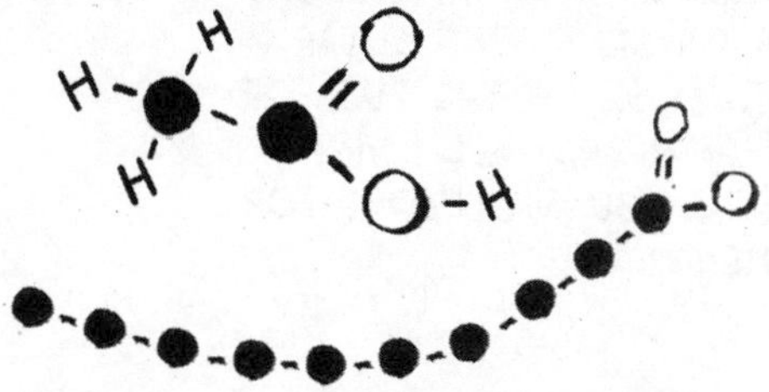

CAPRIC ACID, A 10-CARBON "FATTY ACID" (HYDROGEN OMITTED; SEE NOTE BELOW.)

A **FAT,** OR **TRIGLYCERIDE,** IS A GLYCEROL MOLECULE ATTACHED TO THREE FATTY ACIDS.

GENERALLY, WHEN ONE OR MORE FATTY ACIDS ARE "CAPPED" LIKE THIS, THE RESULTING MOLECULE IS KNOWN AS A

LIPID. AMONG OTHER FUNCTIONS, LIPIDS MAKE WATERPROOF MEMBRANES AND STORE FUEL AS FAT.

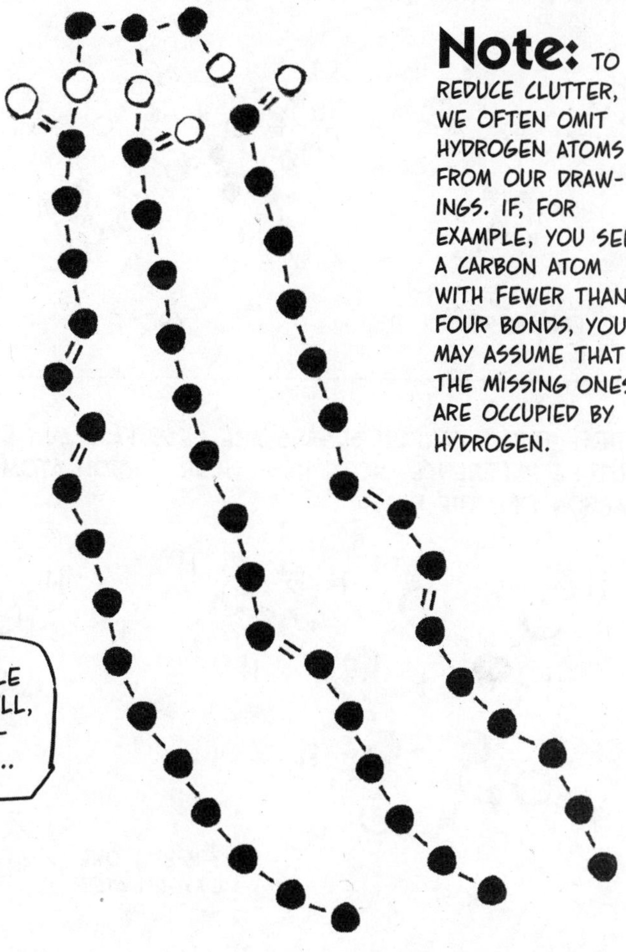

Note: TO REDUCE CLUTTER, WE OFTEN OMIT HYDROGEN ATOMS FROM OUR DRAWINGS. IF, FOR EXAMPLE, YOU SEE A CARBON ATOM WITH FEWER THAN FOUR BONDS, YOU MAY ASSUME THAT THE MISSING ONES ARE OCCUPIED BY HYDROGEN.

A LITTLE TOO WELL, SOME-TIMES...

ADD MORE OXYGEN, GET

SUGARS.

THESE RING-SHAPED MOLECULES USUALLY HAVE TWICE AS MANY HYDROGENS AS OXYGENS. ONE SUGAR, **GLUCOSE,** IS THE FUEL OF LIFE: NEARLY ALL LIVING THINGS BURN GLUCOSE FOR ENERGY.

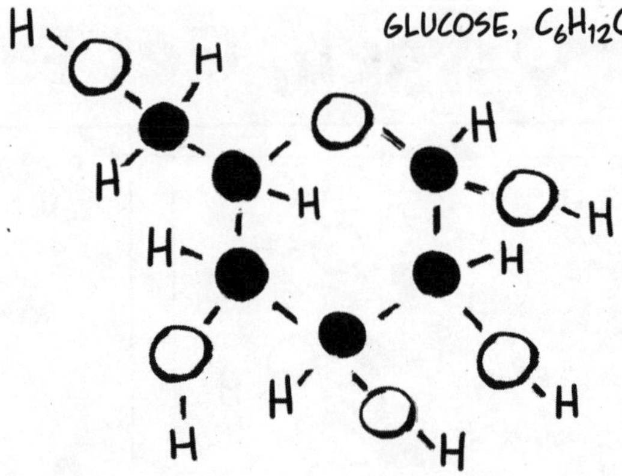

GLUCOSE, $C_6H_{12}O_6$

MOST ORGANISMS ARE ALSO HAPPY TO EAT ANY ONE OF A NUMBER OF OTHER SUGARS. HERE ARE A FEW THAT TURN UP IN HUMAN FOODS LIKE CORN, SUGARCANE, AND COW'S MILK.

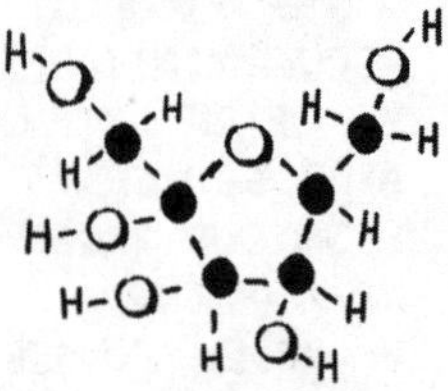

FRUCTOSE

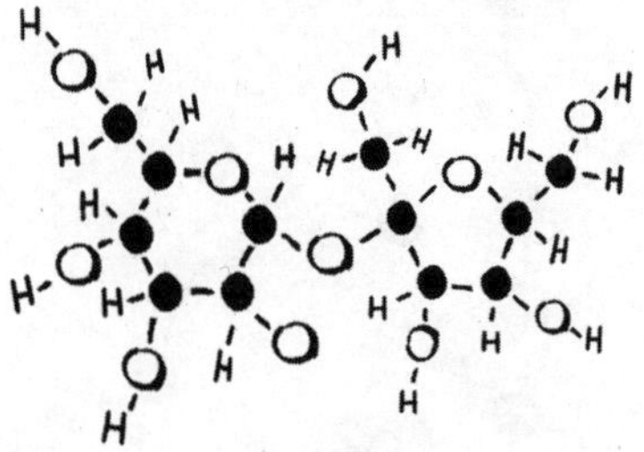

SUCROSE

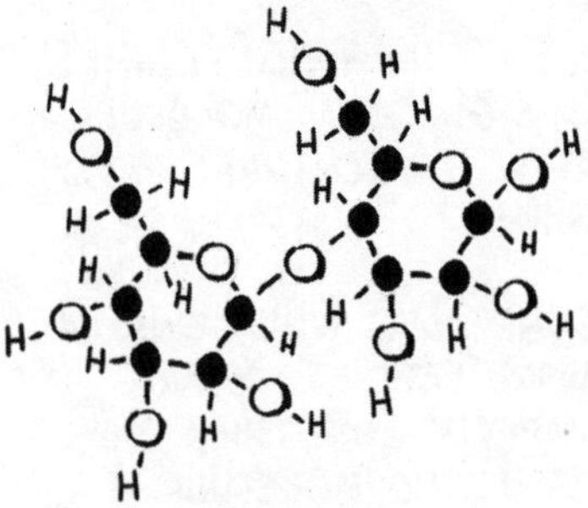

LACTOSE

THESE TWO 5-CARBON SUGARS ARE LESS FAMILIAR, BUT ESSENTIAL TO LIFE. FOR FUTURE REFERENCE, NOTE HOW THEIR CARBON ATOMS ARE NUMBERED. 5′ IS THE CARBON OFF THE RING.

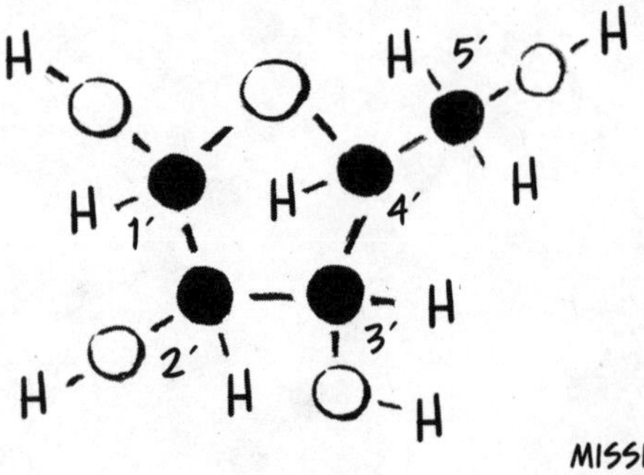

RIBOSE

MISSING ONE OXYGEN ATOM HERE

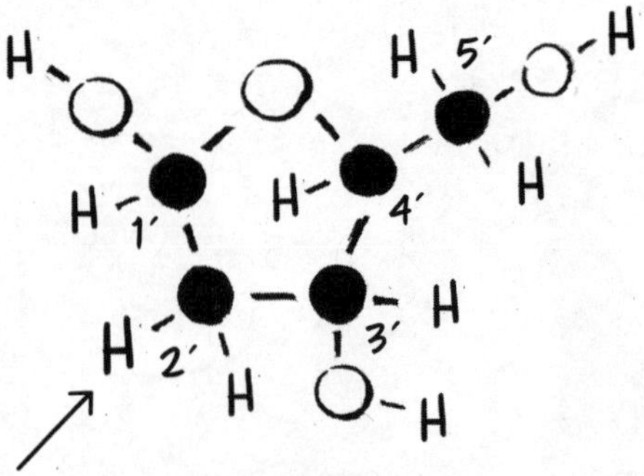

DEOXYRIBOSE

LIKE MANY ORGANIC COMPOUNDS, SUGARS CAN BE STRUNG TOGETHER INTO REPEATING CHAINS, OR **POLYMERS.** THESE **POLYSACCHARIDES** ARE A GOOD WAY TO STORE SUGAR IN THE BODY. (A LONE SUGAR IS SOMETIMES CALLED A MONOSACCHARIDE. DON'T ASK ME WHY.)

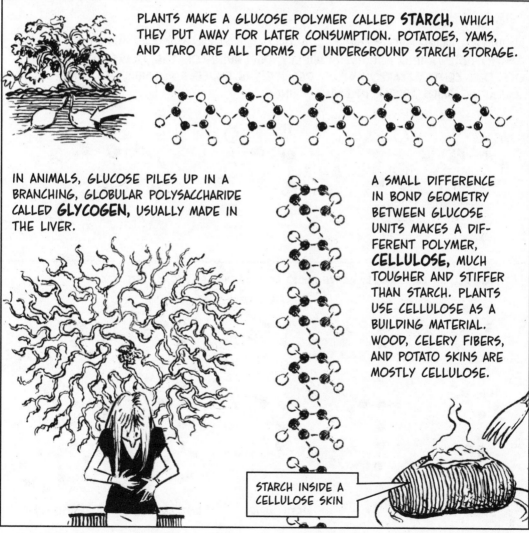

WHEN IT COMES TO NAMES, BIOLOGISTS HAVE INFINITE STORAGE SPACE...

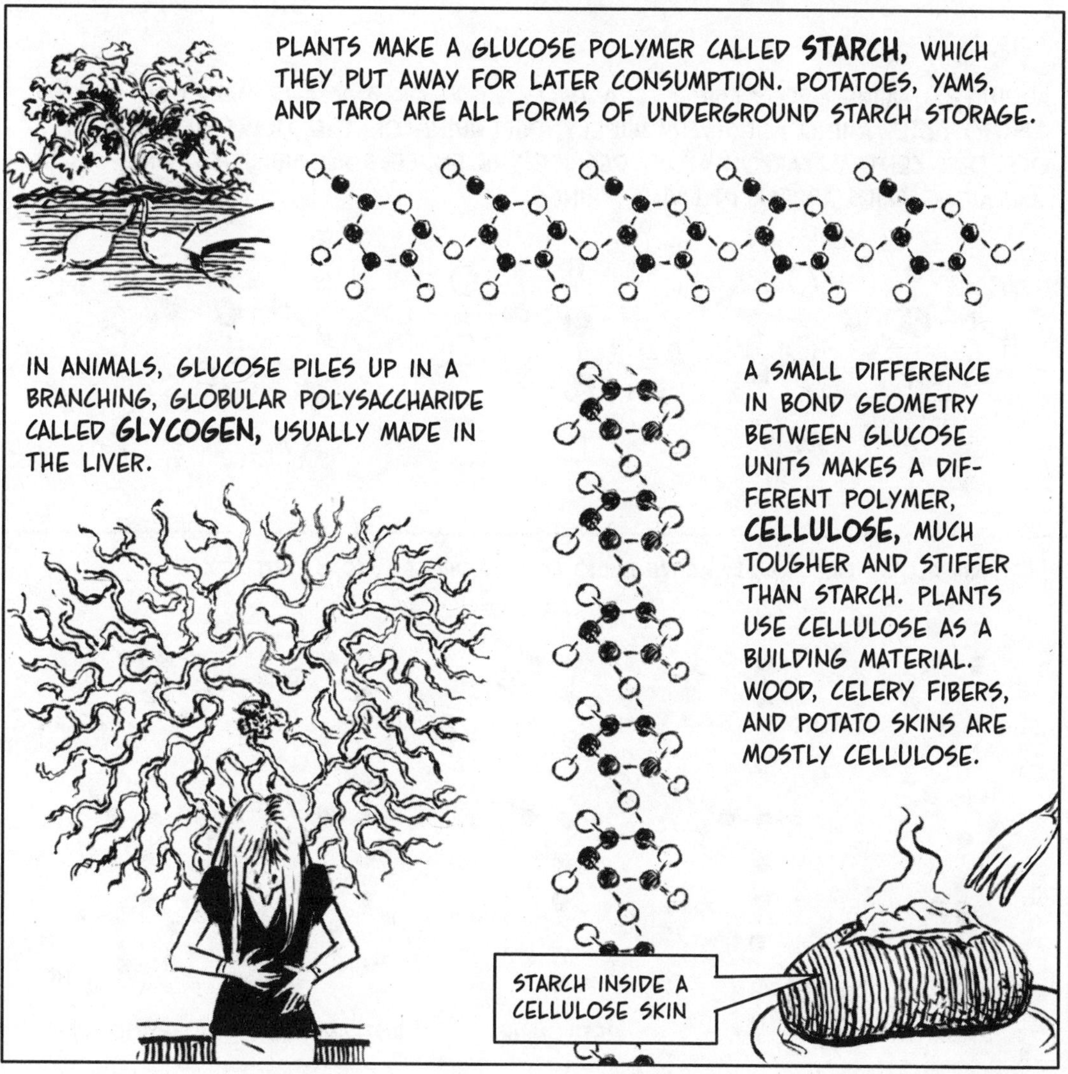
PLANTS MAKE A GLUCOSE POLYMER CALLED STARCH, WHICH THEY PUT AWAY FOR LATER CONSUMPTION. POTATOES, YAMS, AND TARO ARE ALL FORMS OF UNDERGROUND STARCH STORAGE.
IN ANIMALS, GLUCOSE PILES UP IN A BRANCHING, GLOBULAR POLYSACCHARIDE CALLED GLYCOGEN, USUALLY MADE IN THE LIVER.
A SMALL DIFFERENCE IN BOND GEOMETRY BETWEEN GLUCOSE UNITS MAKES A DIFFERENT POLYMER, CELLULOSE, MUCH TOUGHER AND STIFFER THAN STARCH. PLANTS USE CELLULOSE AS A BUILDING MATERIAL. WOOD, CELERY FIBERS, AND POTATO SKINS ARE MOSTLY CELLULOSE.
STARCH INSIDE A CELLULOSE SKIN

Add NITROGEN (plus a pinch of Sulfur), Get PROTEINS.

IN ORGANIC COMPOUNDS, NITROGEN, ATOMIC NUMBER 7, USUALLY BONDS WITH THREE OF ITS OUTER ELECTRONS. NITROGEN COMPOUNDS OFTEN HAVE "AMIDE" OR "AMINE" IN THEIR NAMES (FROM *AMMONIA*, NH_3).

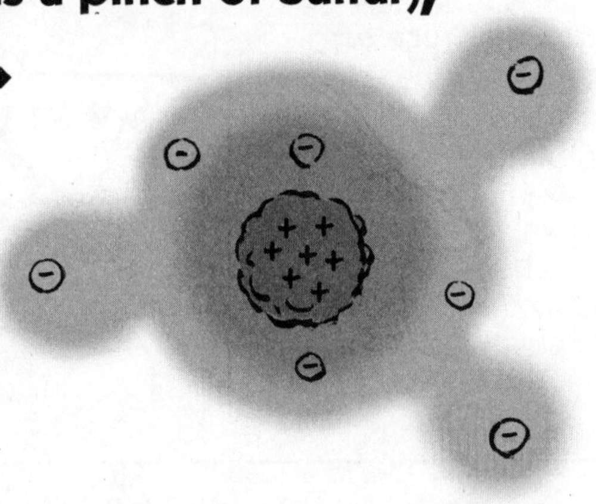

ADDING AN AMINE GROUP (NH_2) TO A TWO-CARBON ORGANIC ACID MAKES AN **AMINO ACID.** AMINO ACIDS VARY WIDELY, DEPENDING ON THE ATOMS HANGING OFF THAT CENTRAL CARBON ATOM. DESPITE THE ENDLESS POSSIBILITIES, ONLY **20** AMINO ACIDS APPEAR IN LIVING THINGS.

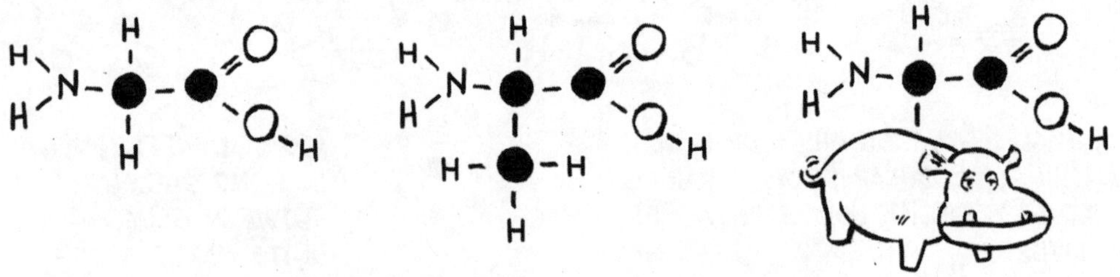

HERE ARE ALL 20 BIOLOGICALLY ACTIVE AMINO ACIDS (HYDROGEN ATOMS OMITTED).

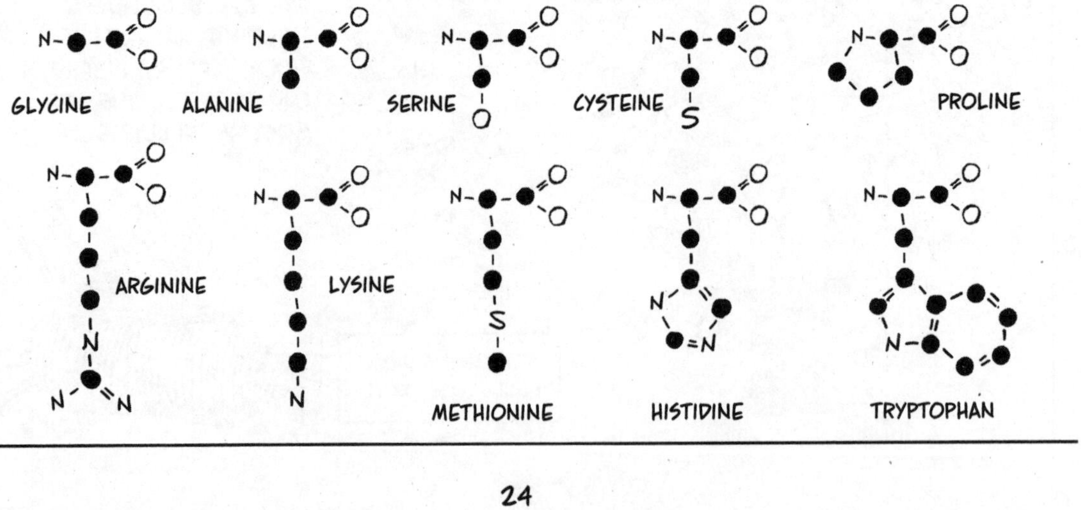

WHAT MAKES AMINO ACIDS SO SPECIAL IS THAT ANY TWO OF THEM WILL JOIN END TO END IN A SO-CALLED **PEPTIDE BOND.**

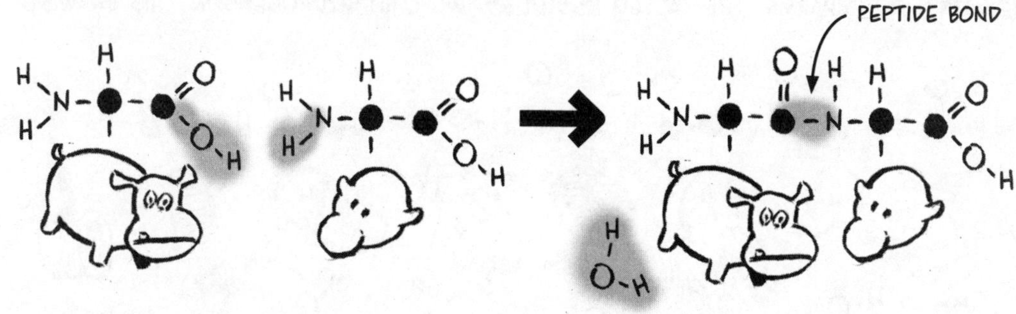

THE TWO-ACID GROUP, OR **DIPEPTIDE,** HAS THE SAME ENDS AS BEFORE, SO A THIRD AMINO ACID CAN EXTEND THE STRING, AND A FOURTH, A FIFTH, A SIXTH...

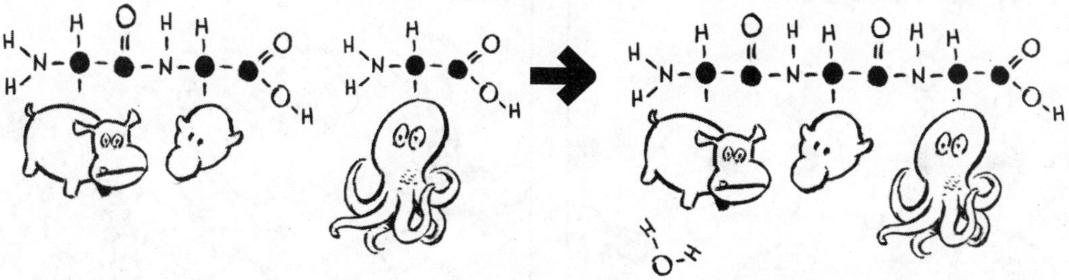

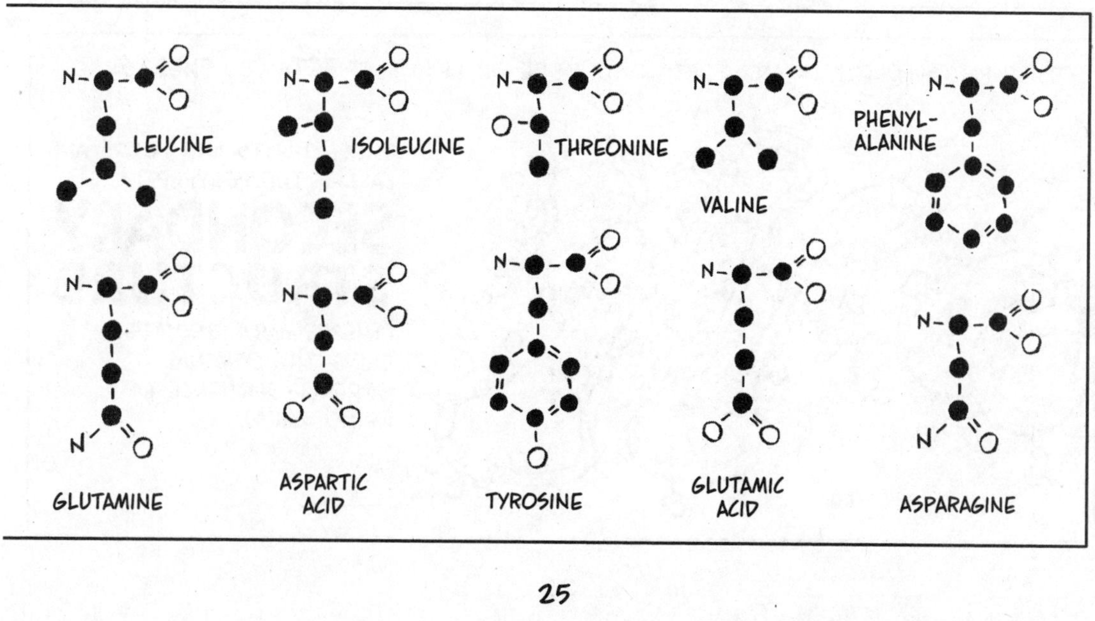

A SINGLE **POLYPEPTIDE** CHAIN CAN GROW TO HUNDREDS, EVEN THOUSANDS, OF AMINO ACIDS, LIKE AN ABSURDLY LONG CHARM BRACELET MADE FROM 20 DIFFERENT CHARMS. IN CHEMISTRY, THE "CHARMS" ARE CALLED **RESIDUES.** WE OMIT HYDROGENS IN THIS DRAWING.

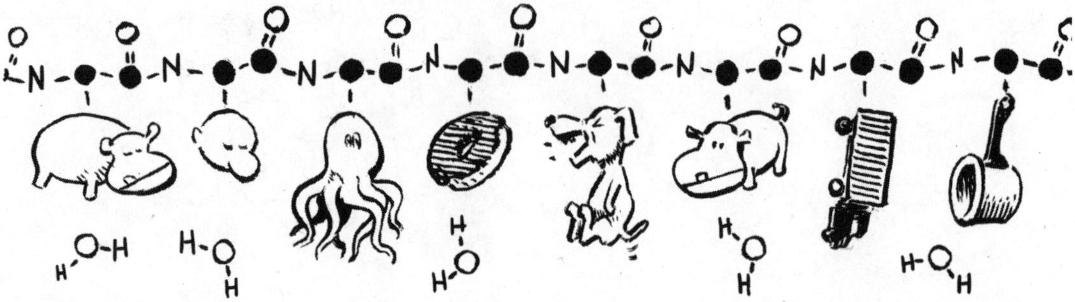

VARIOUS POINTS ALONG THE POLYPEPTIDE HAVE FRACTIONAL ELECTRIC CHARGES THAT ATTRACT AND REPEL EACH OTHER.

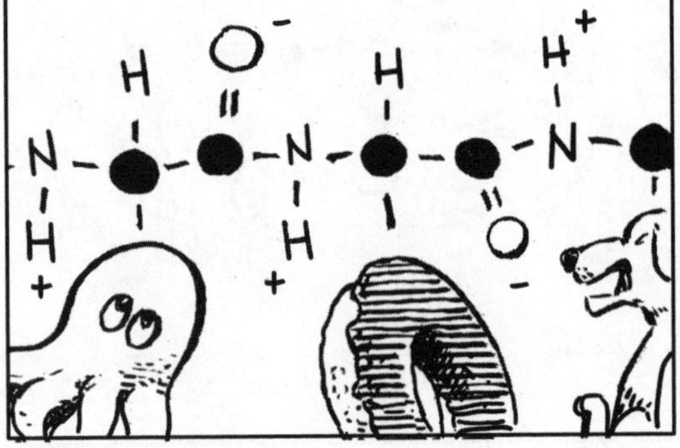

THESE FORCES TWIST PARTS OF THE CHAIN INTO A COIL, THE **ALPHA** (α) **HELIX.**

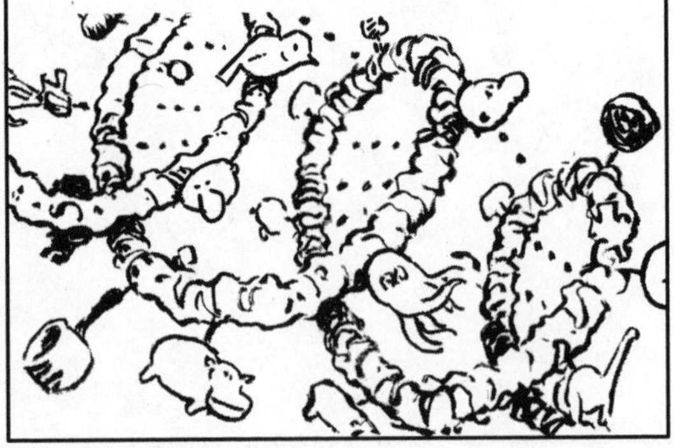

OTHER PARTS FOLD BACK AND FORTH INTO MORE OR LESS FLAT **BETA** (β) **SHEETS.**

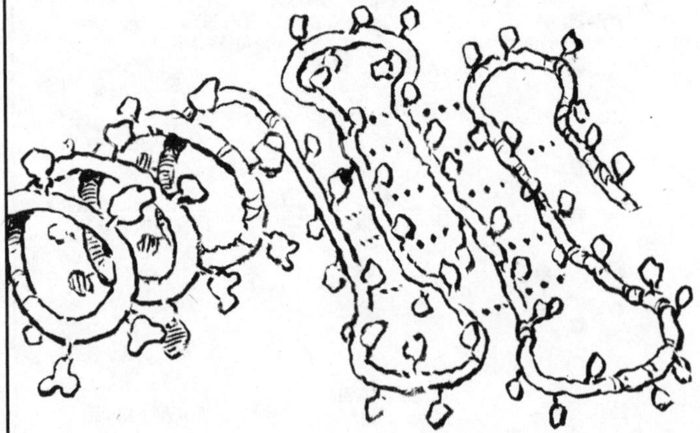

THESE TWISTS AND FOLDS ARE CALLED THE POLYPEPTIDE'S

SECONDARY STRUCTURE

(THE PRIMARY STRUCTURE BEING THE ORIGINAL ORDERED SEQUENCE OF AMINO ACIDS).

AND MORE: SOME OF THE AMINO ACID RESIDUES (THE "CHARMS") ARE **HYDROPHILIC,** AND SOME **HYDROPHOBIC.**

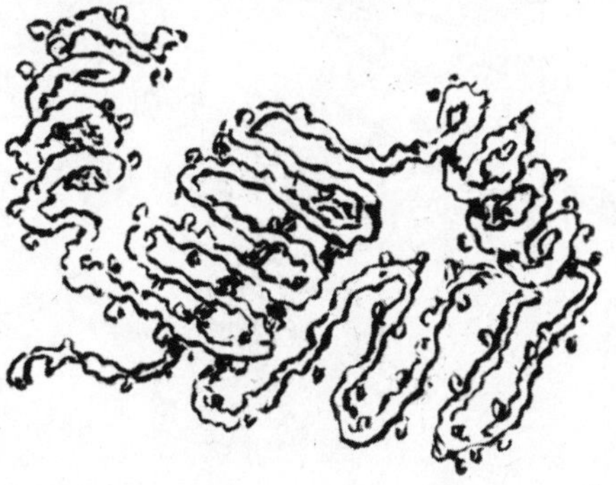

SURROUNDED BY WATER, THE TWISTED, FOLDED POLYPEPTIDE CLOSES UP TIGHTER AS IT TRIES TO TUCK THE HYDROPHOBIC RESIDUES INSIDE, AWAY FROM WATER.

EVENTUALLY, IT PACKS ITSELF INTO A COMPACT MASS, ITS 3-D **TERTIARY STRUCTURE.** OUR POLYPEPTIDE HAS BECOME A

PROTEIN.

SOME PROTEINS CONSIST OF TWO OR MORE POLYPEPTIDE CHAINS FITTED TOGETHER. IN THAT CASE, WE CALL THE FINAL ARRANGEMENT THE PROTEIN'S **QUATERNARY** STRUCTURE.

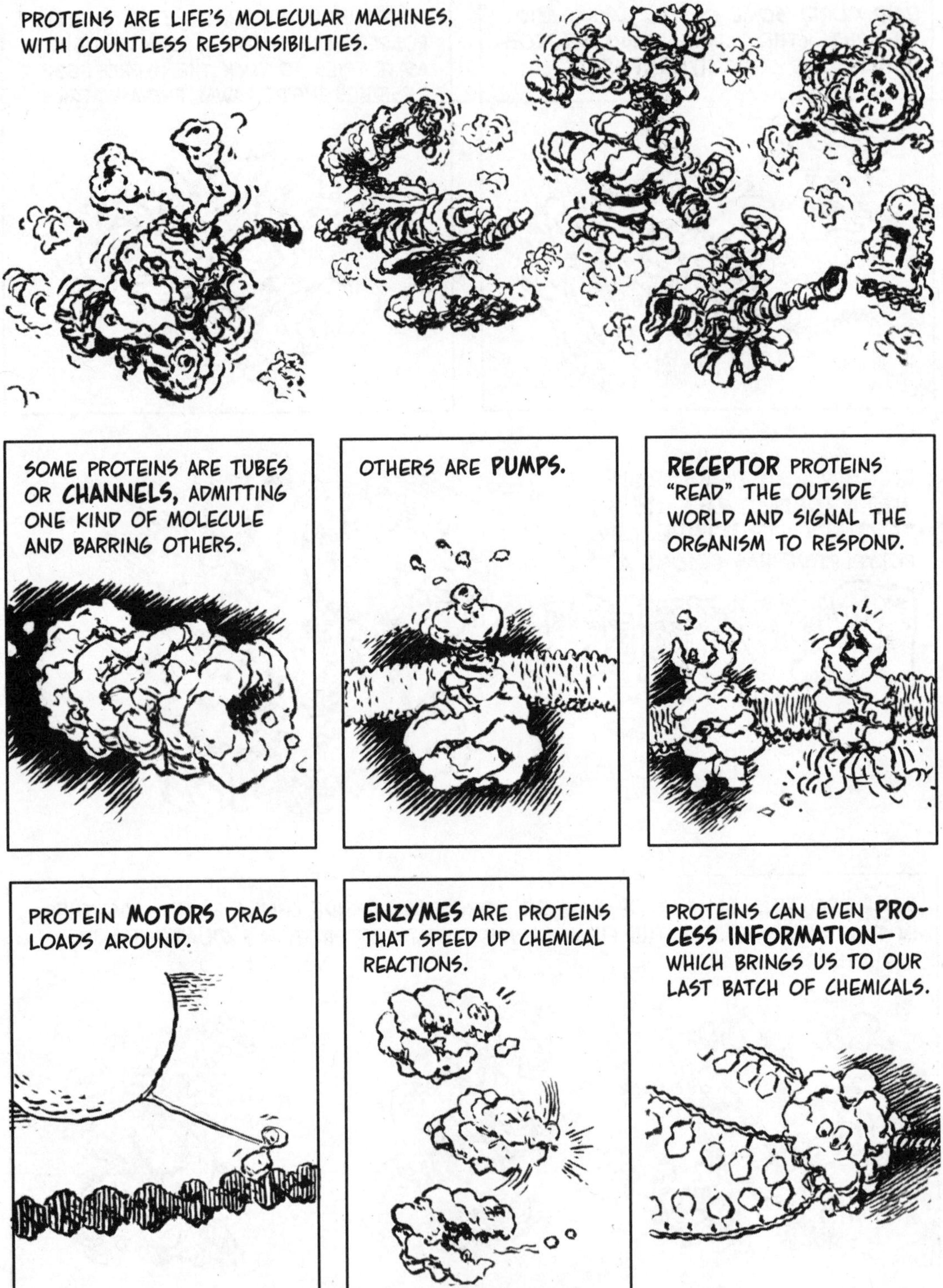
PROTEINS ARE LIFE'S MOLECULAR MACHINES, WITH COUNTLESS RESPONSIBILITIES.
SOME PROTEINS ARE TUBES OR **CHANNELS,** ADMITTING ONE KIND OF MOLECULE AND BARRING OTHERS.
OTHERS ARE **PUMPS.**
RECEPTOR PROTEINS "READ" THE OUTSIDE WORLD AND SIGNAL THE ORGANISM TO RESPOND.
PROTEIN **MOTORS** DRAG LOADS AROUND.
ENZYMES ARE PROTEINS THAT SPEED UP CHEMICAL REACTIONS.
PROTEINS CAN EVEN **PROCESS INFORMATION—** WHICH BRINGS US TO OUR LAST BATCH OF CHEMICALS.

Add PHOSPHORUS, Get Just About Everything Else.

FIVE NITROGENOUS **BASES** FILL OUT OUR LIST OF SIMPLE ORGANIC PARTS. WE'LL OFTEN ABBREVIATE THESE CHARACTERS AS LETTERS OF THE ALPHABET.

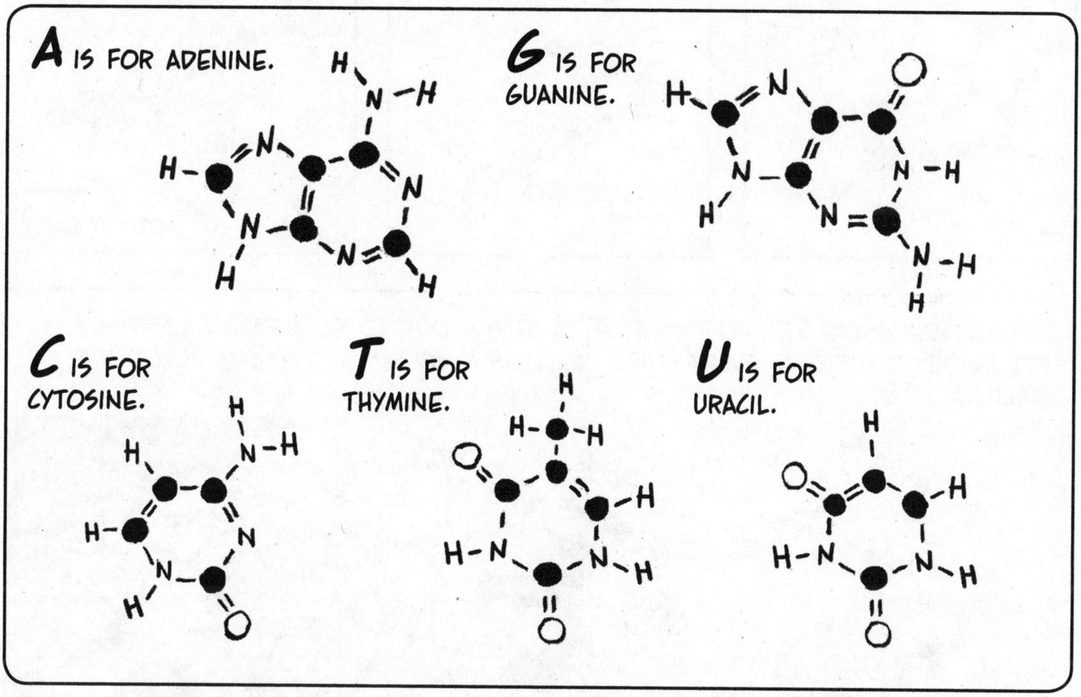

A IS FOR ADENINE.
G IS FOR GUANINE.
C IS FOR CYTOSINE.
T IS FOR THYMINE.
U IS FOR URACIL.

FOUR OF THESE BASES (A, C, G, AND U) HAVE AN AFFINITY FOR THE SUGAR **RIBOSE** (SEE P. 22). WHEN A (FOR EXAMPLE) JOINS RIBOSE, WE GET SOMETHING CALLED **ADENOSINE.**

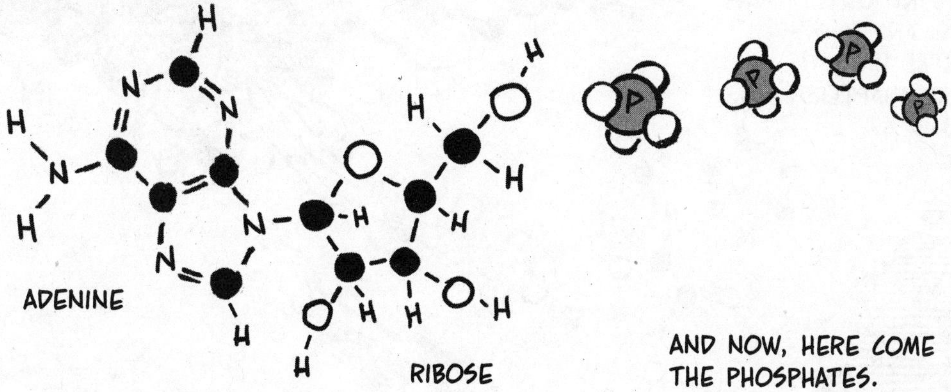

AND NOW, HERE COME THE PHOSPHATES.

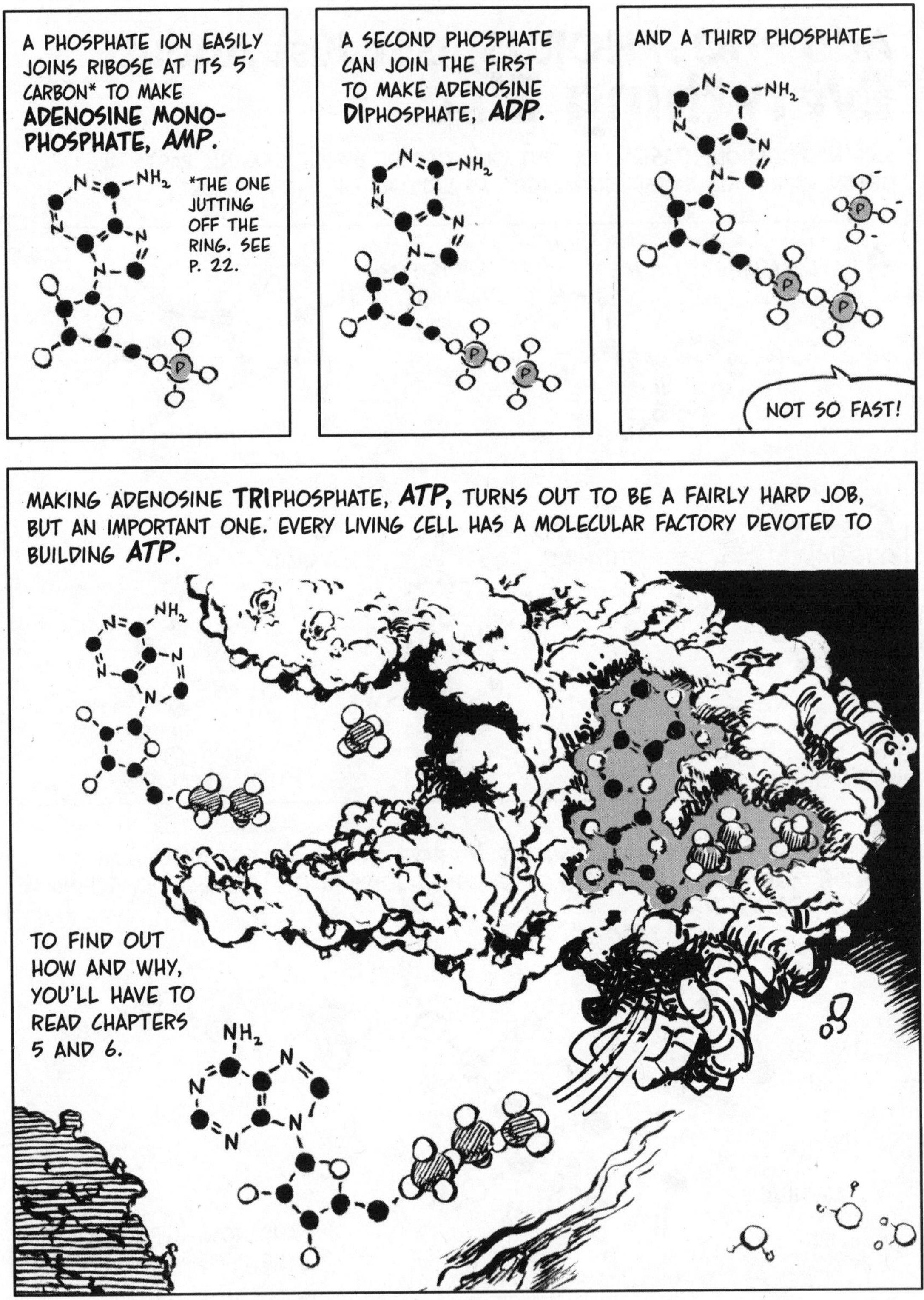
A PHOSPHATE ION EASILY JOINS RIBOSE AT ITS 5′ CARBON* TO MAKE **ADENOSINE MONOPHOSPHATE, AMP.**
*THE ONE JUTTING OFF THE RING. SEE P. 22.
A SECOND PHOSPHATE CAN JOIN THE FIRST TO MAKE ADENOSINE **DI**PHOSPHATE, **ADP.**
AND A THIRD PHOSPHATE—
NOT SO FAST!
MAKING ADENOSINE **TRI**PHOSPHATE, **ATP,** TURNS OUT TO BE A FAIRLY HARD JOB, BUT AN IMPORTANT ONE. EVERY LIVING CELL HAS A MOLECULAR FACTORY DEVOTED TO BUILDING **ATP.**
TO FIND OUT HOW AND WHY, YOU'LL HAVE TO READ CHAPTERS 5 AND 6.

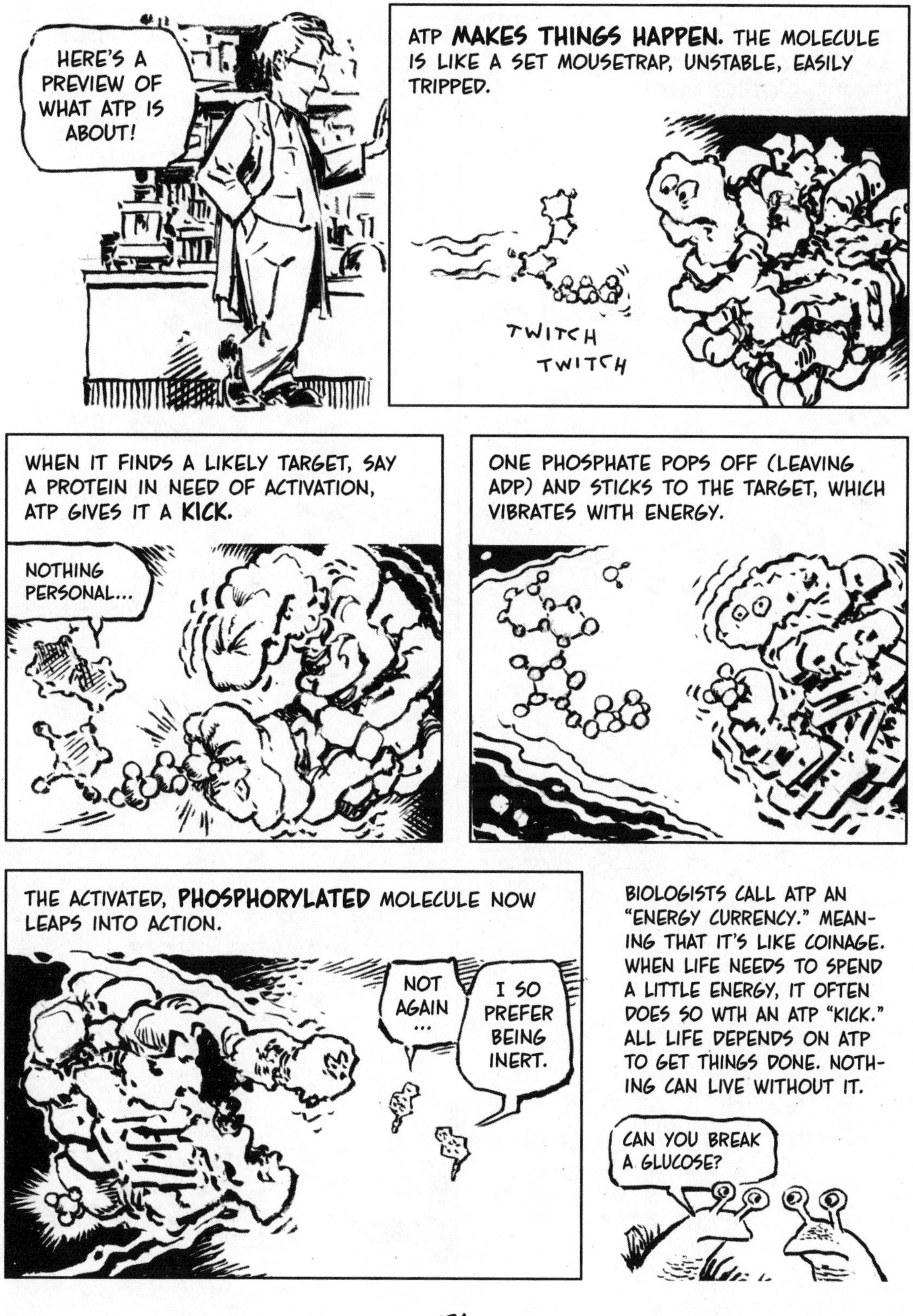
HERE'S A PREVIEW OF WHAT ATP IS ABOUT!
ATP **MAKES THINGS HAPPEN.** THE MOLECULE IS LIKE A SET MOUSETRAP, UNSTABLE, EASILY TRIPPED.
TWITCH TWITCH
WHEN IT FINDS A LIKELY TARGET, SAY A PROTEIN IN NEED OF ACTIVATION, ATP GIVES IT A **KICK.**
NOTHING PERSONAL...
ONE PHOSPHATE POPS OFF (LEAVING ADP) AND STICKS TO THE TARGET, WHICH VIBRATES WITH ENERGY.
THE ACTIVATED, **PHOSPHORYLATED** MOLECULE NOW LEAPS INTO ACTION.
NOT AGAIN ...
I SO PREFER BEING INERT.
BIOLOGISTS CALL ATP AN "ENERGY CURRENCY." MEANING THAT IT'S LIKE COINAGE. WHEN LIFE NEEDS TO SPEND A LITTLE ENERGY, IT OFTEN DOES SO WTH AN ATP "KICK." ALL LIFE DEPENDS ON ATP TO GET THINGS DONE. NOTHING CAN LIVE WITHOUT IT.
CAN YOU BREAK A GLUCOSE?

LIVING SYSTEMS ALSO CARRY SMALLER AMOUNTS OF RIBOPHOSPHATES MADE FROM THE OTHER BASES.

ATP

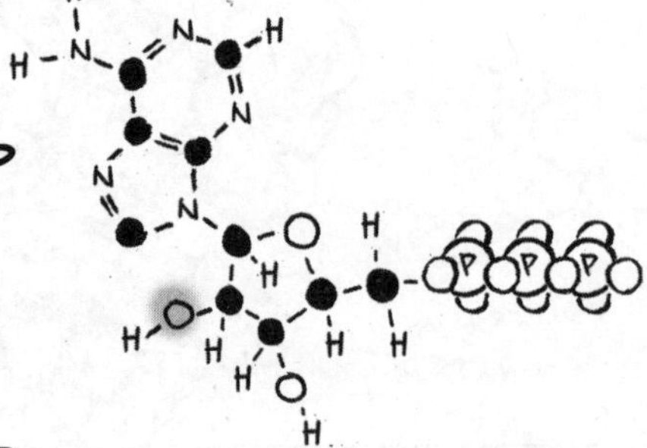

CTP

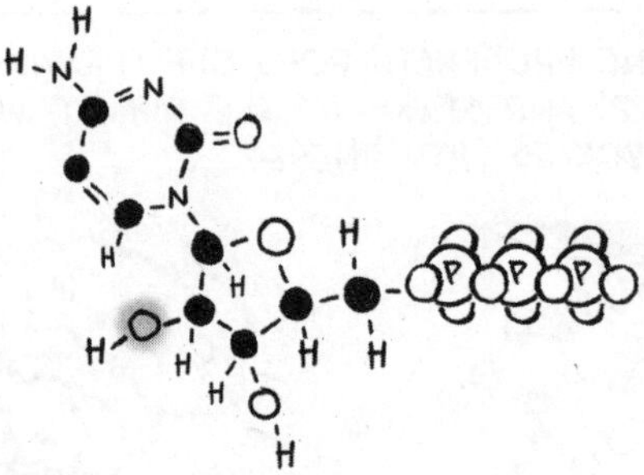

GTP

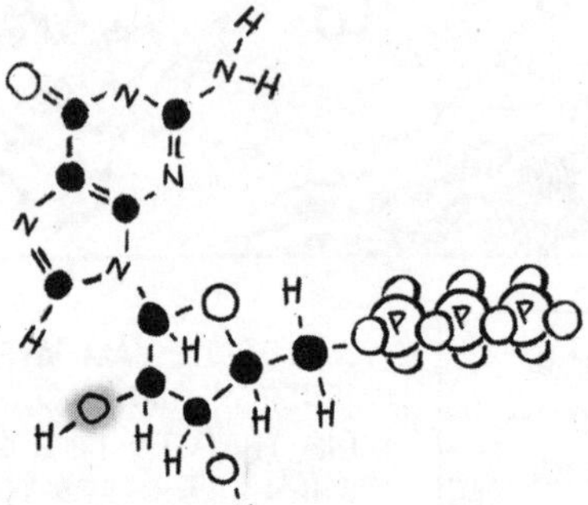

UTP

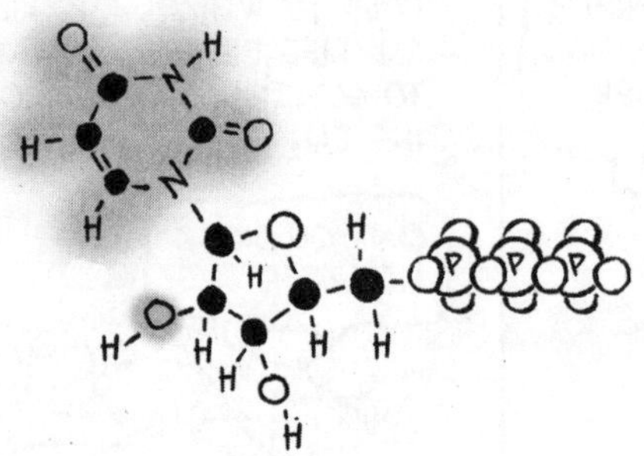

REPLACING RIBOSE WITH DEOXYRIBOSE MAKES SIMILAR MOLECULES. DIFFERENCES ARE HIGHLIGHTED.

dATP

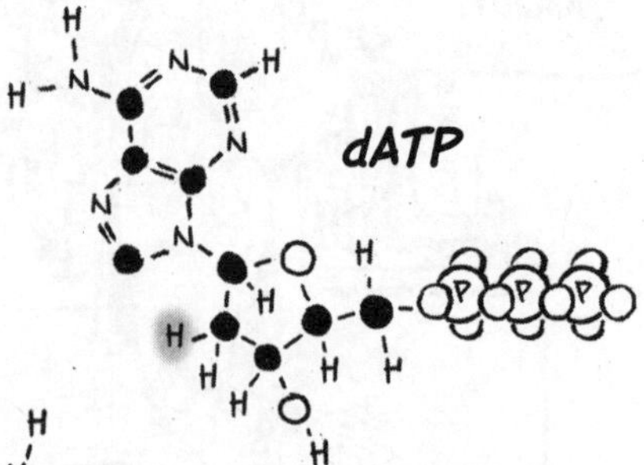

dCTP

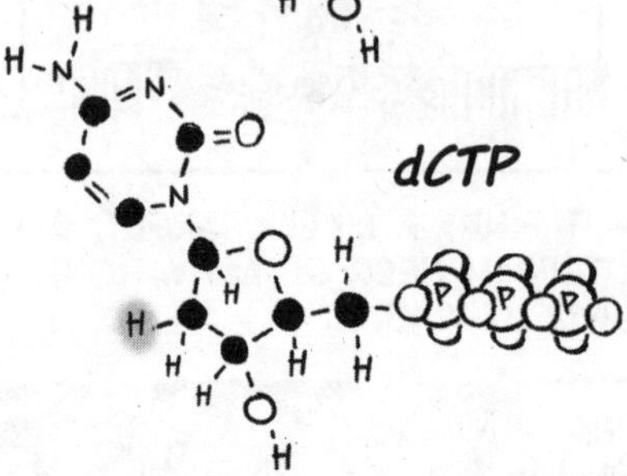

dGTP

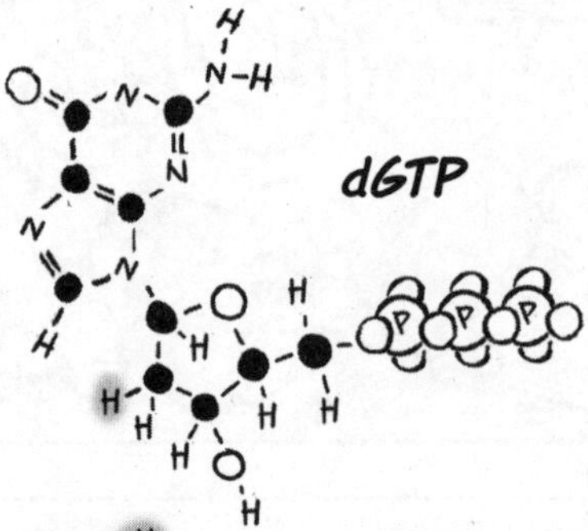

dTTP

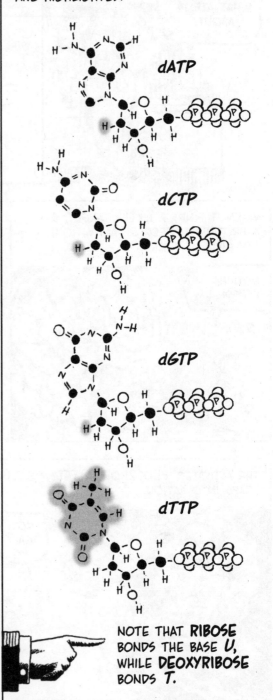

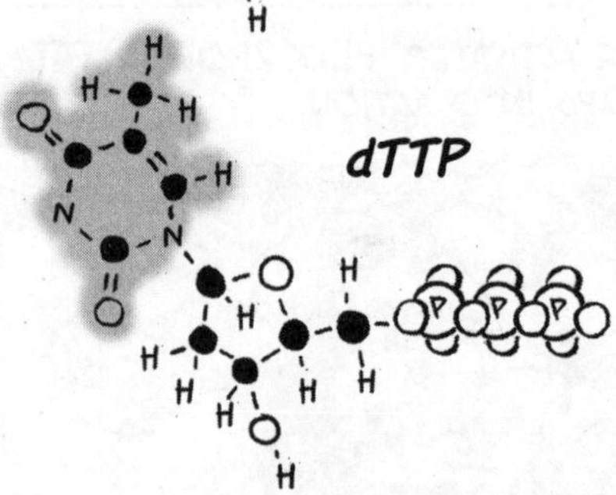

NOTE THAT **RIBOSE** BONDS THE BASE ***U***, WHILE **DEOXYRIBOSE** BONDS ***T***.

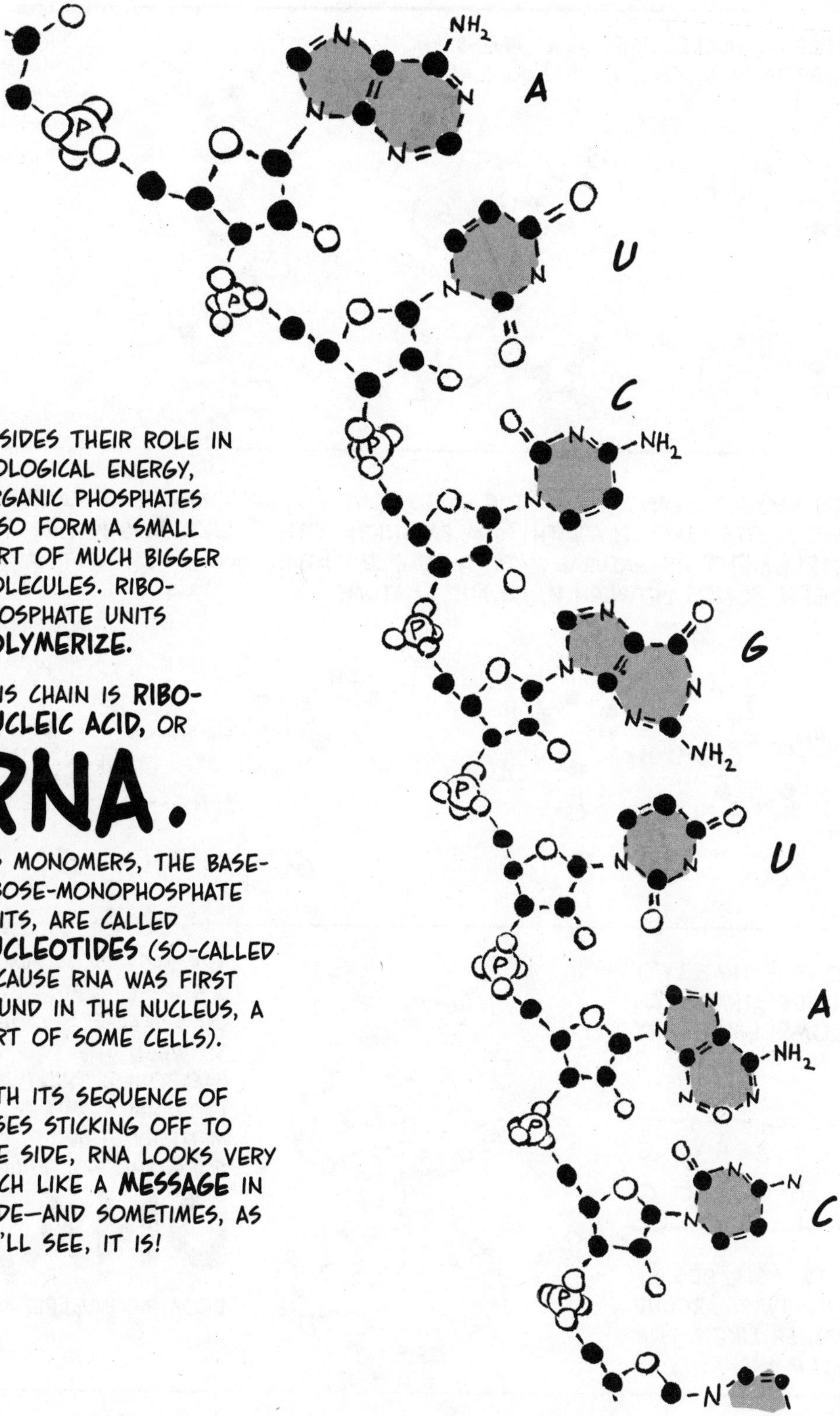

BESIDES THEIR ROLE IN BIOLOGICAL ENERGY, ORGANIC PHOSPHATES ALSO FORM A SMALL PART OF MUCH BIGGER MOLECULES. RIBOPHOSPHATE UNITS **POLYMERIZE.**

THIS CHAIN IS **RIBONUCLEIC ACID,** OR

RNA.

ITS MONOMERS, THE BASE-RIBOSE-MONOPHOSPHATE UNITS, ARE CALLED **NUCLEOTIDES** (SO-CALLED BECAUSE RNA WAS FIRST FOUND IN THE NUCLEUS, A PART OF SOME CELLS).

WITH ITS SEQUENCE OF BASES STICKING OFF TO THE SIDE, RNA LOOKS VERY MUCH LIKE A **MESSAGE** IN CODE—AND SOMETIMES, AS WE'LL SEE, IT IS!

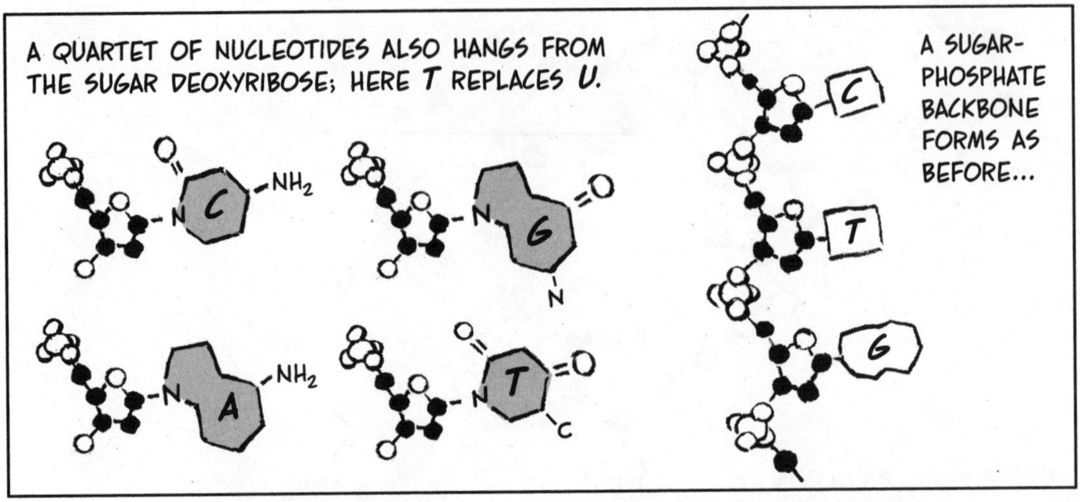
A QUARTET OF NUCLEOTIDES ALSO HANGS FROM THE SUGAR DEOXYRIBOSE; HERE T REPLACES U.
C
G
A
T
A SUGAR-PHOSPHATE BACKBONE FORMS AS BEFORE...
C
T
G

BUT THIS BACKBONE HAS A TWIST: THE BASES A, C, G, AND T CAN FORM **PAIRS.** THE BASE **A** FITS PERFECTLY WITH **T; G** PARTNERS WITH **C.** EACH BASE IS SAID TO **COMPLEMENT** ITS NATURAL MATE. A COMPLEMENTARY PAIR IS CEMENTED BY **HYDROGEN BONDS** BETWEEN **H, N,** AND **O** ATOMS.

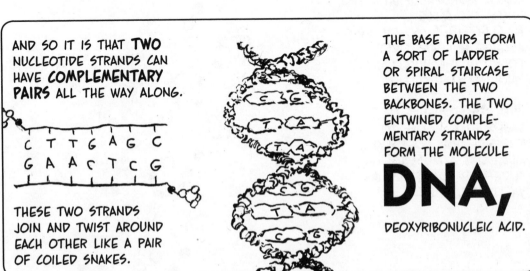

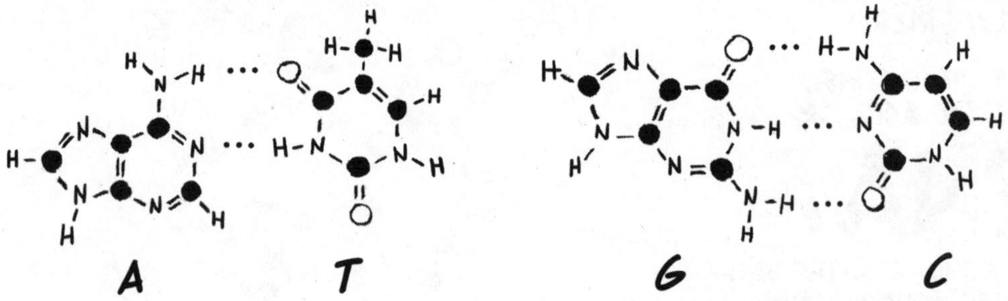

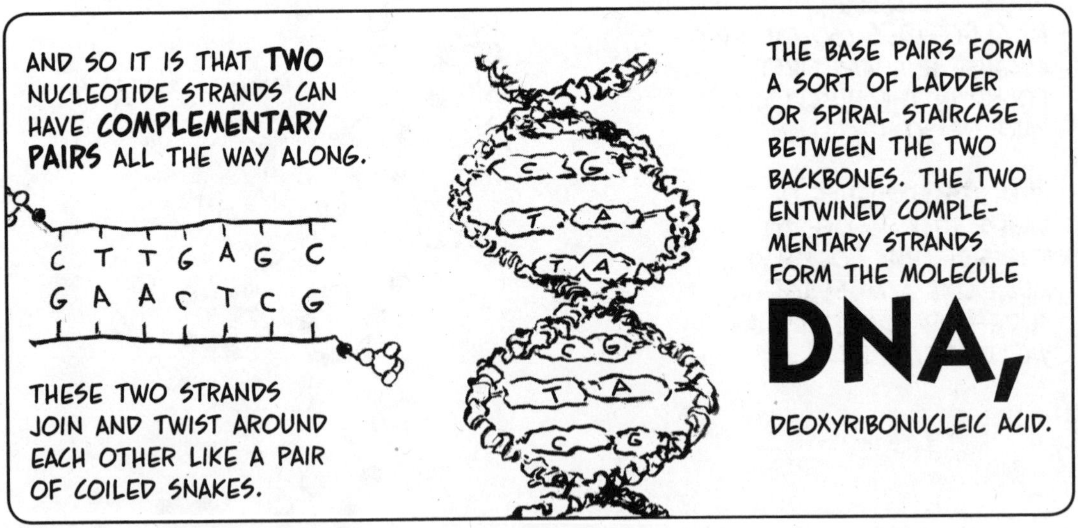
AND SO IT IS THAT TWO NUCLEOTIDE STRANDS CAN HAVE COMPLEMENTARY PAIRS ALL THE WAY ALONG.
C T T G A G C
G A A C T C G
THESE TWO STRANDS JOIN AND TWIST AROUND EACH OTHER LIKE A PAIR OF COILED SNAKES.
THE BASE PAIRS FORM A SORT OF LADDER OR SPIRAL STAIRCASE BETWEEN THE TWO BACKBONES. THE TWO ENTWINED COMPLEMENTARY STRANDS FORM THE MOLECULE
DNA,
DEOXYRIBONUCLEIC ACID.

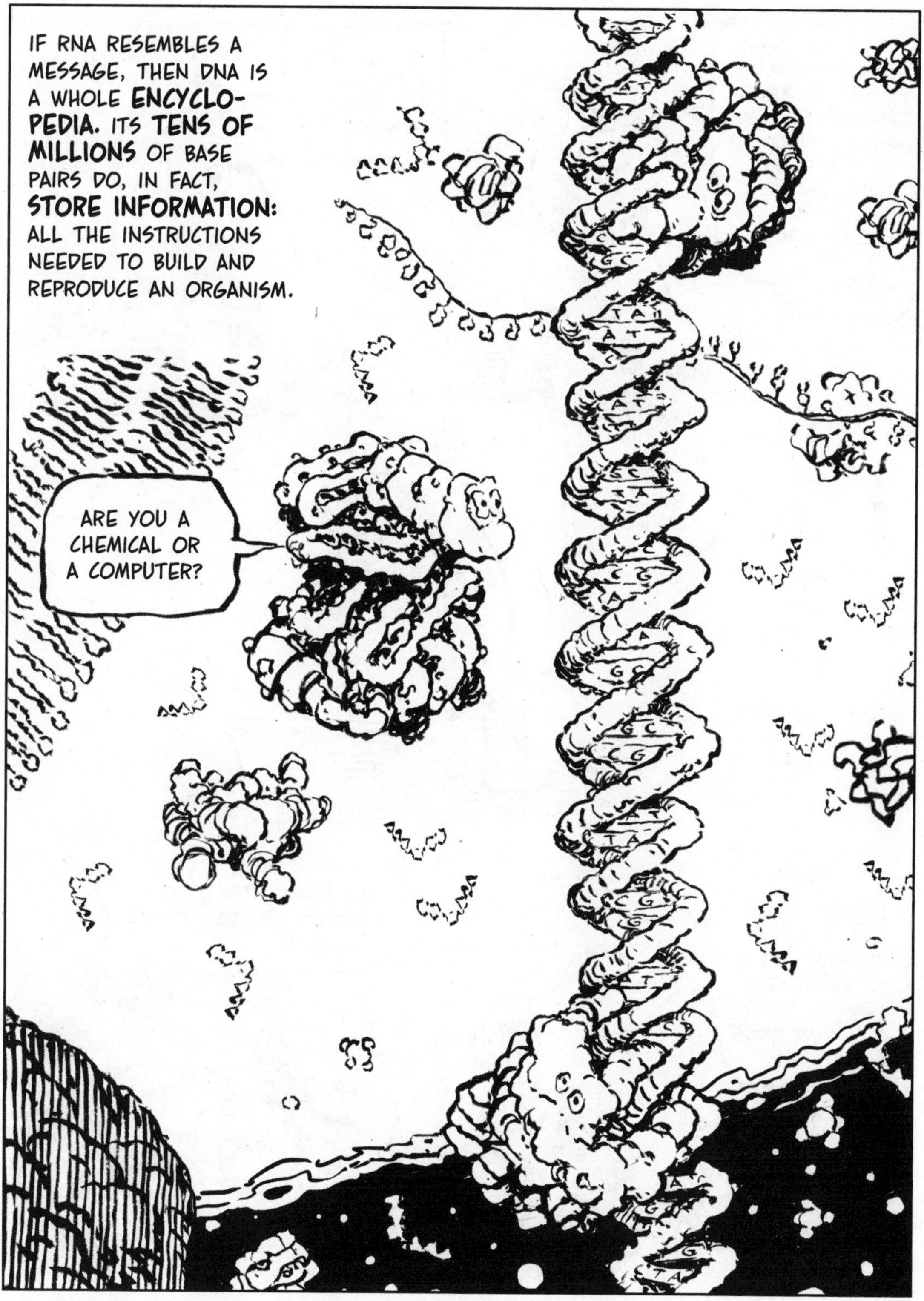
IF RNA RESEMBLES A MESSAGE, THEN DNA IS A WHOLE ENCYCLOPEDIA. ITS TENS OF MILLIONS OF BASE PAIRS DO, IN FACT, STORE INFORMATION: ALL THE INSTRUCTIONS NEEDED TO BUILD AND REPRODUCE AN ORGANISM.
ARE YOU A CHEMICAL OR A COMPUTER?

THESE ARE LIFE'S RAW MATERIALS: LIPIDS, SUGARS AND POLYSACCHARIDES, AMINO ACIDS AND PROTEINS, PHOSPHATES AND NUCLEIC ACIDS—OH, AND WATER!
LET'S POUR 'EM ALL INTO A TUB AND SEE WHAT HAPPENS!
HMMM... NOT MUCH...
NO, TO CREATE LIFE, THESE CHEMICALS HAVE TO BE PACKED TOGETHER AND SEALED UP INSIDE A VERY TIGHT SPACE!
AND PARADOXICALLY, IT'S A PRISON OF THEIR OWN MAKING!

Chapter 4
INTO THE CELL

LIFE INSIDE

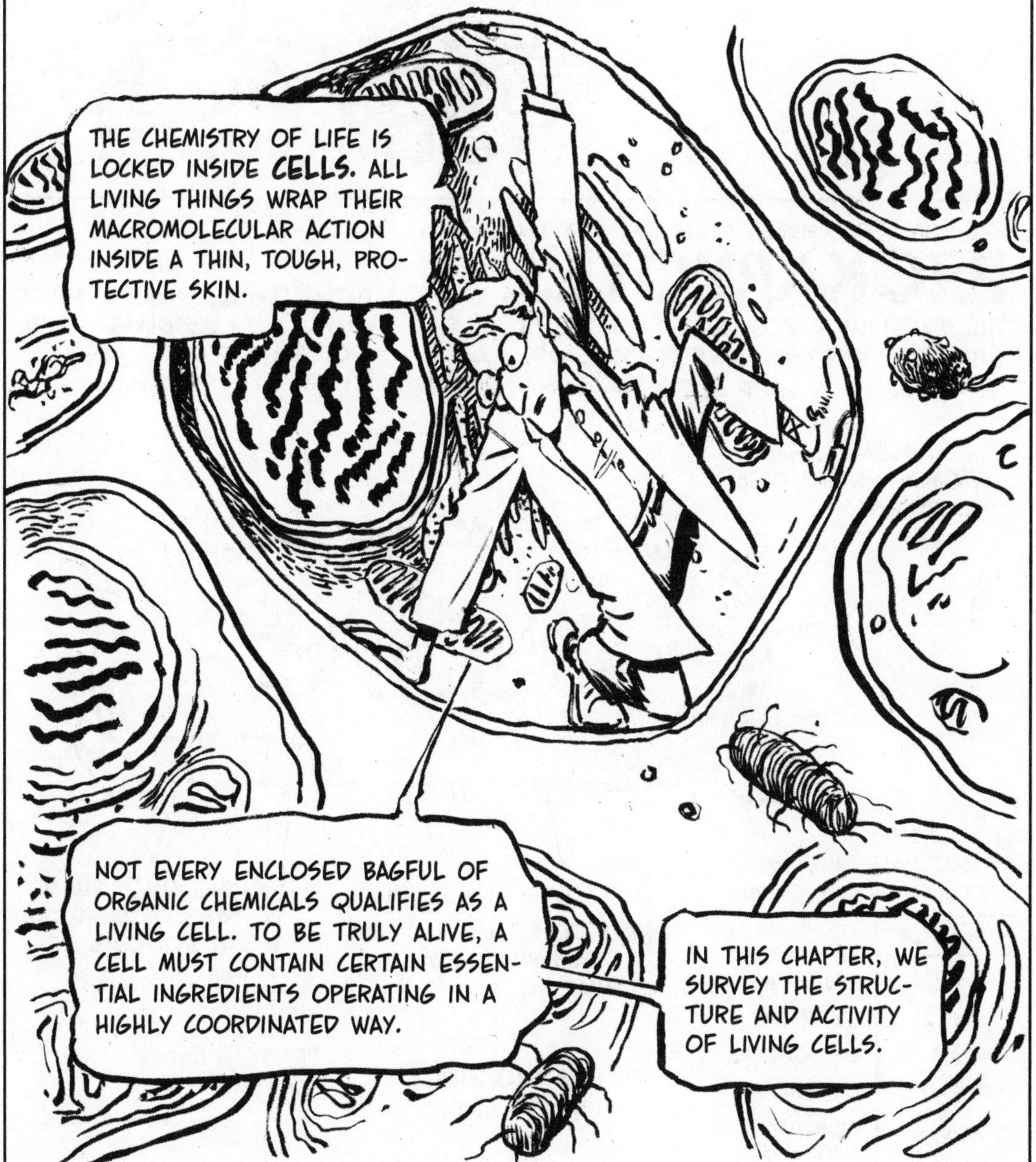
THE CHEMISTRY OF LIFE IS LOCKED INSIDE **CELLS.** ALL LIVING THINGS WRAP THEIR MACROMOLECULAR ACTION INSIDE A THIN, TOUGH, PROTECTIVE SKIN.
NOT EVERY ENCLOSED BAGFUL OF ORGANIC CHEMICALS QUALIFIES AS A LIVING CELL. TO BE TRULY ALIVE, A CELL MUST CONTAIN CERTAIN ESSENTIAL INGREDIENTS OPERATING IN A HIGHLY COORDINATED WAY.
IN THIS CHAPTER, WE SURVEY THE STRUCTURE AND ACTIVITY OF LIVING CELLS.

BIOLOGISTS CLASSIFY CELLS INTO TWO BROAD TYPES, ONE LARGER AND MORE HIGHLY STRUCTURED, THE OTHER SMALLER AND SIMPLER.

1.
2.
THERE ARE TWO KINDS OF BIOLOGISTS: THE KIND THAT CLASSIFIES, AND UNEMPLOYED.

THE SMALLER, SIMPLER CELLS ARE KNOWN AS

PROKARYOTES.

THE BACTERIUM *E. COLI*, WHICH THRIVES IN THE HUMAN GUT (YOU HAVE BILLIONS OF THEM!), MAKES A GOOD MODEL PROKARYOTE.

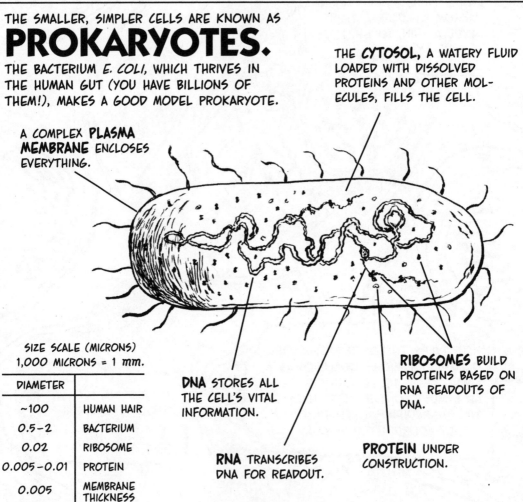

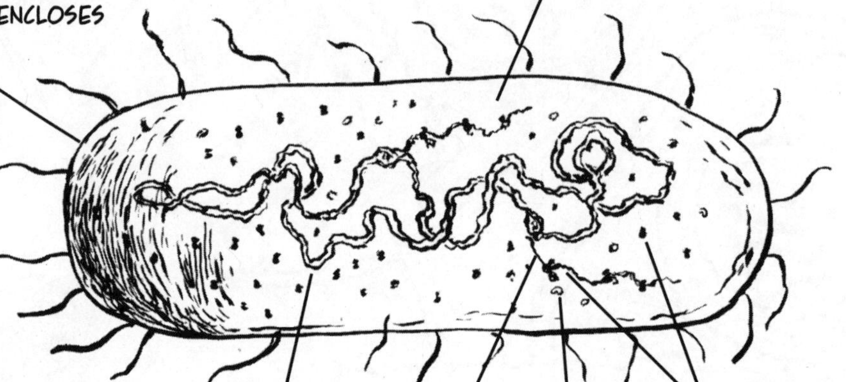
THE **CYTOSOL,** A WATERY FLUID LOADED WITH DISSOLVED PROTEINS AND OTHER MOL-ECULES, FILLS THE CELL.
A COMPLEX **PLASMA MEMBRANE** ENCLOSES EVERYTHING.
DNA STORES ALL THE CELL'S VITAL INFORMATION.
RIBOSOMES BUILD PROTEINS BASED ON RNA READOUTS OF DNA.
PROTEIN UNDER CONSTRUCTION.
RNA TRANSCRIBES DNA FOR READOUT.

SIZE SCALE (MICRONS)
1,000 MICRONS = 1 *mm*.

DIAMETER	
~100	HUMAN HAIR
0.5–2	BACTERIUM
0.02	RIBOSOME
0.005–0.01	PROTEIN
0.005	MEMBRANE THICKNESS

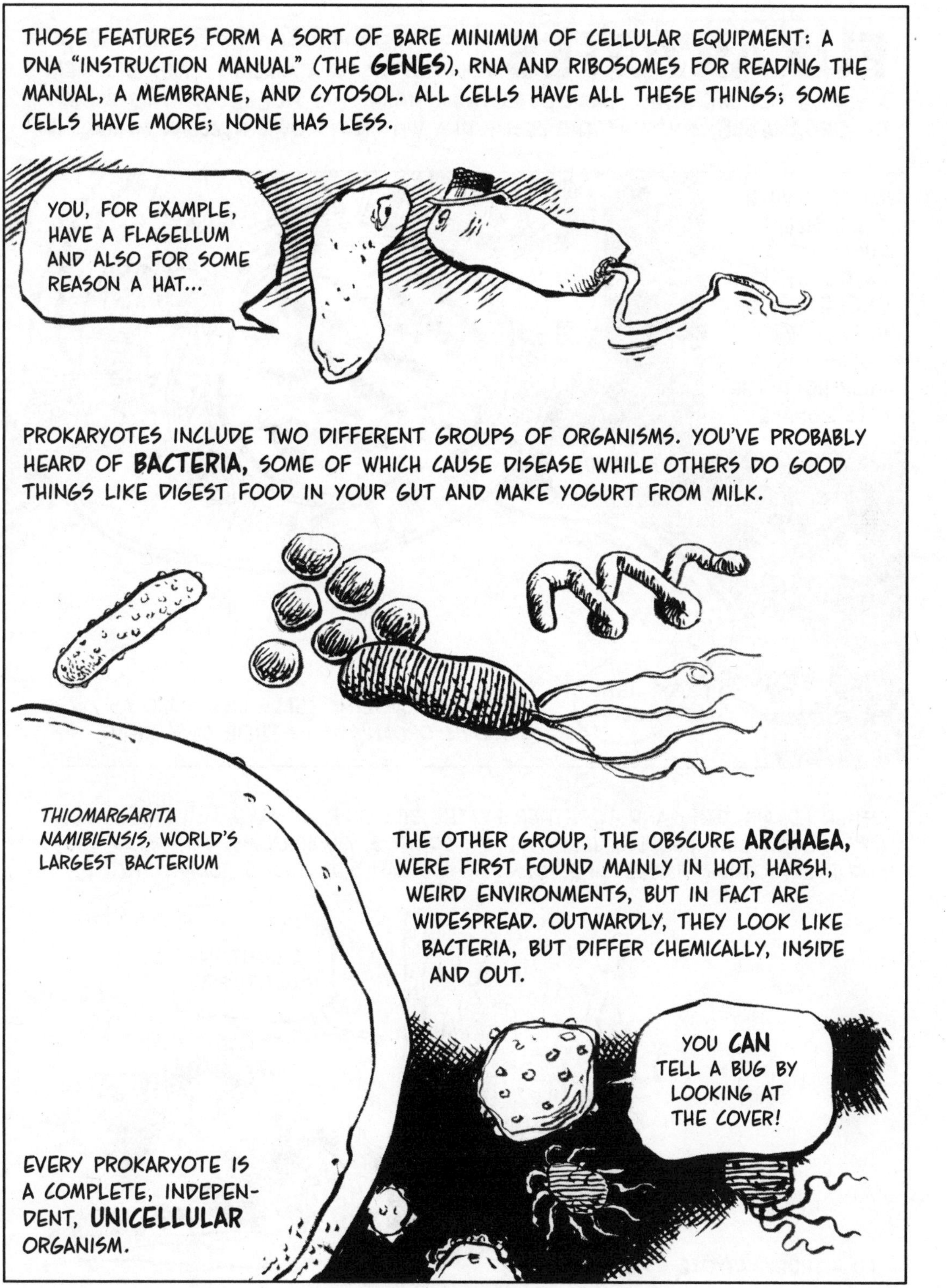
THOSE FEATURES FORM A SORT OF BARE MINIMUM OF CELLULAR EQUIPMENT: A DNA "INSTRUCTION MANUAL" (THE **GENES**), RNA AND RIBOSOMES FOR READING THE MANUAL, A MEMBRANE, AND CYTOSOL. ALL CELLS HAVE ALL THESE THINGS; SOME CELLS HAVE MORE; NONE HAS LESS.
YOU, FOR EXAMPLE, HAVE A FLAGELLUM AND ALSO FOR SOME REASON A HAT...
PROKARYOTES INCLUDE TWO DIFFERENT GROUPS OF ORGANISMS. YOU'VE PROBABLY HEARD OF **BACTERIA,** SOME OF WHICH CAUSE DISEASE WHILE OTHERS DO GOOD THINGS LIKE DIGEST FOOD IN YOUR GUT AND MAKE YOGURT FROM MILK.
THIOMARGARITA NAMIBIENSIS, WORLD'S LARGEST BACTERIUM
THE OTHER GROUP, THE OBSCURE **ARCHAEA,** WERE FIRST FOUND MAINLY IN HOT, HARSH, WEIRD ENVIRONMENTS, BUT IN FACT ARE WIDESPREAD. OUTWARDLY, THEY LOOK LIKE BACTERIA, BUT DIFFER CHEMICALLY, INSIDE AND OUT.
YOU **CAN** TELL A BUG BY LOOKING AT THE COVER!
EVERY PROKARYOTE IS A COMPLETE, INDEPENDENT, **UNICELLULAR** ORGANISM.

EUKARYOTES DWARF THEIR PROKARYOTIC COUSINS, WITH VOLUMES UP TO A MILLION TIMES GREATER. A SINGLE FEATURE GIVES EUKARYOTES THEIR NAME: THE **NUCLEUS.*** THIS BLOB, OR **ORGANELLE,** CONTAINS THE CELL'S DNA WITHIN ITS OWN NUCLEAR MEMBRANE.

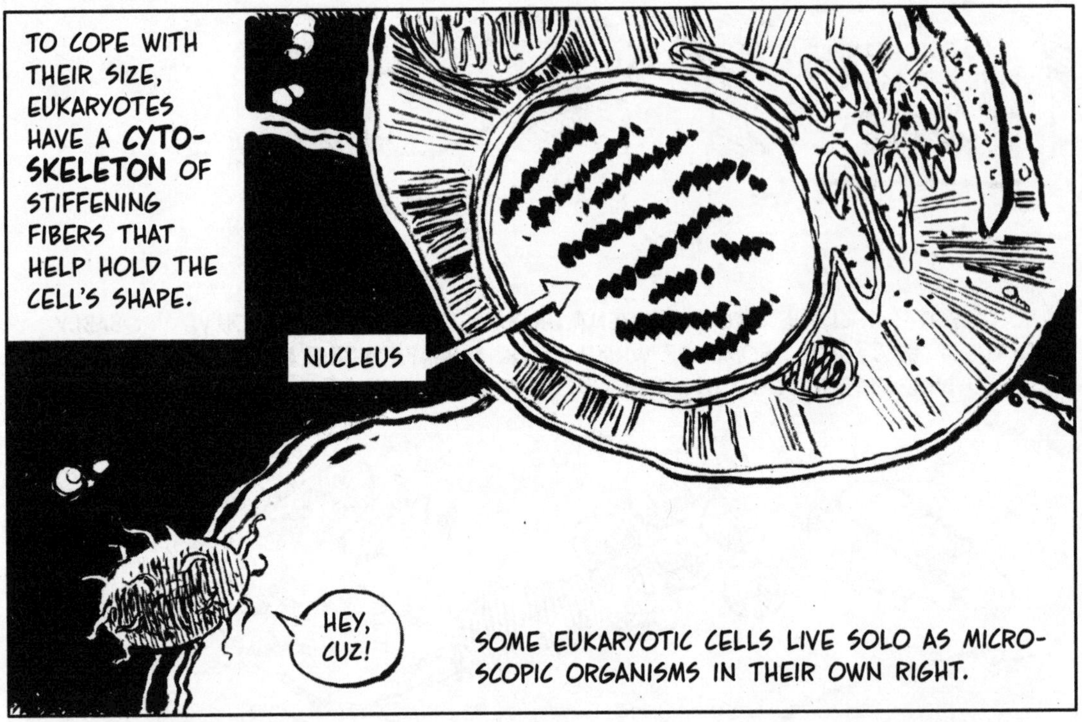
TO COPE WITH THEIR SIZE, EUKARYOTES HAVE A CYTO-SKELETON OF STIFFENING FIBERS THAT HELP HOLD THE CELL'S SHAPE.
NUCLEUS
HEY, CUZ!
SOME EUKARYOTIC CELLS LIVE SOLO AS MICRO-SCOPIC ORGANISMS IN THEIR OWN RIGHT.

OTHER EUKARYOTES BAND TOGETHER BY THE BILLIONS TO MAKE **MULTICELLULAR ORGANISMS** LIKE PEOPLE, JELLYFISH, MOSQUITOS, MUSHROOMS, DOGWOOD TREES, AND EVERY OTHER VISIBLE LIFE FORM. IF YOU CAN SEE IT, IT'S EUKARYOTIC.

I CONTAIN MULTITUDES!

*EU = TRUE. KARYOTE = KERNEL.

EUKARYOTIC CELLS HAVE A CRAZY VARIETY OF SHAPES AND SIZES, EVEN WITHIN A SINGLE BODY. THESE ARE ALL HUMAN CELLS.

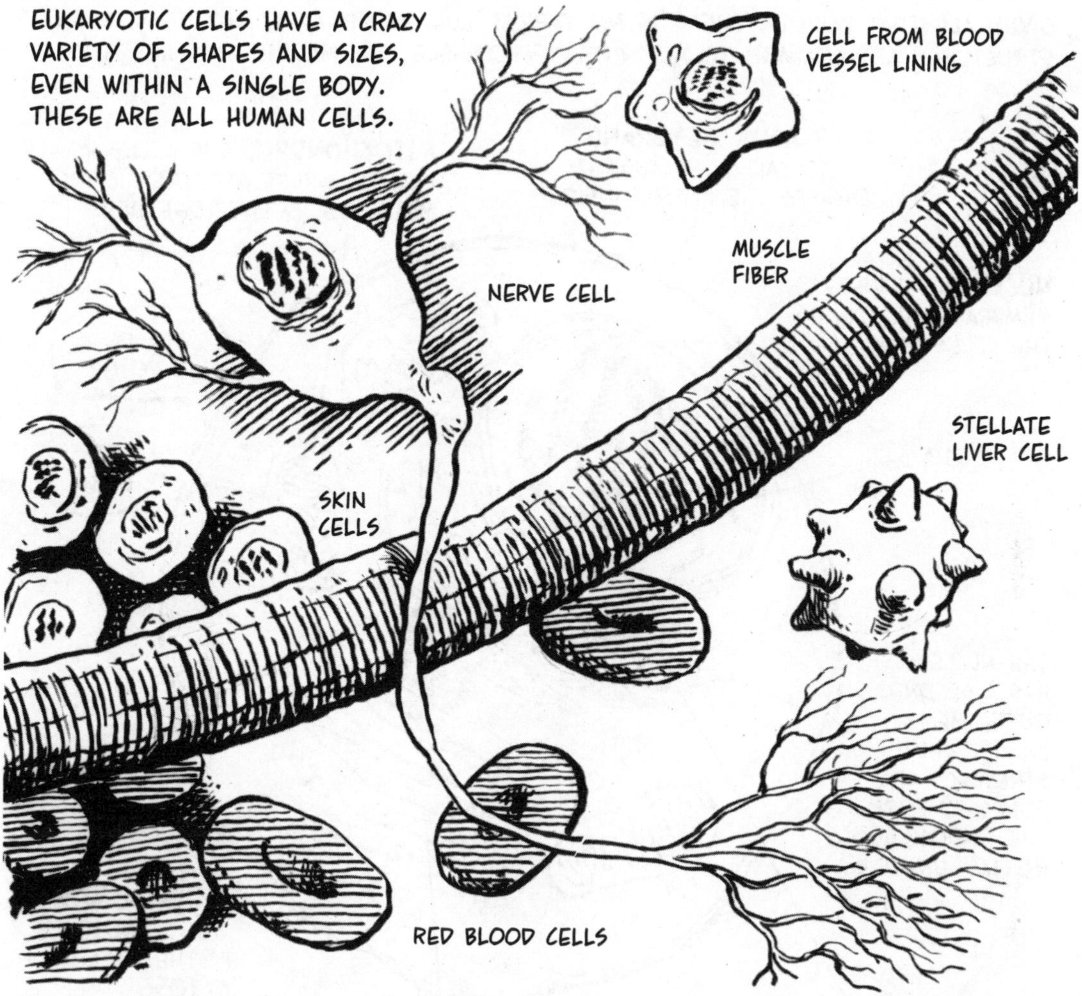

DESPITE THEIR DIFFERENCES, THEY ALL COME FROM A SINGLE SPHERICAL ANCESTOR, THE **FERTILIZED EGG.** THE EGG DIVIDES AGAIN AND AGAIN, AND THE GREAT-GREAT-...GREAT-GRANDDAUGHTERS TAKE ON SPECIALIZED SHAPES AND FUNCTIONS.

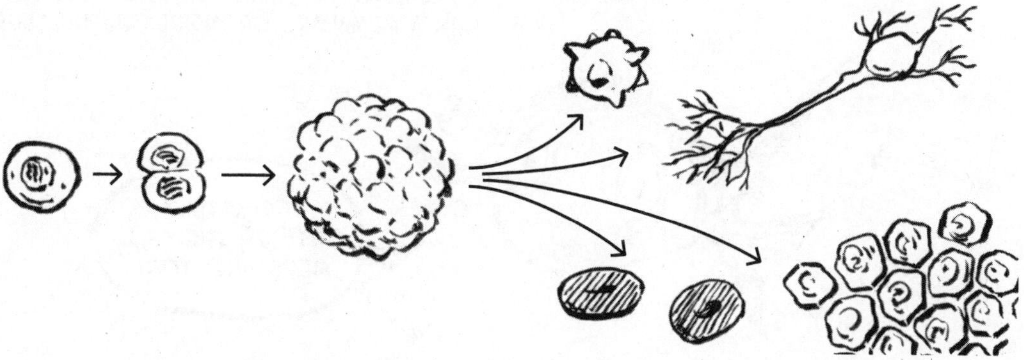

GIVEN ALL THAT VARIETY, THERE IS NO TYPICAL EUKARYOTIC CELL, BUT SOME BASIC STRUCTURES ARE COMMON TO ALL. (TRY, IF POSSIBLE, TO IMAGINE THIS IN 3-D.)

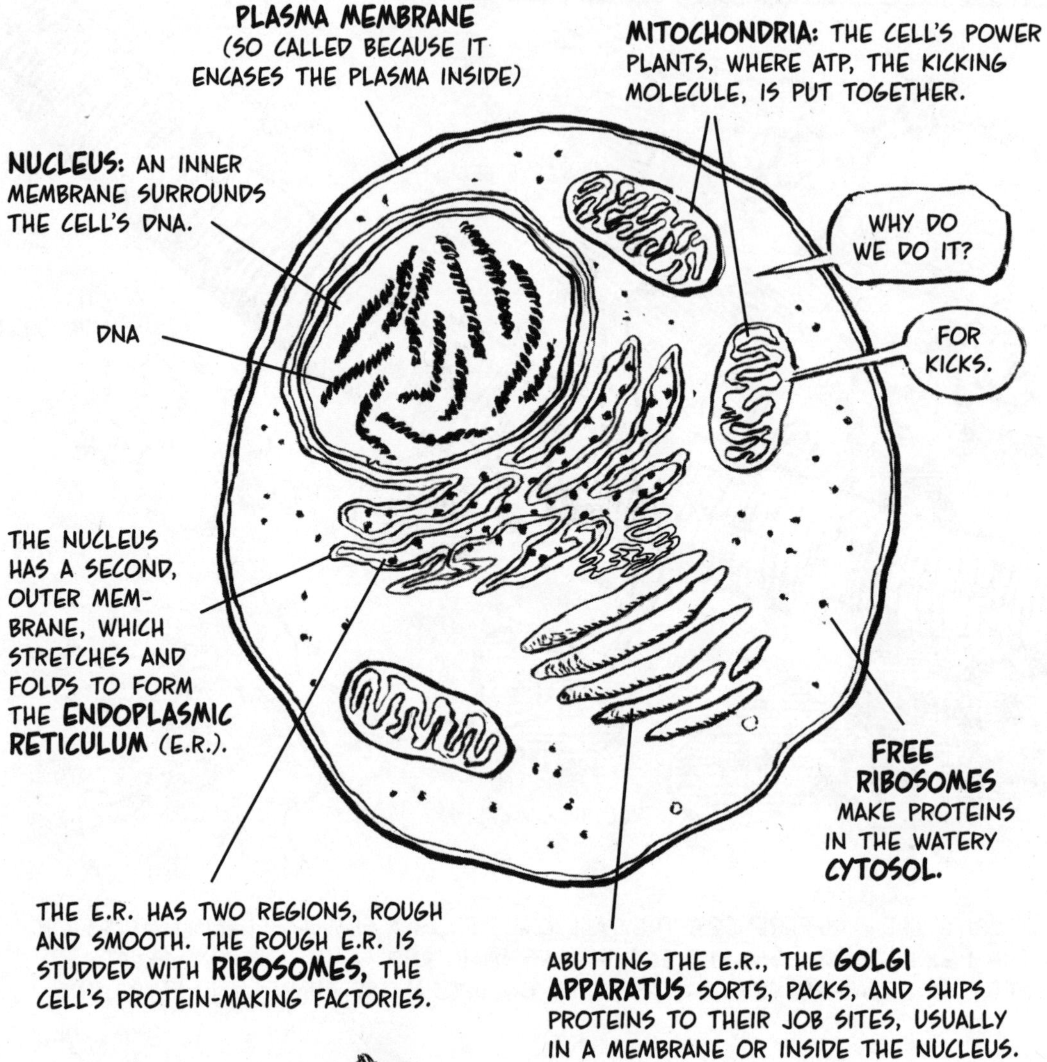

PLASMA MEMBRANE (SO CALLED BECAUSE IT ENCASES THE PLASMA INSIDE)
MITOCHONDRIA: THE CELL'S POWER PLANTS, WHERE ATP, THE KICKING MOLECULE, IS PUT TOGETHER.
NUCLEUS: AN INNER MEMBRANE SURROUNDS THE CELL'S DNA.
WHY DO WE DO IT?
DNA
FOR KICKS.
THE NUCLEUS HAS A SECOND, OUTER MEMBRANE, WHICH STRETCHES AND FOLDS TO FORM THE ENDOPLASMIC RETICULUM (E.R.).
FREE RIBOSOMES MAKE PROTEINS IN THE WATERY CYTOSOL.
THE E.R. HAS TWO REGIONS, ROUGH AND SMOOTH. THE ROUGH E.R. IS STUDDED WITH RIBOSOMES, THE CELL'S PROTEIN-MAKING FACTORIES.
ABUTTING THE E.R., THE GOLGI APPARATUS SORTS, PACKS, AND SHIPS PROTEINS TO THEIR JOB SITES, USUALLY IN A MEMBRANE OR INSIDE THE NUCLEUS.

IT'S WATERY OUTSIDE THE CELL MEMBRANE, TOO!

PLANTS, OUR PARTNERS IN (AND FOR) LIFE, HAVE A FEW EXTRA FEATURES THAT REFLECT PLANTS' SPECIAL ROLE AND UNIQUE CHALLENGES.

FOR EXAMPLE, UNLIKE ANIMALS, PLANTS CAN'T WALK TO WATER AND TAKE A DRINK. PLANTS FACE A WATER PROBLEM.

SO THEIR CELLS STORE WATER IN THE SPACE BETWEEN THEIR PLASMA MEMBRANE AND A SEMI-RIGID **CELL WALL** MADE OF CELLULOSE OR OTHER POLYSACCHARIDES.

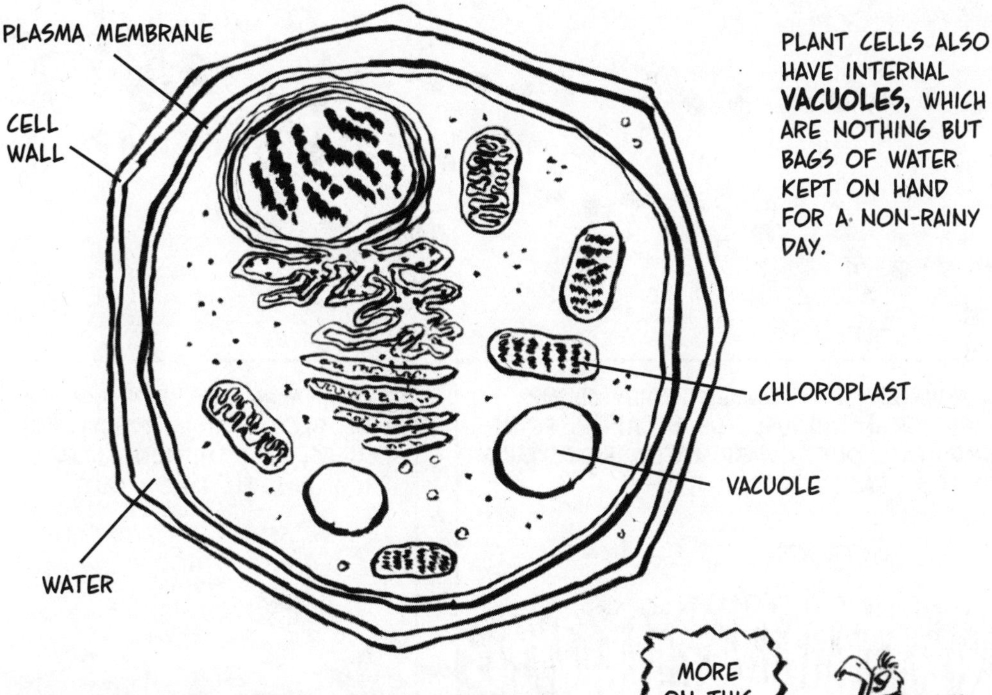

PLANT CELLS ALSO HAVE INTERNAL **VACUOLES,** WHICH ARE NOTHING BUT BAGS OF WATER KEPT ON HAND FOR A NON-RAINY DAY.

PLANTS ALSO HAVE A SPECIAL ABILITY: THEY CAN **EAT AIR.** MORE SPECIFICALLY, PLANTS TAKE IN **CARBON DIOXIDE GAS** AND USE IT TO MAKE GLUCOSE FUEL. BRIGHT GREEN **CHLOROPLASTS** HOLD THE EQUIPMENT FOR STARTING THAT PROCESS.

YOU MAY HAVE NOTICED HOW OFTEN PROTEINS WERE MENTIONED IN THOSE LAST FEW PAGES. THE CELL BUILDS THEM, MOVES THEM, AND USES THEM FOR—WHAT, EXACTLY?
NEARLY EVERYTHING.

BEGIN WITH THE PLASMA MEMBRANE. MOST OF THE MEMBRANE IS MADE OF FATTY, WATER-RESISTANT LIPIDS—BUT THESE LIPIDS HAVE ONE PHOSPHATE END ATTACHED TO A POLAR HEAD. **PHOSPHOLIPIDS** ARE PART WATER-FRIENDLY, PART WATER-HATING.

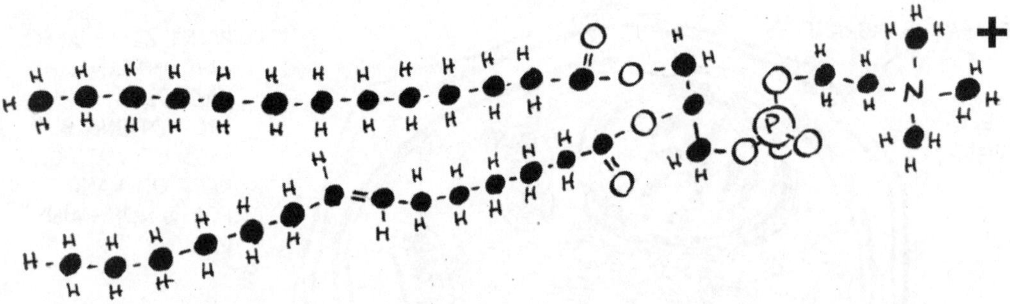

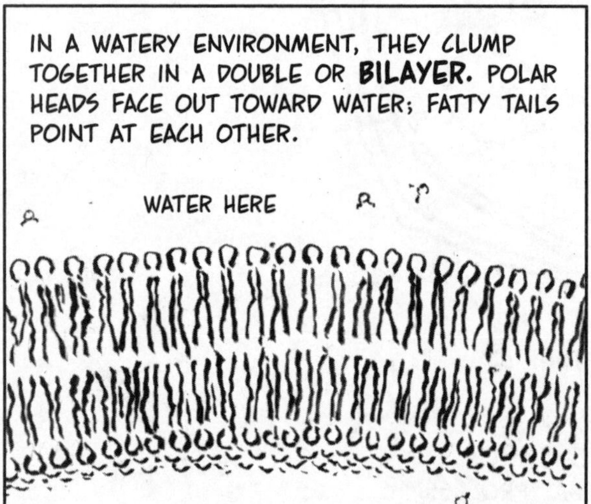
IN A WATERY ENVIRONMENT, THEY CLUMP TOGETHER IN A DOUBLE OR **BILAYER.** POLAR HEADS FACE OUT TOWARD WATER; FATTY TAILS POINT AT EACH OTHER.
WATER HERE
WATER HERE

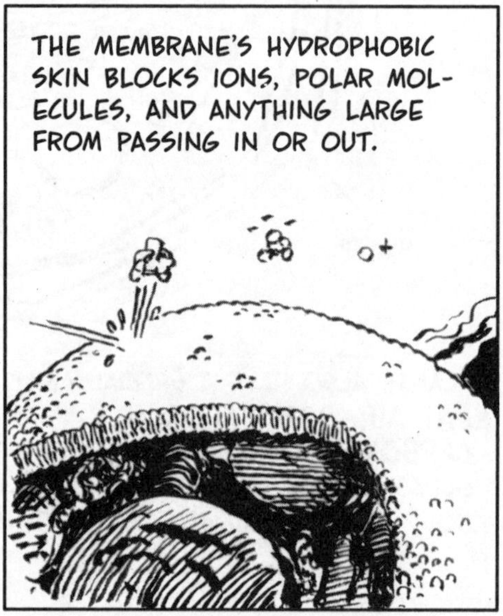
THE MEMBRANE'S HYDROPHOBIC SKIN BLOCKS IONS, POLAR MOLECULES, AND ANYTHING LARGE FROM PASSING IN OR OUT.

MANY PROTEINS DRIFT AROUND IN THE BILAYER. EACH PROTEIN HAS A SPECIFIC JOB: ATTACHING TO OTHER CELLS, SENSING CHEMICAL MESSAGES, MOVING MATERIAL IN AND OUT, AND MANY MORE.

MEMBRANE PROTEINS ACTIVELY MANAGE THE CELL'S CHEMISTRY TO PRESERVE **HOMEOSTASIS,** A STABLE YET DYNAMIC INTERIOR STATE BEST SUITED TO LIFE.

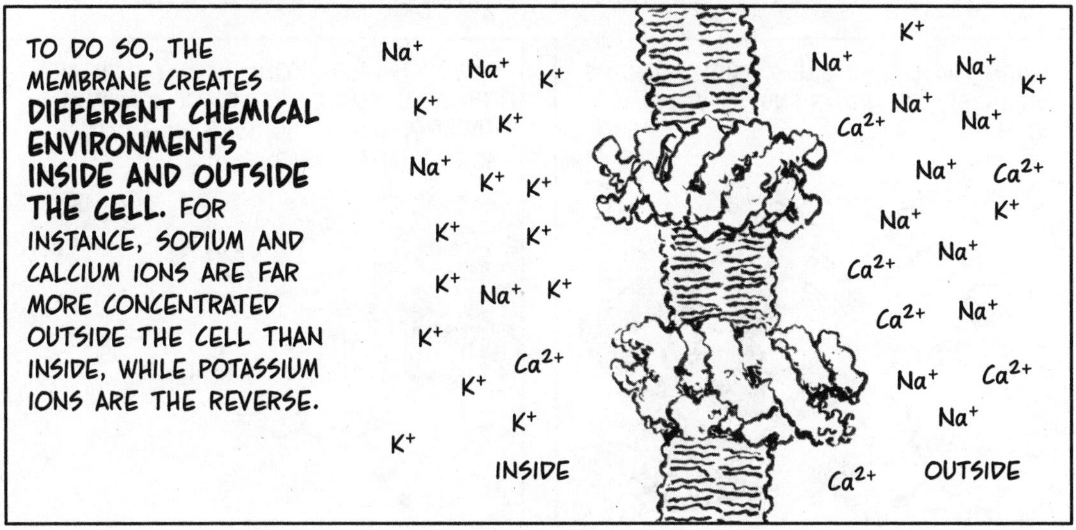
TO DO SO, THE MEMBRANE CREATES **DIFFERENT CHEMICAL ENVIRONMENTS INSIDE AND OUTSIDE THE CELL.** FOR INSTANCE, SODIUM AND CALCIUM IONS ARE FAR MORE CONCENTRATED OUTSIDE THE CELL THAN INSIDE, WHILE POTASSIUM IONS ARE THE REVERSE.
Na^+ Na^+ K^+ K^+ K^+ Na^+ K^+ K^+ K^+ K^+ K^+ Na^+ K^+ K^+ Ca^{2+} K^+ K^+ K^+
INSIDE
K^+ Na^+ Na^+ K^+ Na^+ Ca^{2+} Na^+ Na^+ Ca^{2+} Na^+ K^+ Ca^{2+} Na^+ Ca^{2+} Na^+ Na^+ Ca^{2+} Na^+ Ca^{2+}
OUTSIDE

CHANNELS AND PUMPS

SOME MEMBRANE PROTEINS ARE LIKE PIPES THAT ARE OPEN TO JUST **ONE KIND** OF ION OR MOLECULE. **AQUAPORINS,** FOR EXAMPLE, LET WATER AND ONLY WATER FLOW FREELY IN AND OUT.

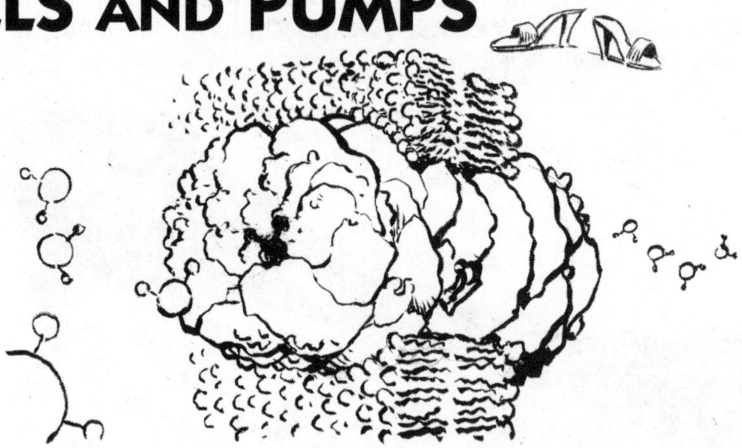

ION CHANNELS, BY CONTRAST, ARE PLUGGED OR **GATED** WITH A LID THAT IS USUALLY CLOSED.
Na+

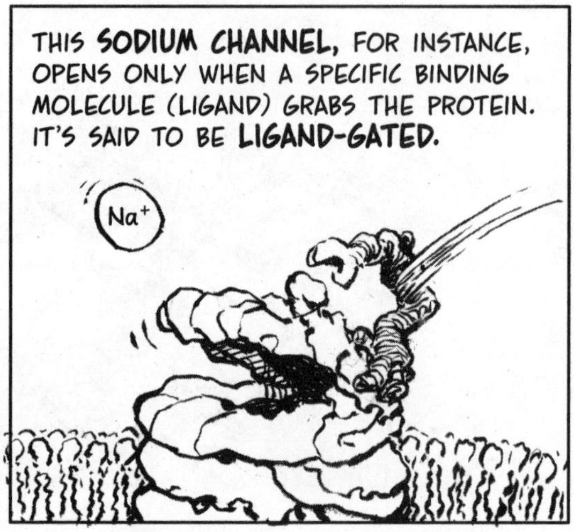
THIS **SODIUM CHANNEL,** FOR INSTANCE, OPENS ONLY WHEN A SPECIFIC BINDING MOLECULE (LIGAND) GRABS THE PROTEIN. IT'S SAID TO BE **LIGAND-GATED.**
Na+

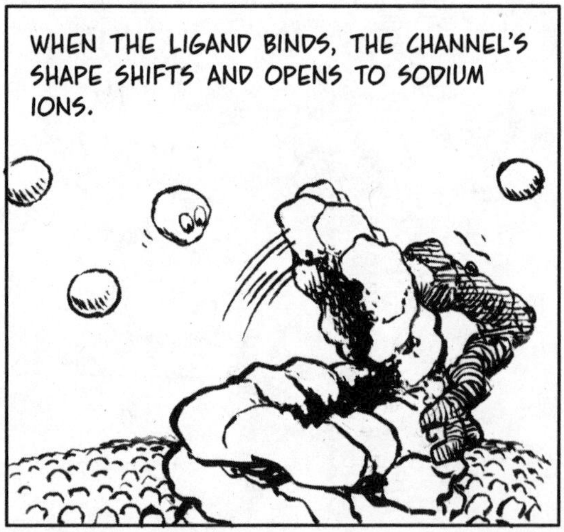
WHEN THE LIGAND BINDS, THE CHANNEL'S SHAPE SHIFTS AND OPENS TO SODIUM IONS.

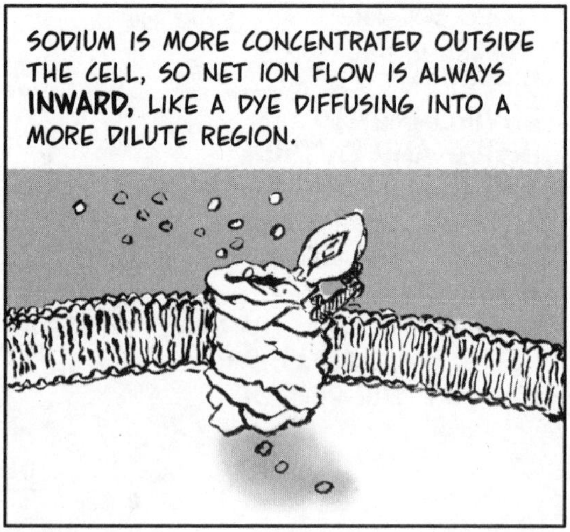
SODIUM IS MORE CONCENTRATED OUTSIDE THE CELL, SO NET ION FLOW IS ALWAYS **INWARD,** LIKE A DYE DIFFUSING INTO A MORE DILUTE REGION.

SODIUM-POTASSIUM PUMP

SODIUM IONS WON'T FLOW OUTWARD ON THEIR OWN, JUST AS DYE WON'T "UNDIFFUSE." THE CELL NEEDS TO **PUMP** SODIUM "UPHILL" AGAINST ITS "CONCENTRATION GRADIENT." THIS TAKES **WORK.**

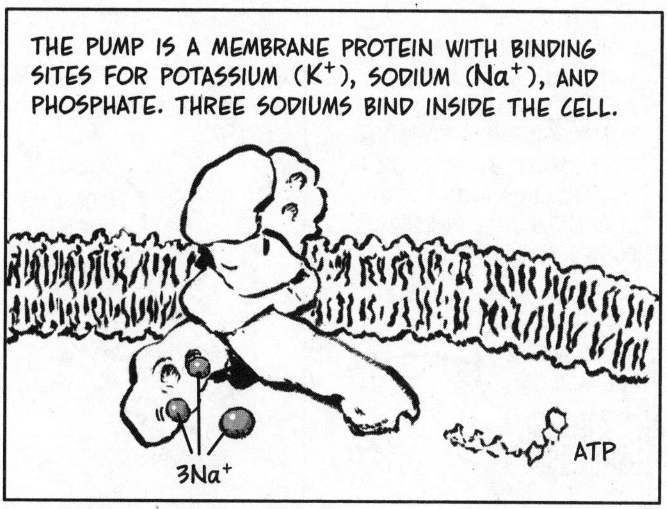
THE PUMP IS A MEMBRANE PROTEIN WITH BINDING SITES FOR POTASSIUM (K^+), SODIUM (Na^+), AND PHOSPHATE. THREE SODIUMS BIND INSIDE THE CELL.
$3Na^+$
ATP

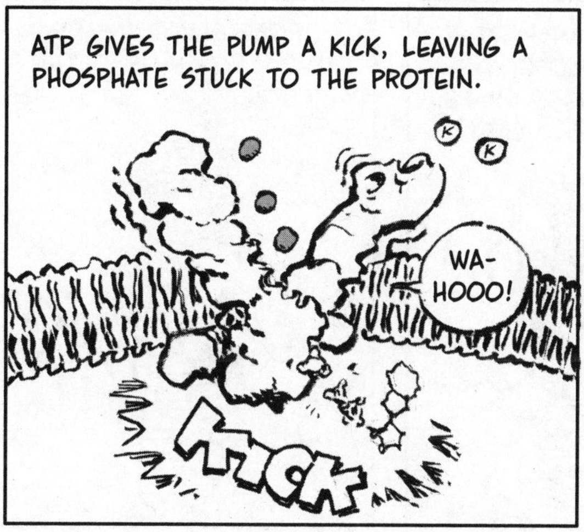

ATP GIVES THE PUMP A KICK, LEAVING A PHOSPHATE STUCK TO THE PROTEIN.

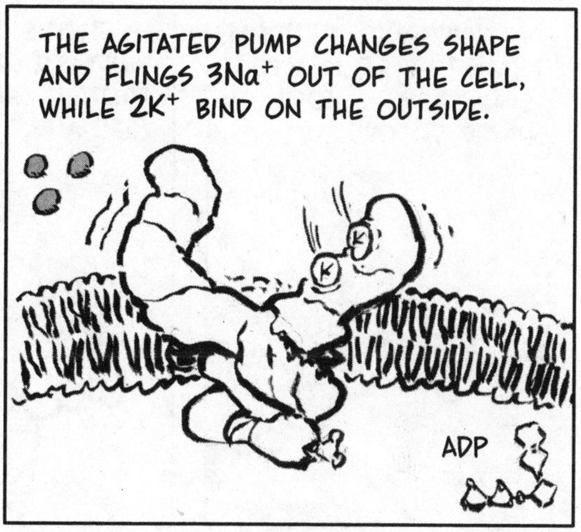
THE AGITATED PUMP CHANGES SHAPE AND FLINGS $3Na^+$ OUT OF THE CELL, WHILE $2K^+$ BIND ON THE OUTSIDE.
ADP

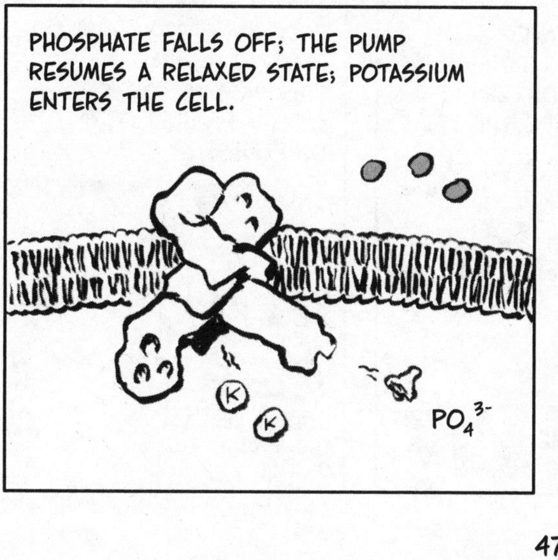
PHOSPHATE FALLS OFF; THE PUMP RESUMES A RELAXED STATE; POTASSIUM ENTERS THE CELL.
PO_4^{3-}

BECAUSE $3Na^+$ GO OUT FOR EVERY $2K^+$ THAT COME IN, THIS PUMP CAUSES A **CHARGE IMBALANCE** ACROSS THE MEMBRANE.

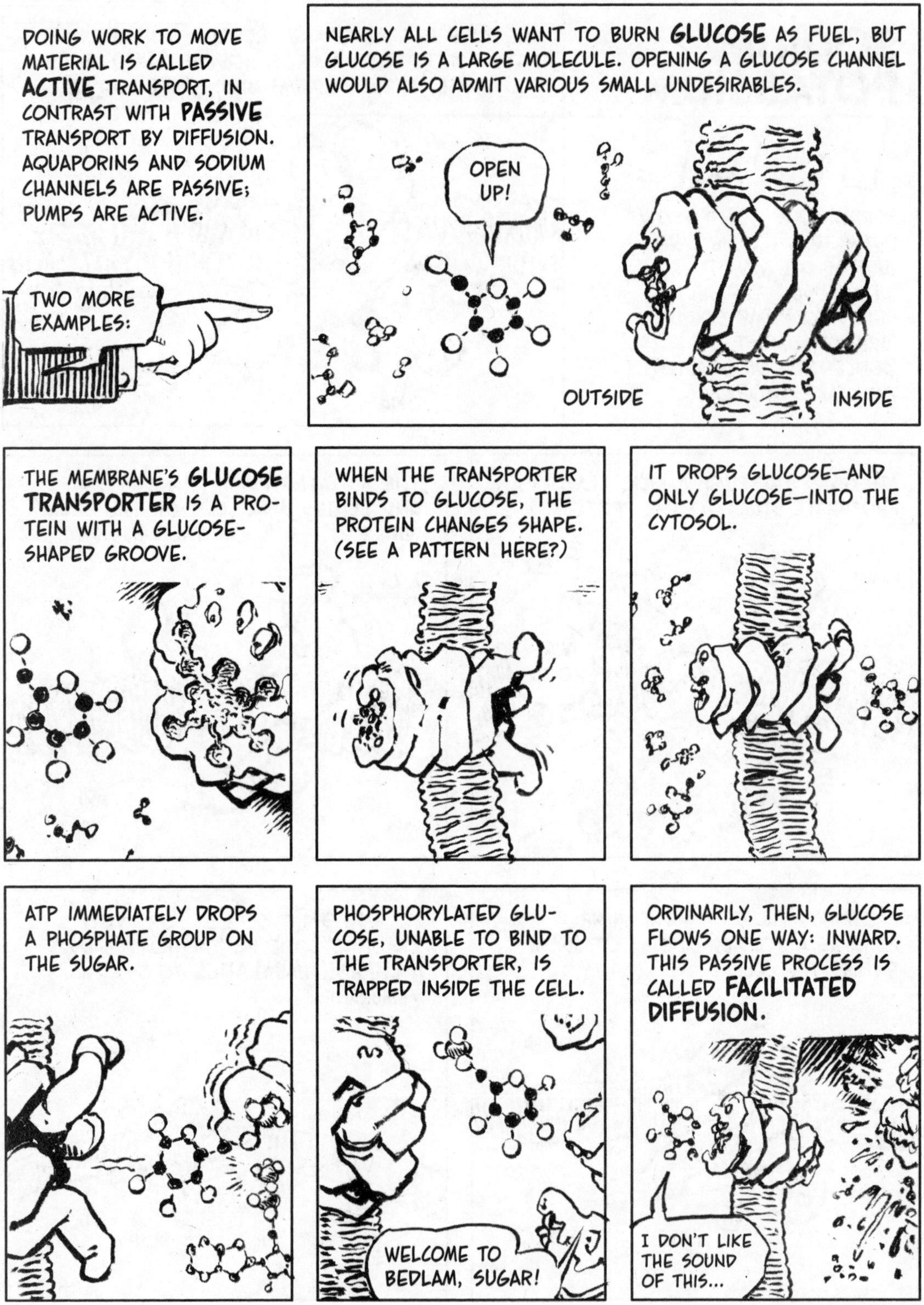
DOING WORK TO MOVE MATERIAL IS CALLED ACTIVE TRANSPORT, IN CONTRAST WITH PASSIVE TRANSPORT BY DIFFUSION. AQUAPORINS AND SODIUM CHANNELS ARE PASSIVE; PUMPS ARE ACTIVE.
TWO MORE EXAMPLES:
NEARLY ALL CELLS WANT TO BURN GLUCOSE AS FUEL, BUT GLUCOSE IS A LARGE MOLECULE. OPENING A GLUCOSE CHANNEL WOULD ALSO ADMIT VARIOUS SMALL UNDESIRABLES.
OPEN UP!
OUTSIDE
INSIDE
THE MEMBRANE'S GLUCOSE TRANSPORTER IS A PROTEIN WITH A GLUCOSE-SHAPED GROOVE.
WHEN THE TRANSPORTER BINDS TO GLUCOSE, THE PROTEIN CHANGES SHAPE. (SEE A PATTERN HERE?)
IT DROPS GLUCOSE—AND ONLY GLUCOSE—INTO THE CYTOSOL.
ATP IMMEDIATELY DROPS A PHOSPHATE GROUP ON THE SUGAR.
PHOSPHORYLATED GLUCOSE, UNABLE TO BIND TO THE TRANSPORTER, IS TRAPPED INSIDE THE CELL.
WELCOME TO BEDLAM, SUGAR!
ORDINARILY, THEN, GLUCOSE FLOWS ONE WAY: INWARD. THIS PASSIVE PROCESS IS CALLED FACILITATED DIFFUSION.
I DON'T LIKE THE SOUND OF THIS...

Endocytosis and Exocytosis

ARE HIGHLY ACTIVE TRANSPORT PROCESSES. THE NAMES MEAN "INTO CELL" AND "OUT OF CELL."

BIOLOGISTS AND THEIR DICTIONARIES!

ENDOCYTOSIS BEGINS WHEN WORKER PROTEINS, POWERED BY ATP KICKS, PULL A REGION OF THE PLASMA MEMBRANE INWARD.

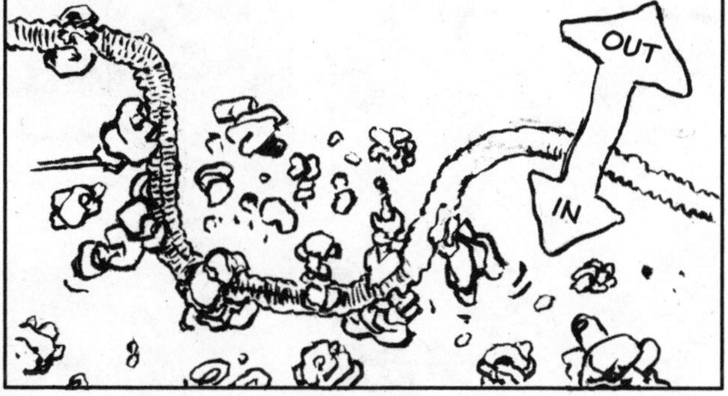
OUT
IN

WHATEVER WAS BOUND TO THE OUTER SURFACE THERE IS SEALED INSIDE A PACKAGE, OR **VESICLE.**

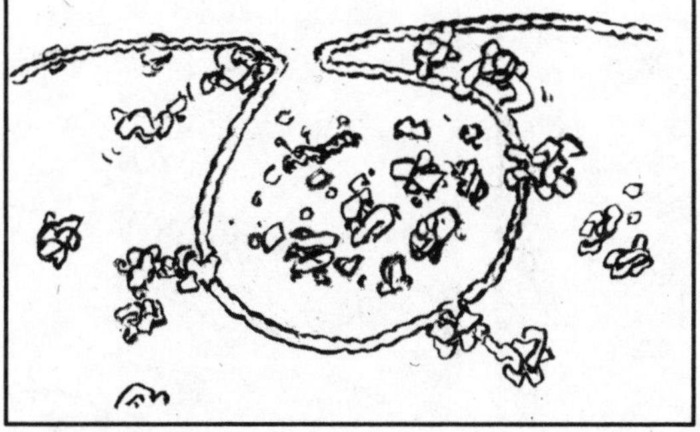

THE VESICLE BREAKS FREE OF THE MEMBRANE AND TRAVELS INTO THE CELL.

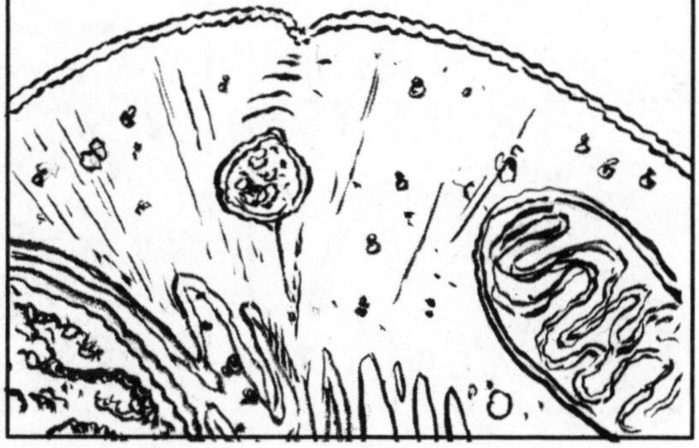

DEEP INSIDE, THE VESICLE'S MEMBRANE IS BROKEN UP AND ITS CONTENTS MADE AVAILABLE.

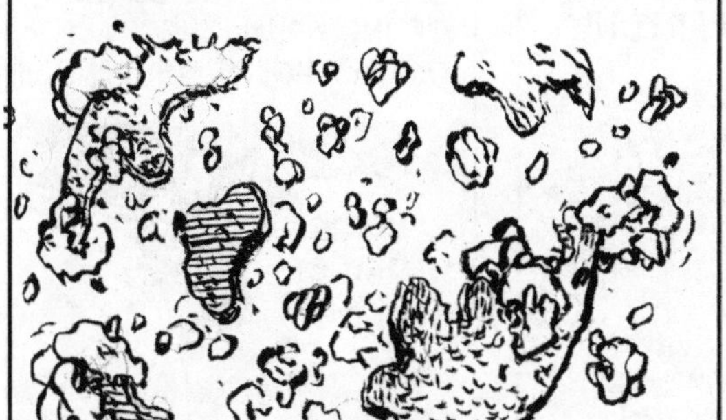

EXOCYTOSIS IS THE REVERSE PROCESS: INSIDE THE CELL, A VESICLE CARRIES CARGO TO THE PLASMA MEMBRANE AND RELEASES THE CARGO OUTSIDE THE CELL.

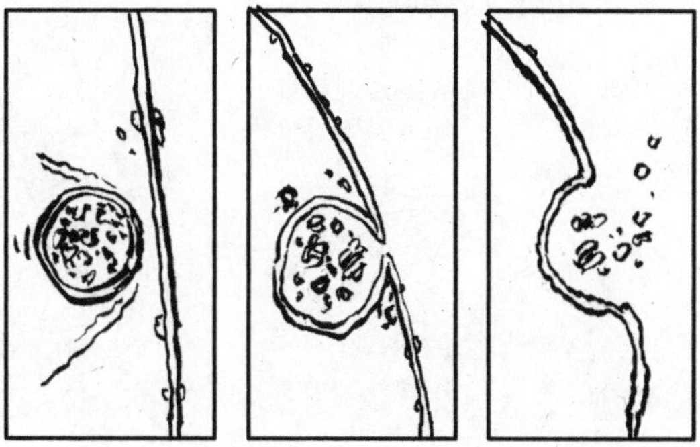

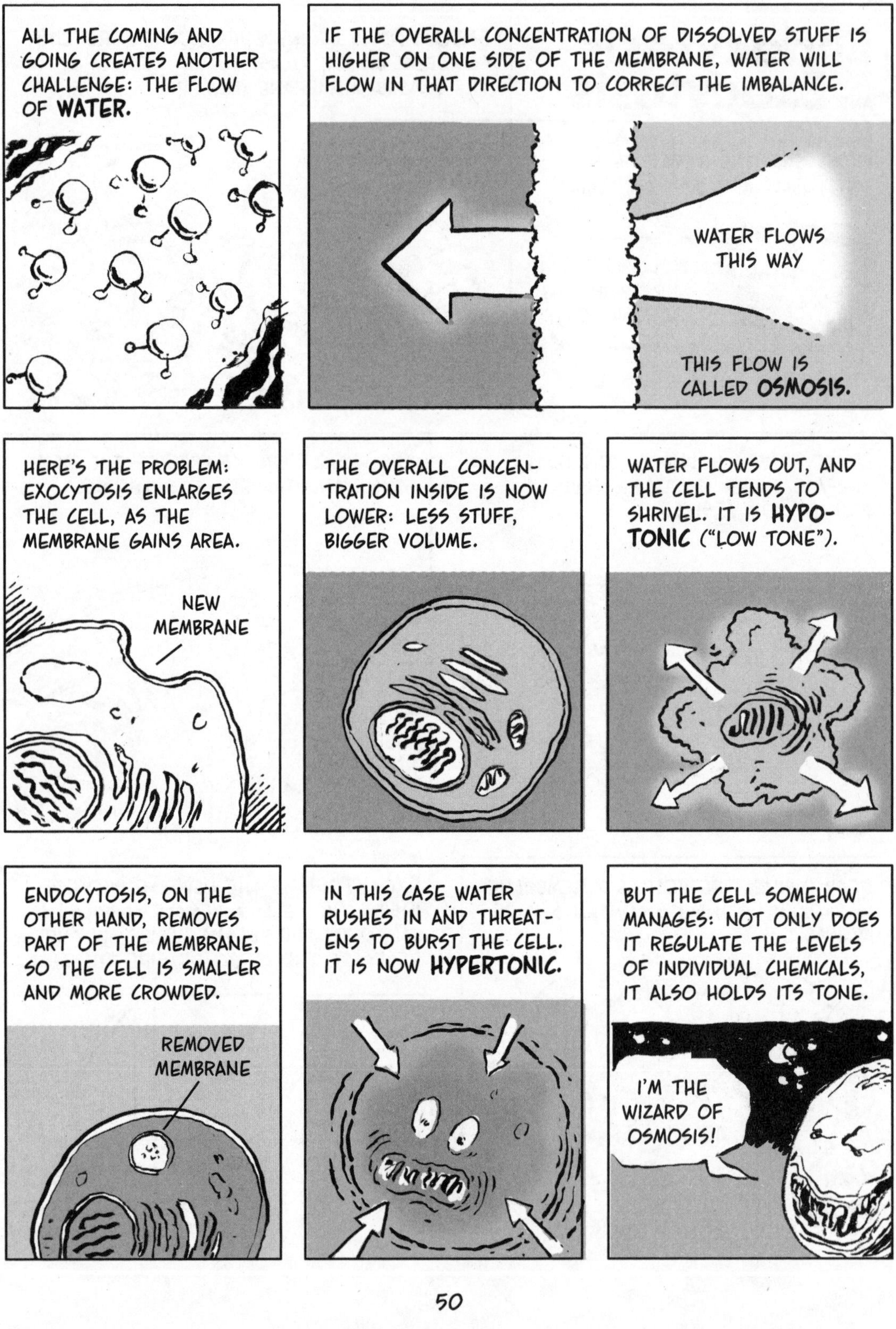
ALL THE COMING AND GOING CREATES ANOTHER CHALLENGE: THE FLOW OF WATER.
IF THE OVERALL CONCENTRATION OF DISSOLVED STUFF IS HIGHER ON ONE SIDE OF THE MEMBRANE, WATER WILL FLOW IN THAT DIRECTION TO CORRECT THE IMBALANCE.
WATER FLOWS THIS WAY
THIS FLOW IS CALLED OSMOSIS.
HERE'S THE PROBLEM: EXOCYTOSIS ENLARGES THE CELL, AS THE MEMBRANE GAINS AREA.
NEW MEMBRANE
THE OVERALL CONCENTRATION INSIDE IS NOW LOWER: LESS STUFF, BIGGER VOLUME.
WATER FLOWS OUT, AND THE CELL TENDS TO SHRIVEL. IT IS HYPOTONIC ("LOW TONE").
ENDOCYTOSIS, ON THE OTHER HAND, REMOVES PART OF THE MEMBRANE, SO THE CELL IS SMALLER AND MORE CROWDED.
REMOVED MEMBRANE
IN THIS CASE WATER RUSHES IN AND THREATENS TO BURST THE CELL. IT IS NOW HYPERTONIC.
BUT THE CELL SOMEHOW MANAGES: NOT ONLY DOES IT REGULATE THE LEVELS OF INDIVIDUAL CHEMICALS, IT ALSO HOLDS ITS TONE.
I'M THE WIZARD OF OSMOSIS!

MOLECULES COMING INTO THE CELL BREAK UP AND REASSEMBLE IN NEW WAYS. THE OUTPUT OF ONE REACTION BECOMES THE INPUT FOR OTHERS. TAKEN TOGETHER, THIS WEB OF CHEMISTRY IS CALLED THE CELL'S

METABOLISM.

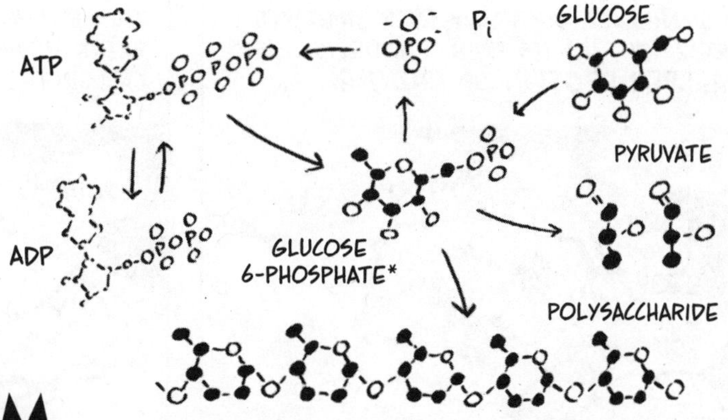
ATP
ADP
P_i
GLUCOSE
PYRUVATE
GLUCOSE 6-PHOSPHATE*
POLYSACCHARIDE

*THE "6" TELLS US WHICH OF GLUCOSE'S CARBON ATOMS IS ATTACHED TO THE PHOSPHATE GROUP.

JUST AS YOUR BODY BREAKS DOWN FOOD WHEN YOU EAT IT, A CELL BREAKS DOWN NEWLY ARRIVED LARGE MOLECULES INTO BITS. REACTIONS THAT TAKE THINGS APART ARE CALLED **CATABOLIC.**

LET'S GET CRACKING!
OOPS!

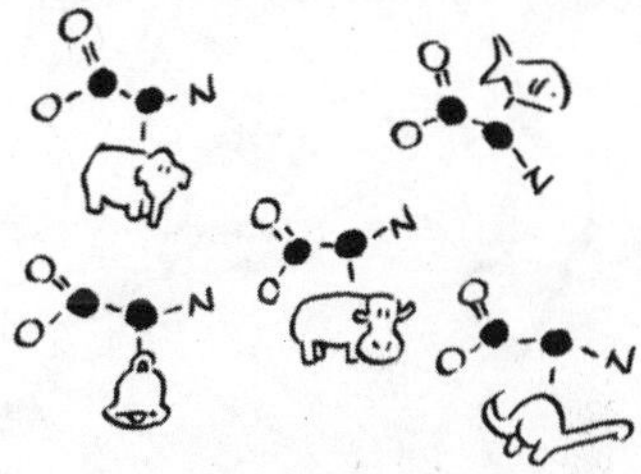

INGESTED PROTEINS BREAK DOWN TO AMINO ACIDS.

THE CELL ALSO BUILDS LARGER MOLECULES FROM SMALL STUFF: THESE REACTIONS ARE CALLED **ANABOLIC.**

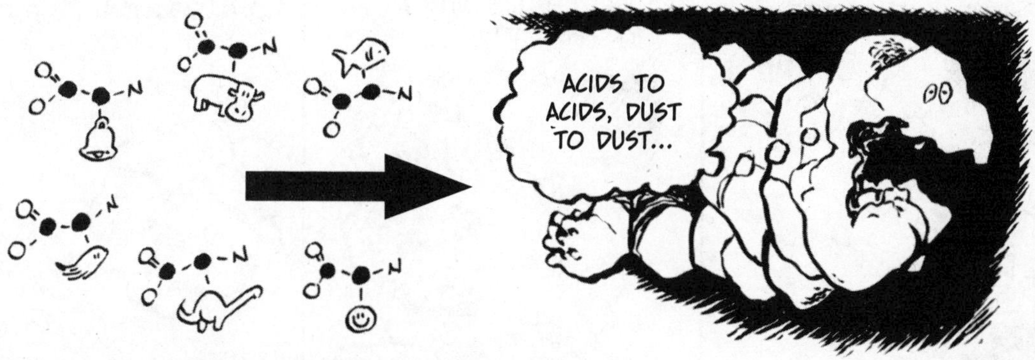
ACIDS TO ACIDS, DUST TO DUST...

AMINO ACIDS ASSEMBLE INTO NEW PROTEINS.

REMARKABLY, VIRTUALLY EVERY METABOLIC REACTION HAS ITS OWN UNIQUE, SPECIFIC **HELPER PROTEIN** OR **ENZYME**.

THIS ENZYME, FOR EXAMPLE, HELPS TO BREAK DOWN ONE PARTICULAR MOLECULE, CALLED THE ENZYME'S **SUBSTRATE**.

BINDING THE SUBSTRATE, THE ENZYME CHANGES SHAPE AND CRACKS THE SMALL MOLECULE INTO PIECES.

THESE FRAGMENTS SWIM AWAY, AND THE ENZYME REVERTS TO ITS ORIGINAL SHAPE UNTIL ANOTHER SUBSTRATE MOLECULE COMES ALONG.

ANABOLIC ENZYMES HAVE SITES TO BIND MULTIPLE MOLECULES AT ONCE.

THE SUBSTRATES BIND, AND THE PROTEIN COMBINES THEM, USUALLY WITH A KICK FROM ATP.

ENZYMES ARE ALL-IMPORTANT. NO ENZYMES, NO METABOLISM!

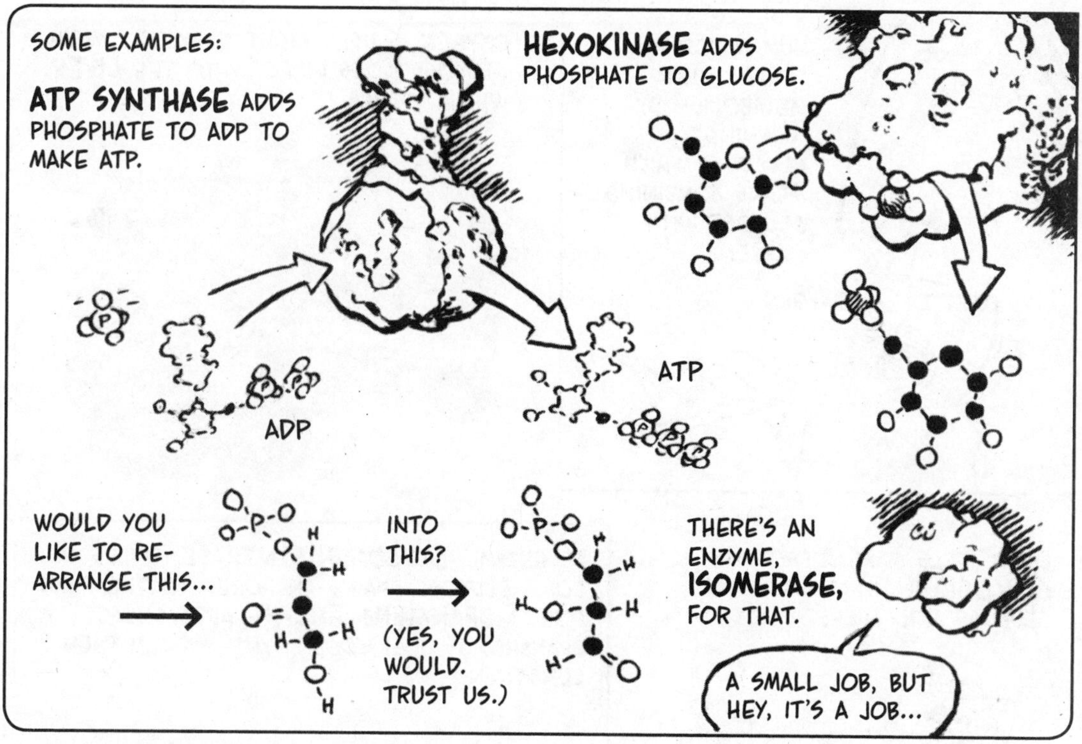
SOME EXAMPLES:
ATP SYNTHASE ADDS PHOSPHATE TO ADP TO MAKE ATP.
ADP
ATP
HEXOKINASE ADDS PHOSPHATE TO GLUCOSE.
WOULD YOU LIKE TO RE-ARRANGE THIS...
INTO THIS?
(YES, YOU WOULD. TRUST US.)
THERE'S AN ENZYME, ISOMERASE, FOR THAT.
A SMALL JOB, BUT HEY, IT'S A JOB...

PEPSIN BREAKS PROTEIN CHAINS. ONE OF THE FIRST ENZYMES TO BE DISCOVERED—IN 1836, BEFORE ANYONE THOUGHT OF PUTTING "ASE" AT THE END OF EVERY ENZYME'S NAME—PEPSIN IS SECRETED INTO THE STOMACH BY SPECIALIZED **GASTRIC CHIEF CELLS.**

CANNIBAL!
SNAP!

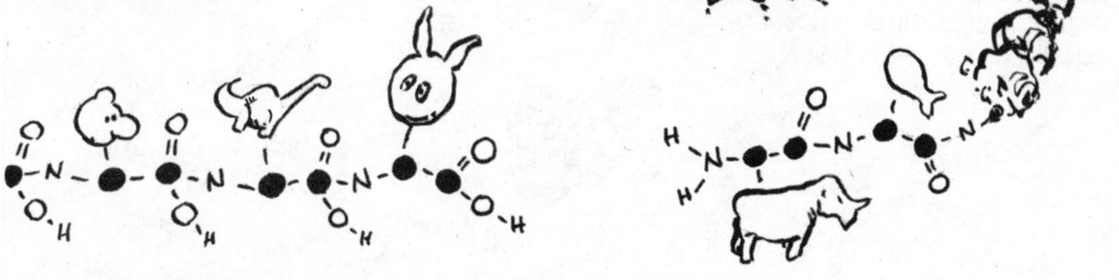

THE **PEPTIDE BOND** BETWEEN AMINO ACIDS WAS NAMED AFTER PEPSIN.

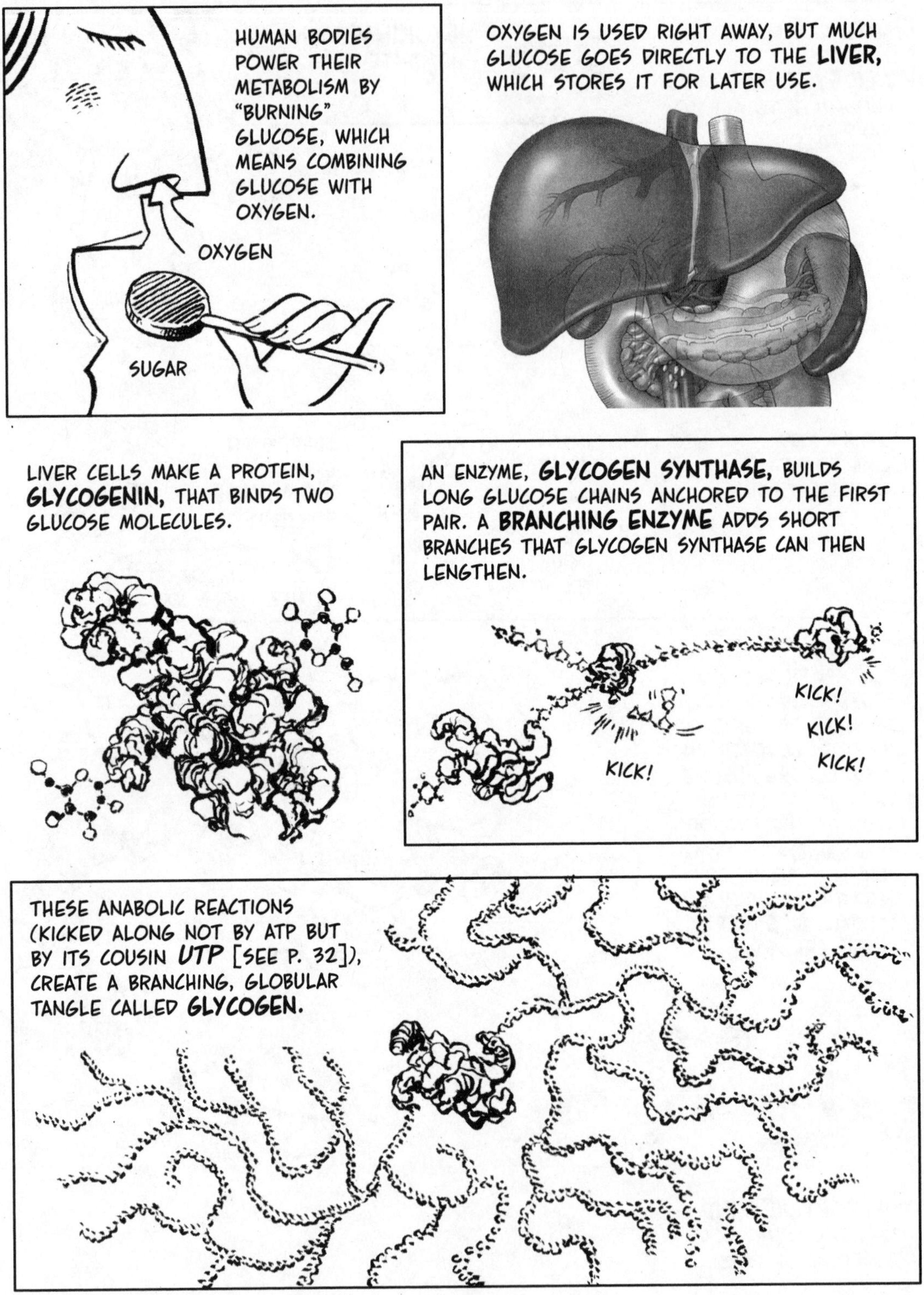
HUMAN BODIES POWER THEIR METABOLISM BY "BURNING" GLUCOSE, WHICH MEANS COMBINING GLUCOSE WITH OXYGEN.
OXYGEN
SUGAR
OXYGEN IS USED RIGHT AWAY, BUT MUCH GLUCOSE GOES DIRECTLY TO THE **LIVER,** WHICH STORES IT FOR LATER USE.
LIVER CELLS MAKE A PROTEIN, **GLYCOGENIN,** THAT BINDS TWO GLUCOSE MOLECULES.
AN ENZYME, **GLYCOGEN SYNTHASE,** BUILDS LONG GLUCOSE CHAINS ANCHORED TO THE FIRST PAIR. A **BRANCHING ENZYME** ADDS SHORT BRANCHES THAT GLYCOGEN SYNTHASE CAN THEN LENGTHEN.
KICK!
KICK!
KICK!
KICK!
THESE ANABOLIC REACTIONS (KICKED ALONG NOT BY ATP BUT BY ITS COUSIN ***UTP*** [SEE P. 32]), CREATE A BRANCHING, GLOBULAR TANGLE CALLED **GLYCOGEN.**

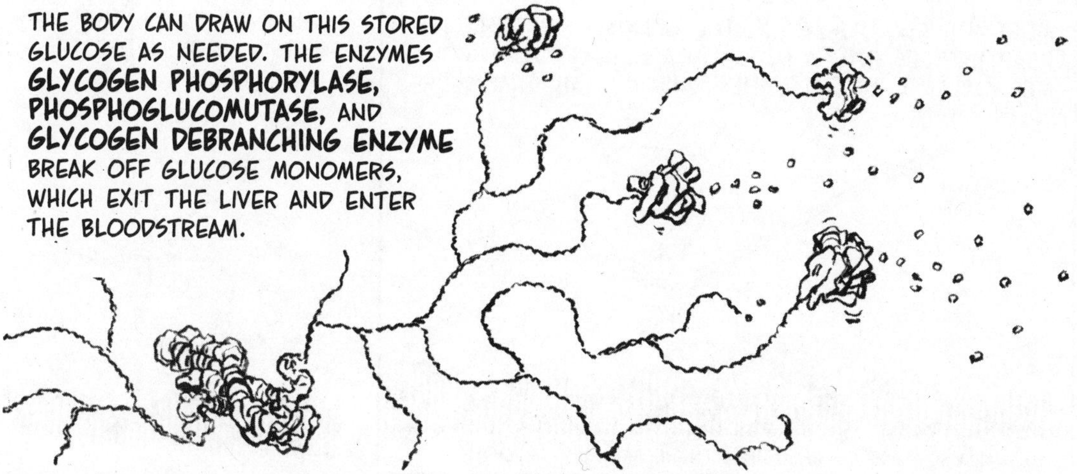
THE BODY CAN DRAW ON THIS STORED GLUCOSE AS NEEDED. THE ENZYMES **GLYCOGEN PHOSPHORYLASE, PHOSPHOGLUCOMUTASE,** AND **GLYCOGEN DEBRANCHING ENZYME** BREAK OFF GLUCOSE MONOMERS, WHICH EXIT THE LIVER AND ENTER THE BLOODSTREAM.

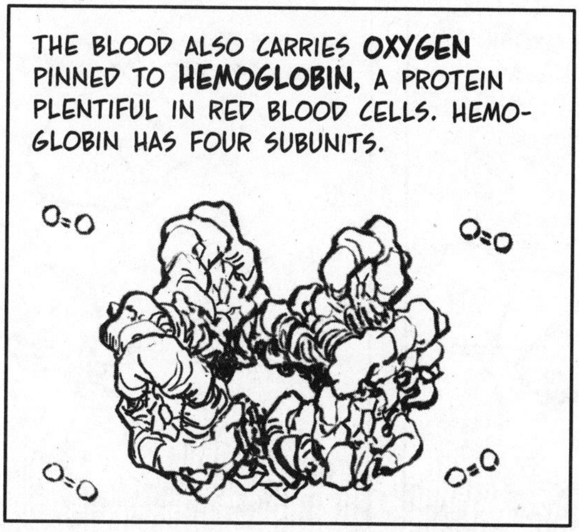
THE BLOOD ALSO CARRIES **OXYGEN** PINNED TO **HEMOGLOBIN,** A PROTEIN PLENTIFUL IN RED BLOOD CELLS. HEMOGLOBIN HAS FOUR SUBUNITS.
O=O
O=O
O=O
O=O

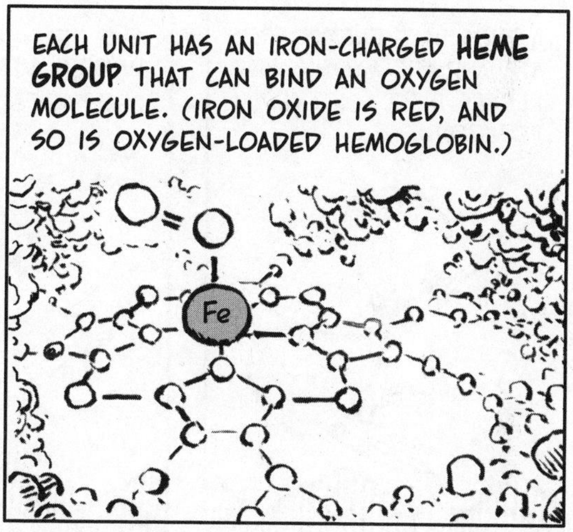
EACH UNIT HAS AN IRON-CHARGED **HEME GROUP** THAT CAN BIND AN OXYGEN MOLECULE. (IRON OXIDE IS RED, AND SO IS OXYGEN-LOADED HEMOGLOBIN.)
Fe

THE BLOOD CARRIES OXYGEN AND SUGAR TO THE HUNGRY CELLS THAT NEED THEM.

WE'LL DESCRIBE GLUCOSE OXIDATION IN CHAPTER 6. FOR NOW, LET'S LOOK AT HOW CELLS TURN ENERGY INTO **DIRECTED MOVEMENT.**
YES!!

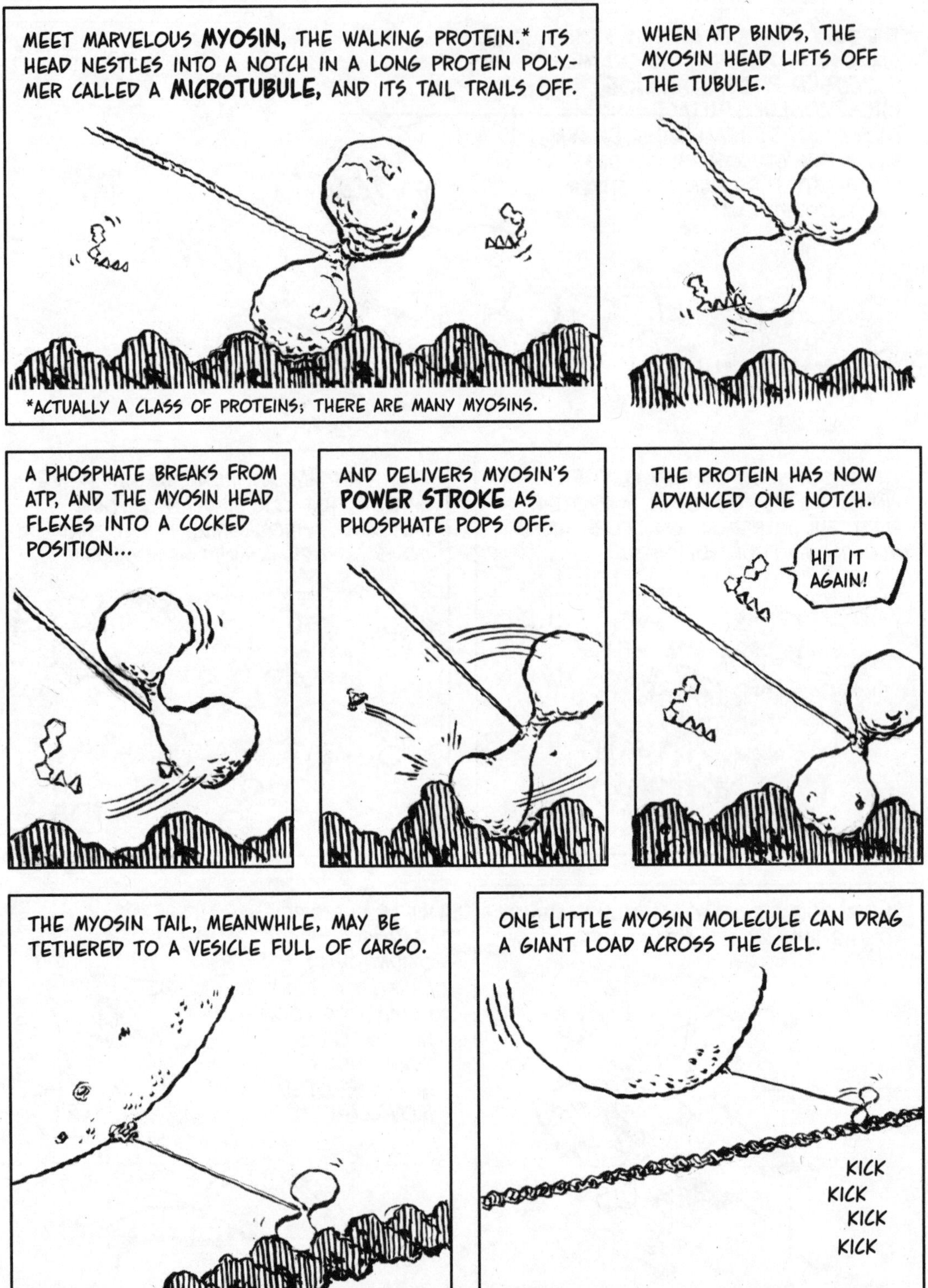
MEET MARVELOUS **MYOSIN,** THE WALKING PROTEIN.* ITS HEAD NESTLES INTO A NOTCH IN A LONG PROTEIN POLYMER CALLED A **MICROTUBULE,** AND ITS TAIL TRAILS OFF.
*ACTUALLY A CLASS OF PROTEINS; THERE ARE MANY MYOSINS.
WHEN ATP BINDS, THE MYOSIN HEAD LIFTS OFF THE TUBULE.
A PHOSPHATE BREAKS FROM ATP, AND THE MYOSIN HEAD FLEXES INTO A COCKED POSITION...
AND DELIVERS MYOSIN'S **POWER STROKE** AS PHOSPHATE POPS OFF.
THE PROTEIN HAS NOW ADVANCED ONE NOTCH.
HIT IT AGAIN!
THE MYOSIN TAIL, MEANWHILE, MAY BE TETHERED TO A VESICLE FULL OF CARGO.
ONE LITTLE MYOSIN MOLECULE CAN DRAG A GIANT LOAD ACROSS THE CELL.
KICK
KICK
KICK
KICK

IN MUSCLES, MANY MYOSIN HEADS SIT ON FIBERS MADE OF ANOTHER PROTEIN, **ACTIN.** BOTH ACTIN AND MYOSIN ARE ANCHORED TO A PLATE-LIKE STRUCTURE AT THEIR ENDS, AND THE ENTIRE ARRANGEMENT FACES A SECOND VERSION OF ITSELF.

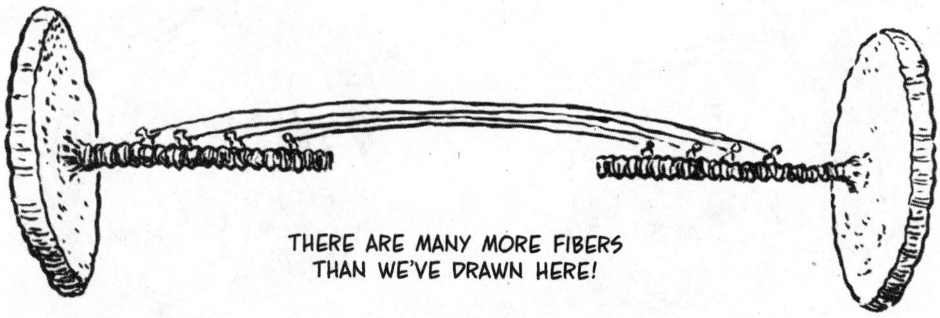

THERE ARE MANY MORE FIBERS THAN WE'VE DRAWN HERE!

THE MYOSIN TAILS BRAID TOGETHER INTO A THICK CABLE THAT JOINS THE TWO SIDES AND IMMOBILIZES THE MYOSIN. INSTEAD, ITS POWER STROKES **MOVE** THE **ACTIN.** THE TWO PLATES ARE PULLED TOGETHER, AND THE MUSCLE **GETS SHORTER.**

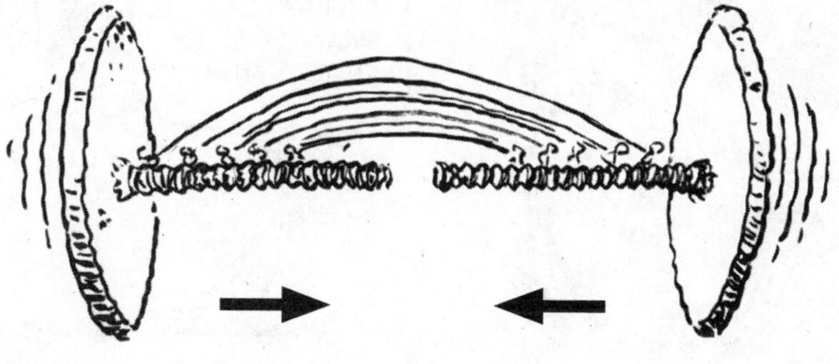

WE'VE ALL SEEN HOW A MUSCLE CONTRACTS IN ONE DIRECTION WHEN IT'S USED.

ALL OUR MUSCLE MOVEMENTS, FROM FLEXING BICEPS TO RAISING EYEBROWS, WORK THIS WAY.

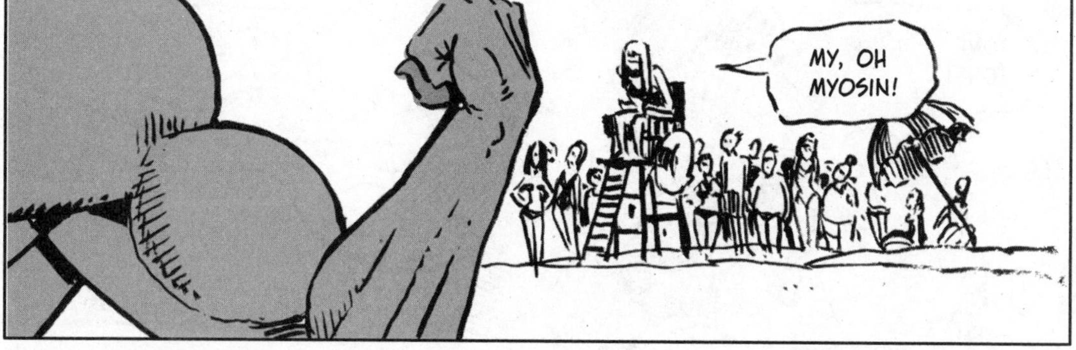

MY, OH MYOSIN!

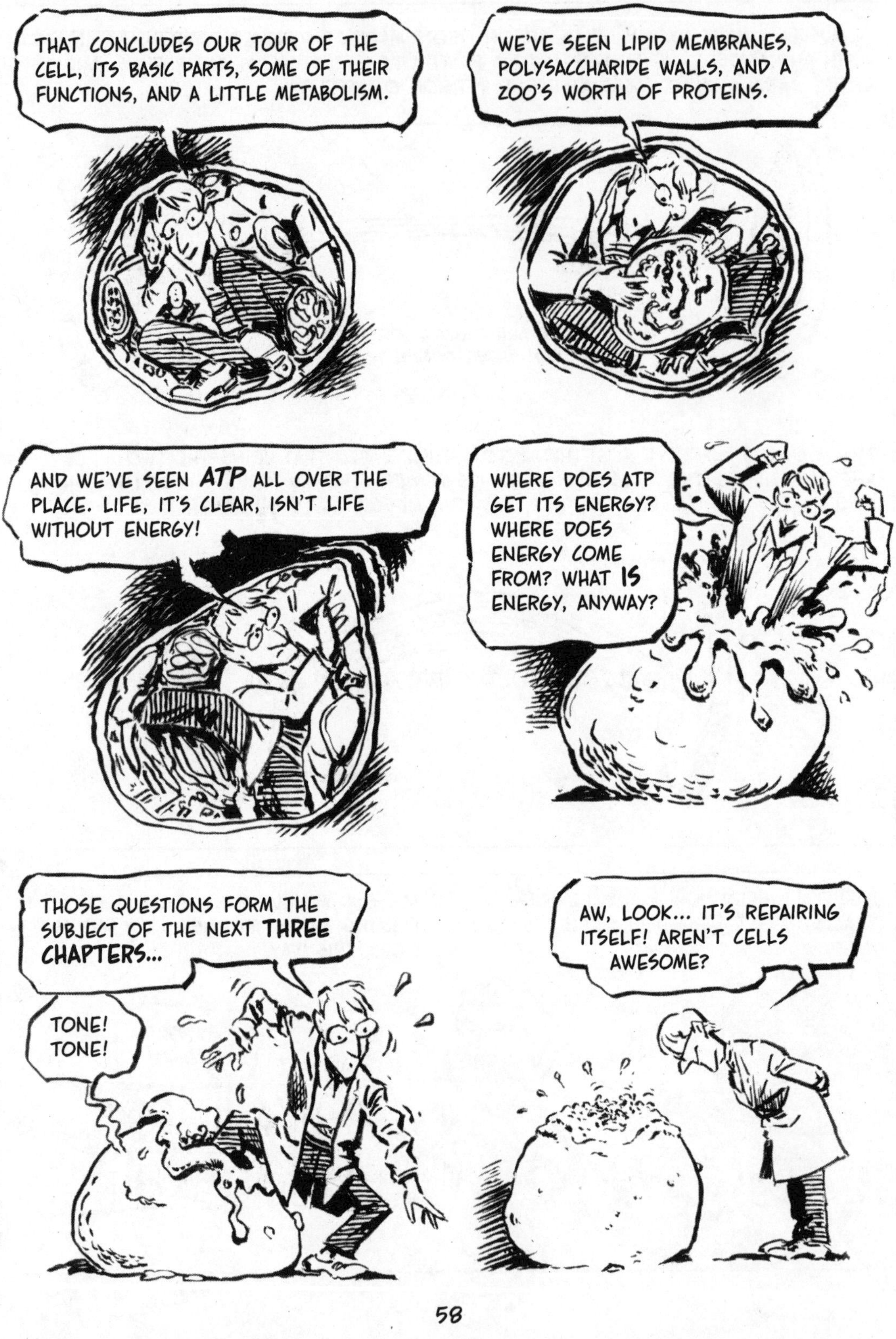
THAT CONCLUDES OUR TOUR OF THE CELL, ITS BASIC PARTS, SOME OF THEIR FUNCTIONS, AND A LITTLE METABOLISM.
WE'VE SEEN LIPID MEMBRANES, POLYSACCHARIDE WALLS, AND A ZOO'S WORTH OF PROTEINS.
AND WE'VE SEEN ATP ALL OVER THE PLACE. LIFE, IT'S CLEAR, ISN'T LIFE WITHOUT ENERGY!
WHERE DOES ATP GET ITS ENERGY? WHERE DOES ENERGY COME FROM? WHAT IS ENERGY, ANYWAY?
THOSE QUESTIONS FORM THE SUBJECT OF THE NEXT THREE CHAPTERS...
TONE! TONE!
AW, LOOK... IT'S REPAIRING ITSELF! AREN'T CELLS AWESOME?

Chapter 5
ENERGY

YOU HAVE IT, DESPITE HOW YOU MAY FEEL AT THE MOMENT.

TECHNICALLY, ENERGY EQUALS **WORK,** AND WORK IS **FORCE** ACTING OVER A **DISTANCE.** WHEN A RHINOCEROS PUSHES A RAILCAR DOWN A TRACK, THE RHINO EXPENDS **MECHANICAL ENERGY.**

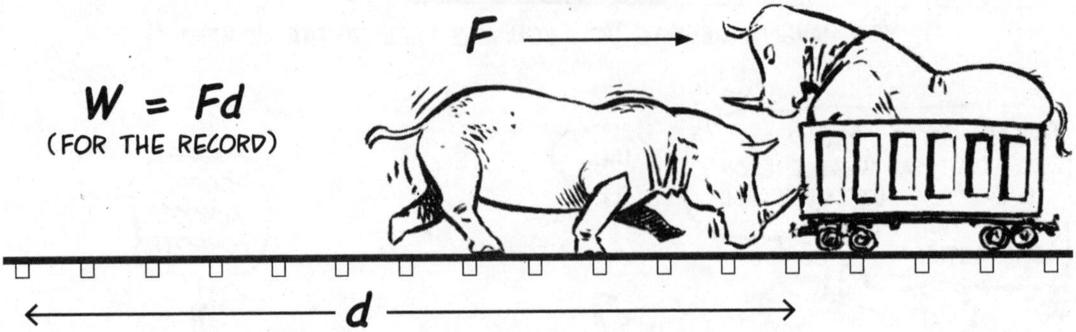

NOW SUPPOSE THE RHINO **STOPS** PUSHING (AND THE TRACK IS FRICTIONLESS). THE CAR WILL **KEEP ROLLING** AT CONSTANT SPEED. THE CAR AND ITS CONTENTS NOW HAVE **KINETIC ENERGY,** THE ENERGY OF MOTION.

$$K.E. = \frac{1}{2}mv^2$$

THE IDEA HERE IS THAT THE RHINO'S WORK GIVES THE CAR ENERGY THAT APPEARS IN A NEW FORM, **KINETIC** ENERGY.

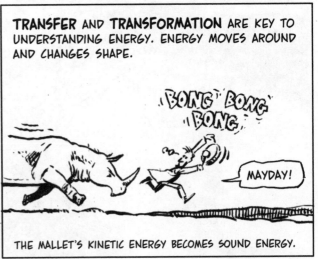

TRANSFER AND **TRANSFORMATION** ARE KEY TO UNDERSTANDING ENERGY. ENERGY MOVES AROUND AND CHANGES SHAPE.

THE MALLET'S KINETIC ENERGY BECOMES SOUND ENERGY.

MORE TRANSFORMATIONS: WHEN TWO THINGS ARE RUBBED TOGETHER, THEY WARM UP. MECHANICAL ENERGY BECOMES **HEAT.**

REASON: TEMPERATURE IS A MEASURE OF MOLECULES' **KINETIC ENERGY** AS THEY JIGGLE. RUBBING APPLIES FORCE TO MOLECULES AND SPEEDS THEM UP.

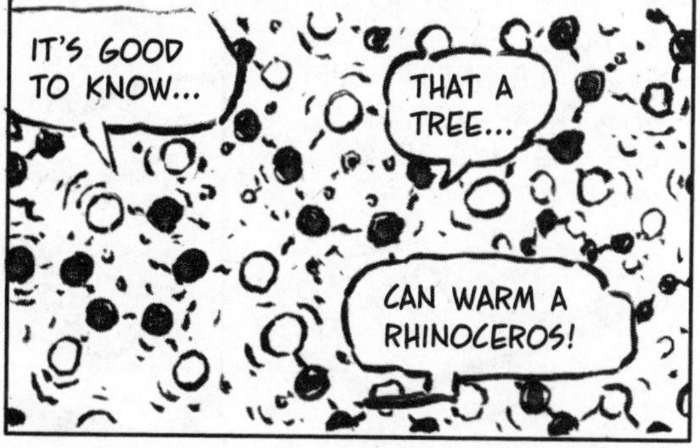

IT'S GOOD TO KNOW...
THAT A TREE...
CAN WARM A RHINOCEROS!

HEAT IS CONVERTED TO MECHANICAL ENERGY IN A **STEAM ENGINE.** HEAT BOILS WATER; EXPANDING STEAM DRIVES THE WORKS.

HEAT ALSO DOES WORK AT THE ATOMIC LEVEL, AS THE SHAKING ACTION OPPOSES ATTRACTIVE FORCES BEWEEN MOLECULES. (BREAKING HYDROGEN BONDS IN WATER MAKES STEAM!)

ALL OF ENERGY'S MANY GUISES CAN BE TURNED TO **MECHANICAL ENERGY.** THAT'S WHY MOST TEXTBOOKS DEFINE ENERGY AS THE "CAPACITY" TO DO WORK, WHICH SOUNDS LIKE SCOLDING TO US.

GET A JOB!

ENERGY NEVER POPS INTO EXISTENCE OR VANISHES, BUT ONLY MOVES AROUND AND CHANGES FORM. THE TOTAL AMOUNT OF ENERGY IN THE UNIVERSE IS CONSTANT, SAYS THE **FIRST LAW OF THERMODYNAMICS.**

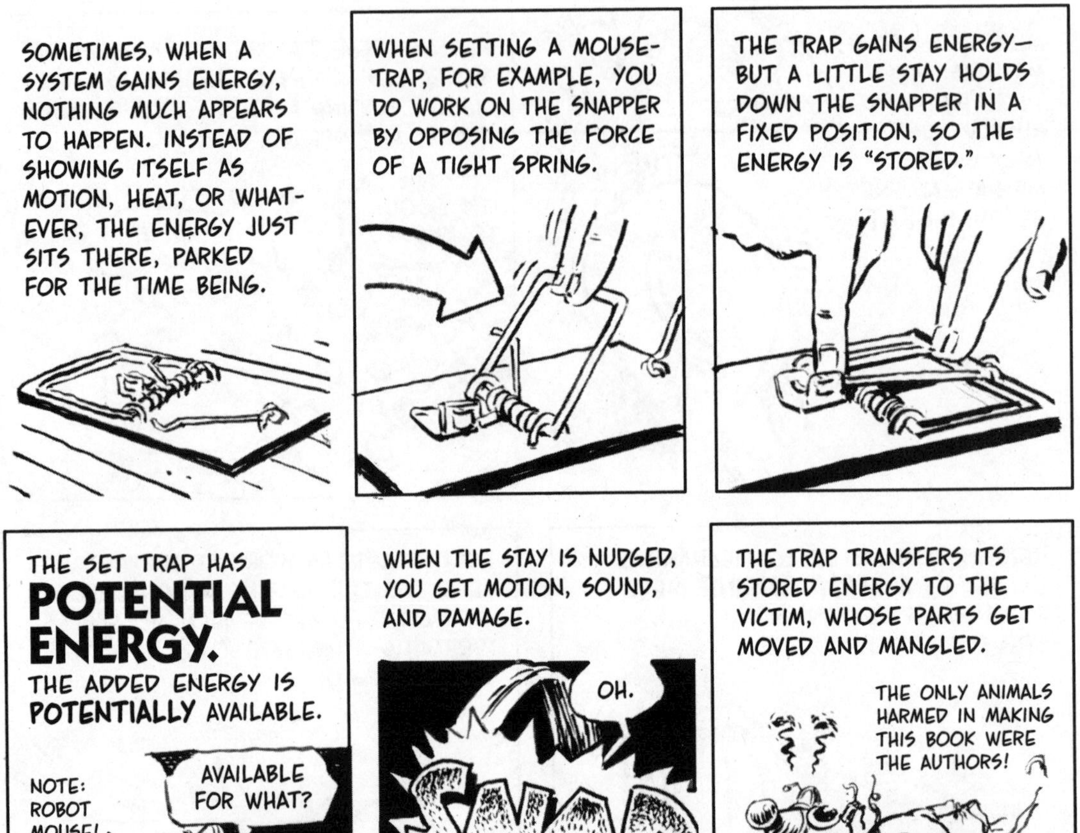
SOMETIMES, WHEN A SYSTEM GAINS ENERGY, NOTHING MUCH APPEARS TO HAPPEN. INSTEAD OF SHOWING ITSELF AS MOTION, HEAT, OR WHATEVER, THE ENERGY JUST SITS THERE, PARKED FOR THE TIME BEING.
WHEN SETTING A MOUSETRAP, FOR EXAMPLE, YOU DO WORK ON THE SNAPPER BY OPPOSING THE FORCE OF A TIGHT SPRING.
THE TRAP GAINS ENERGY— BUT A LITTLE STAY HOLDS DOWN THE SNAPPER IN A FIXED POSITION, SO THE ENERGY IS "STORED."
THE SET TRAP HAS **POTENTIAL ENERGY.**
THE ADDED ENERGY IS **POTENTIALLY** AVAILABLE.
NOTE: ROBOT MOUSE!
AVAILABLE FOR WHAT?
WHEN THE STAY IS NUDGED, YOU GET MOTION, SOUND, AND DAMAGE.
OH.
SNAP
THE TRAP TRANSFERS ITS STORED ENERGY TO THE VICTIM, WHOSE PARTS GET MOVED AND MANGLED.
THE ONLY ANIMALS HARMED IN MAKING THIS BOOK WERE THE AUTHORS!
ON PAGE 47, WE SAW **ELECTRIC POTENTIAL** ACROSS A PLASMA MEMBRANE, WITH A NET POSITIVE CHARGE OUTSIDE THE CELL.
WHEN CHANNELS OPEN, IONS FLOW IN; POTENTIAL ENERGY BECOMES KINETIC.
THE **ELECTRICAL ENERGY** OF FLOWING IONS DRIVES THE ANIMAL BRAIN AND NERVOUS SYSTEM.

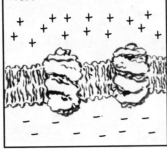

Chemical Energy

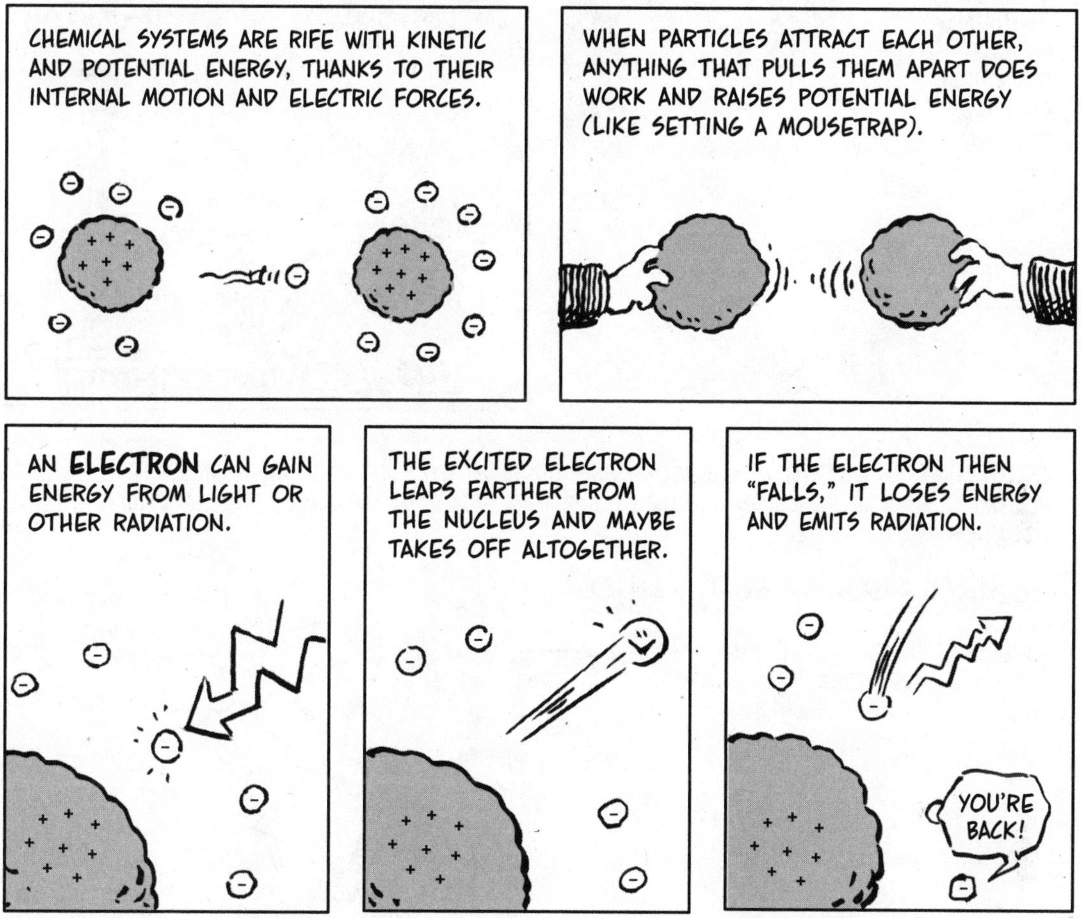
CHEMICAL SYSTEMS ARE RIFE WITH KINETIC AND POTENTIAL ENERGY, THANKS TO THEIR INTERNAL MOTION AND ELECTRIC FORCES.
WHEN PARTICLES ATTRACT EACH OTHER, ANYTHING THAT PULLS THEM APART DOES WORK AND RAISES POTENTIAL ENERGY (LIKE SETTING A MOUSETRAP).
AN **ELECTRON** CAN GAIN ENERGY FROM LIGHT OR OTHER RADIATION.
THE EXCITED ELECTRON LEAPS FARTHER FROM THE NUCLEUS AND MAYBE TAKES OFF ALTOGETHER.
IF THE ELECTRON THEN "FALLS," IT LOSES ENERGY AND EMITS RADIATION.
YOU'RE BACK!

IN CHEMICAL REACTIONS, ATOMS AND ELECTRONS MOVE; OLD BONDS BREAK AND NEW ONES FORM. ENERGY MOVES **WITHIN** THE REACTION SYSTEM, AND ENERGY MAY ALSO SPREAD TO THE **OUTSIDE WORLD.** IN THIS REACTION, COMBINING METHANE WITH OXYGEN **RELEASES ENERGY.** (THIS REACTION MAKES THE FLAME ON A GAS STOVE.)

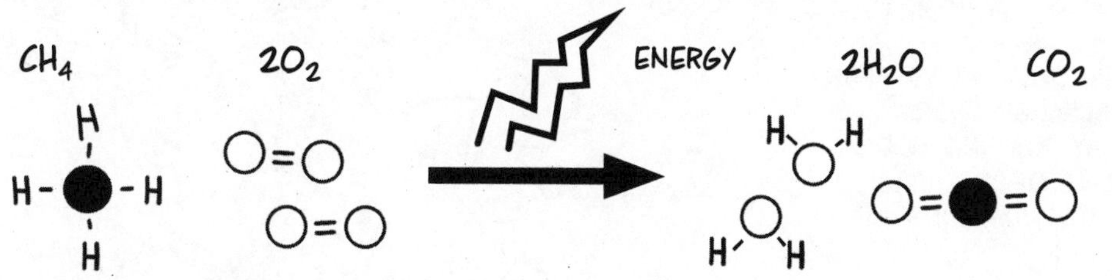

ANY ENERGY-RELEASING REACTION IS CALLED **EXERGONIC.**

EXERGONIC EXAMPLES: **BURNING WOOD** OBVIOUSLY RELEASES ENERGY.

HOT STUFF!

WE'VE ALREADY SEEN THAT FIRE'S ENERGY CAN DO WORK.

OXIDIZING GLUCOSE IS EXERGONIC. WOOD IS MOSTLY CELLULOSE, A GLUCOSE POLYMER, SO THE REACTION IS REALLY BURNING WOOD IN DISGUISE. THE REACTION EQUATION IS:

$$C_6H_{12}O_6 + 6O_2 \longrightarrow 6CO_2 + 6H_2O$$

WE'LL SEE IN THE NEXT CHAPTER WHAT KIND OF WORK THIS ENERGY CAN DO.

IS DISSOLVING TEETH A KIND OF WORK?

SPLITTING ATP CAN KICK OUT ENERGY WITHIN THE CELL, BECAUSE THE REACTION

$$ATP \longrightarrow ADP + P_i$$

IS EXERGONIC. WE'VE SEEN THE RELEASED ENERGY DO WORK BY PUMPING SODIUM IONS OUT OF THE CELL AND POTASSIUM IONS IN.

(SOUND EFFECTS AND DIALOG NOT NECESSARILY RECORDED VERBATIM.)

THE UNIVERSE FAVORS EXERGONIC REACTIONS, BECAUSE ENERGY TENDS TO SPREAD.* THAT'S WHY CHEMISTS CALL THESE REACTIONS **SPONTANEOUS**. EVEN SO, EXERGONIC REACTIONS MAY NOT **START** SPONTANEOUSLY BUT INSTEAD REQUIRE SOME **ACTIVATION**.

IT TAKES A FIRE TO START A FIRE!
AND THEN IT'S %$#%*& HARD TO PUT IT OUT!

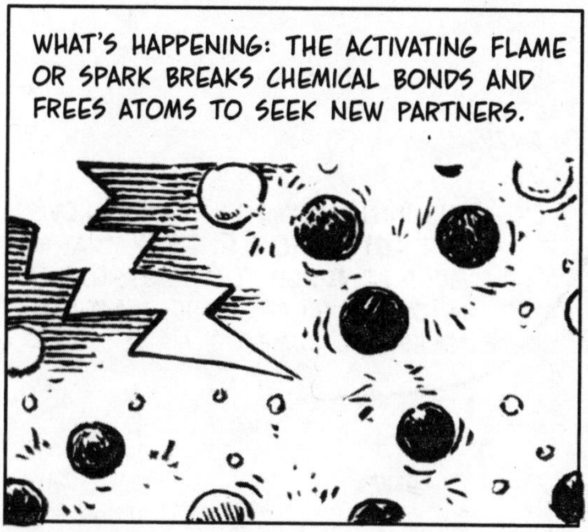
WHAT'S HAPPENING: THE ACTIVATING FLAME OR SPARK BREAKS CHEMICAL BONDS AND FREES ATOMS TO SEEK NEW PARTNERS.

FROM THAT POINT ON, THE REACTION'S **OWN ENERGY OUTPUT** KEEPS THINGS GOING.
A FIRE FEEDS ITSELF!

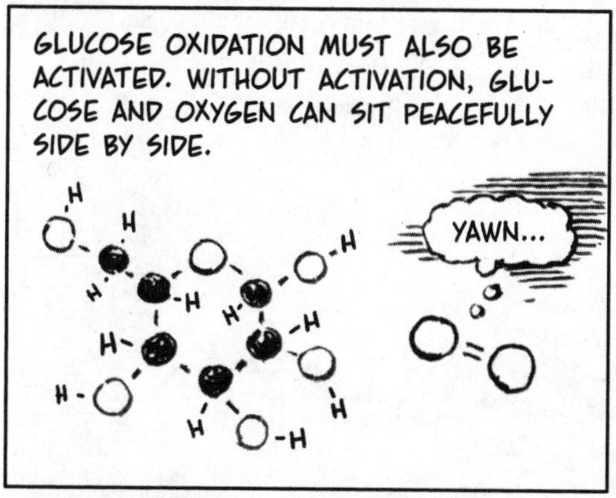
GLUCOSE OXIDATION MUST ALSO BE ACTIVATED. WITHOUT ACTIVATION, GLUCOSE AND OXYGEN CAN SIT PEACEFULLY SIDE BY SIDE.
YAWN...

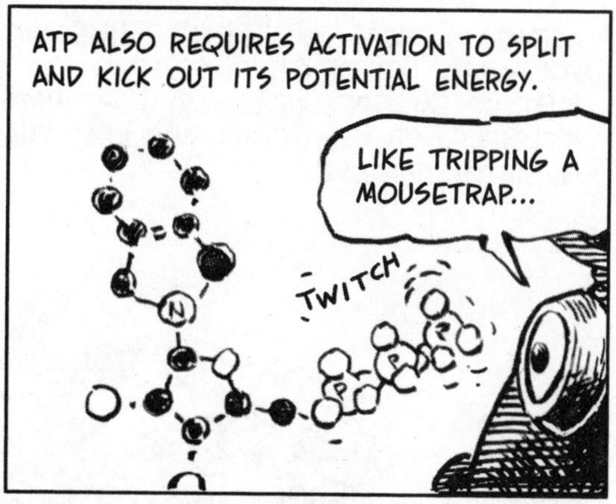
ATP ALSO REQUIRES ACTIVATION TO SPLIT AND KICK OUT ITS POTENTIAL ENERGY.
LIKE TRIPPING A MOUSETRAP...
TWITCH

*ENERGY SPREADS BECAUSE SPREAD-OUT STATES ARE VASTLY MORE PROBABLE THAN CONCENTRATED STATES. ENERGY'S INEXORABLE SPREADING IS CALLED THE **SECOND LAW OF THERMODYNAMICS**.

REACTION ENERGY TRANSFERRED TO THE OUTSIDE WORLD IS CALLED **FREE ENERGY,** WRITTEN **G,** AFTER J. WILLARD **GIBBS,** WHO FIRST ANALYZED IT THOROUGHLY.

$$\Delta G = \Delta H - T \Delta S$$

J. WILLARD GIBBS
1839–1903

ΔG IS THE FREE ENERGY **CHANGE** OF THE **REACTION.** AN **EXERGONIC** REACTION HAS **NEGATIVE** $\Delta G.$*

$$REACTANTS + \Delta G \rightarrow PRODUCTS$$
$$\Delta G < 0$$

THE WORLD'S GAIN IS THE REACTION'S LOSS!

*Δ, DELTA, IS NOT A NUMBER BUT STANDARD SCIENTIFIC SHORTHAND FOR "THE CHANGE IN."

THE GRAPH OF AN EXERGONIC REACTION'S FREE ENERGY OVER TIME LOOKS LIKE THIS. THE ENERGY OF THE REACTION SYSTEM FALLS AS THE REACTION PROCEEDS.

ACTIVATION ENERGY

ΔG

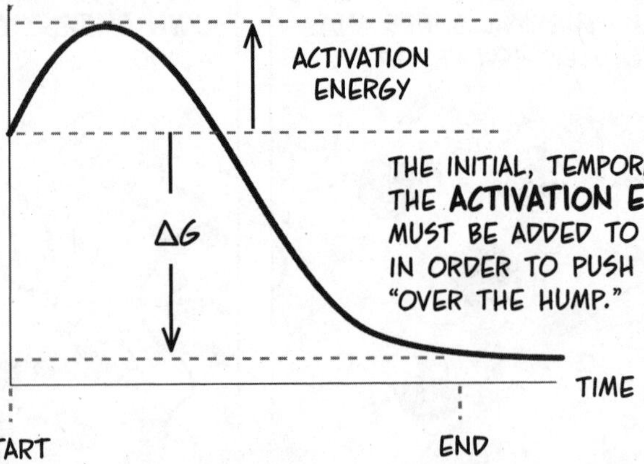

THE INITIAL, TEMPORARY RISE SHOWS THE **ACTIVATION ENERGY** THAT MUST BE ADDED TO THE SYSTEM IN ORDER TO PUSH THE REACTION "OVER THE HUMP."

TIME

START

END

ACTIVATION ENERGY IS A **GOOD THING.** WITHOUT IT, YOU—AND A LOT OF OTHER STUFF—WOULD BURN, RUST, OR EXPLODE. WE'RE VERY FORTUNATE THAT MOLECULES STICK TOGETHER AND KEEP THEIR POTENTIAL ENERGY!

THANKS, UNIVERSE!!

HOW DO LIVING SYSTEMS OVERCOME ACTIVATION ENERGY? HOW DO THEY MAKE REACTIONS RUN? THEY CAN'T EXACTLY LIGHT A MATCH TO THEMSELVES!

YOU **COULD** SIT ON A VOLCANO...
UM... NO.

ANSWER: WITH **ENZYMES.** AS SHOWN ON P. 52, EVERY METABOLIC REACTION COMES WITH ITS OWN SPECIFIC ENZYME. THE ENZYME'S ACTIVE SITE OR SITES PUTS REACTANTS IN FAVORABLE POSITIONS, SO THEY FIND EACH OTHER EASILY RATHER THAN FLYING AROUND RANDOMLY.

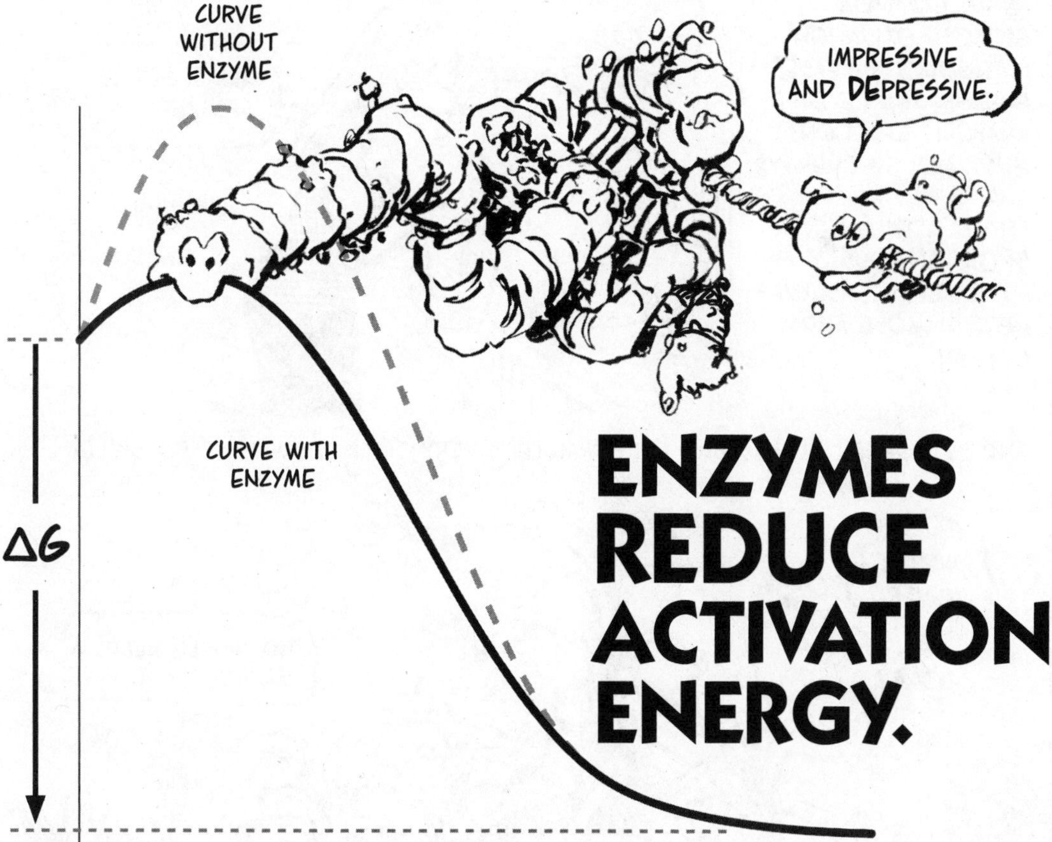
CURVE WITHOUT ENZYME
IMPRESSIVE AND **DE**PRESSIVE.
CURVE WITH ENZYME
ΔG

ENZYMES REDUCE ACTIVATION ENERGY.

AS A GENERAL RULE, **CATABOLIC** REACTIONS, WHICH BREAK UP LARGER MOLECULES, ARE **EXERGONIC,** WHILE **ANABOLIC** REACTIONS, WHICH BUILD LARGER MOLECULES, ARE **ENDERGONIC,** MEANING THAT THEY **ABSORB FREE ENERGY.**

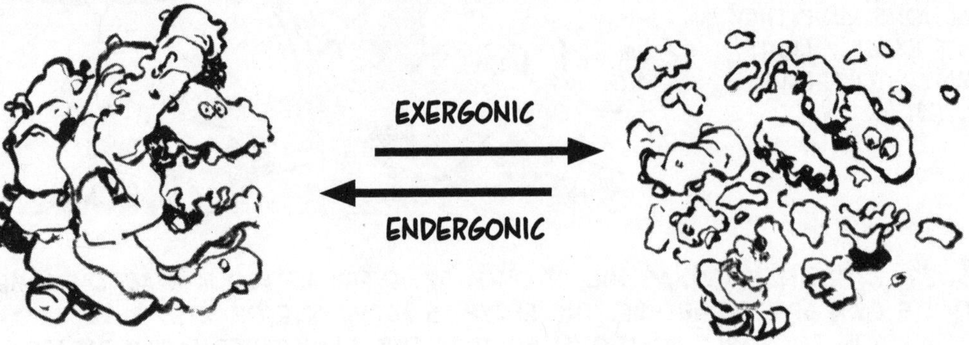

EXERGONIC
ENDERGONIC

AS AN EXAMPLE, GLUCOSE OXIDATION (CATABOLIC) RELEASES ENERGY. THE REVERSE, ANABOLIC REACTION MUST ABSORB ENERGY. THAT'S WHY WATER AND CO_2 WILL NEVER **SPONTANEOUSLY** ASSEMBLE THEMSELVES INTO GLUCOSE AND OXYGEN.
PSSSHT
ALL I GET IS SELTZER!
AND YET, GREEN PLANTS CAN PERFORM THAT VERY TRICK. HOW IS IT POSSIBLE?
WE'LL NEVER TELL!
THE WORLD NEEDS A TALKING PLANT, DAVE.

THE UNIVERSE, REMEMBER, FAVORS EXERGONIC REACTIONS. ENERGY "WANTS" TO SPREAD OUT. HOW, THEN, CAN AN ENDERGONIC REACTION EVER HAPPEN? HOW CAN IT PULL IN ENERGY FROM THE REST OF THE UNIVERSE?

SOUNDS FUTILE.

FOR INSTANCE, BUILDING **GLYCOGEN** FROM GLUCOSE IS AN ANABOLIC, ENDERGONIC REACTION. GLUCOSE MONOMERS WILL VIRTUALLY **NEVER** JOIN THE CHAIN SPONTANEOUSLY.

COME ON...

BUT GIVE IT A KICK FROM ***UTP*** (WITH AN ENZYME ASSIST), AND THE SUGAR HOOKS UP!

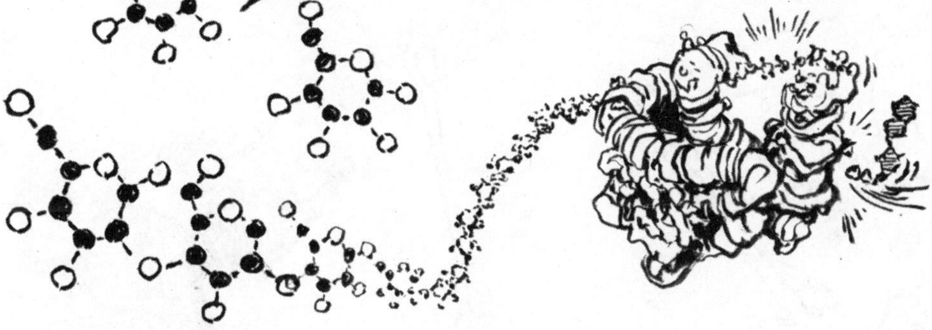

THE ENDERGONIC REACTION IS ***COUPLED*** WITH AN EXERGONIC REACTION, NAMELY THE SPLITTING OF ***UTP.*** WHEN ALL ATOMIC REARRANGEMENTS ARE ACCOUNTED FOR, THE OVERALL REACTION **RELEASES** ENERGY.

$$\text{GLYCOGEN + GLUCOSE + UTP} \longrightarrow \text{LONGER GLYCOGEN} + \text{UDP} + P_i$$

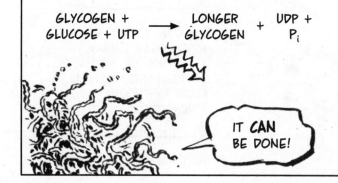

IT **CAN** BE DONE!

THIS EXPLAINS THE EFFECTIVENESS OF KICKS FROM ATP (AND UTP AND GTP); THE EXERGONIC

$$\textbf{ATP} \longrightarrow \textbf{ADP} + \textbf{P}_i$$

PUTS OUT **MORE THAN ENOUGH ENERGY** TO DRIVE THE ENDERGONIC PART OF THE REACTION.

OF COURSE, SPLITTING ATP (OR GTP OR UTP) LEAVES THE CELL WITH A PROBLEM: HOW TO REPLENISH IT? THE CELL DESPERATELY NEEDS ITS ATP, BUT PUTTING IT BACK TOGETHER, I.E., THE REACTION

$$\text{ADP} + \text{P}_i \rightarrow \text{ATP}$$

IS ENDERGONIC.

THE UNIVERSE IS A CRUEL PLACE TO LIVE...

MAKING ATP IS LIKE SETTING A MOUSETRAP. ENERGY MUST COME FROM SOMEWHERE ELSE. WHEN YOU SET A MOUSETRAP BY HAND, YOU COUPLE **YOUR OWN METABOLISM** TO THE TRAP. YOU SPLIT ATP TO MOVE YOUR MUSCLES TO PUSH THE SNAPPER!

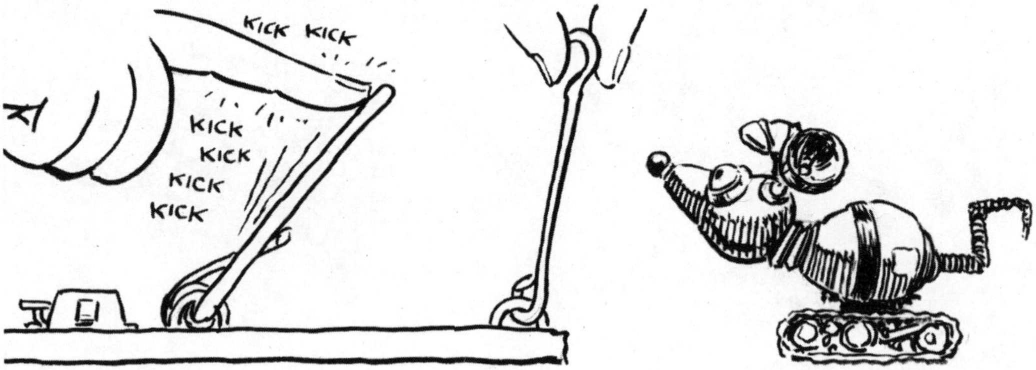

WHAT EXERGONIC REACTION CAN THE CELL COUPLE TO ADP AND P_i TO PUSH THEM TOGETHER?

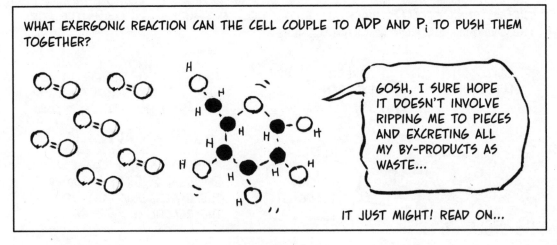

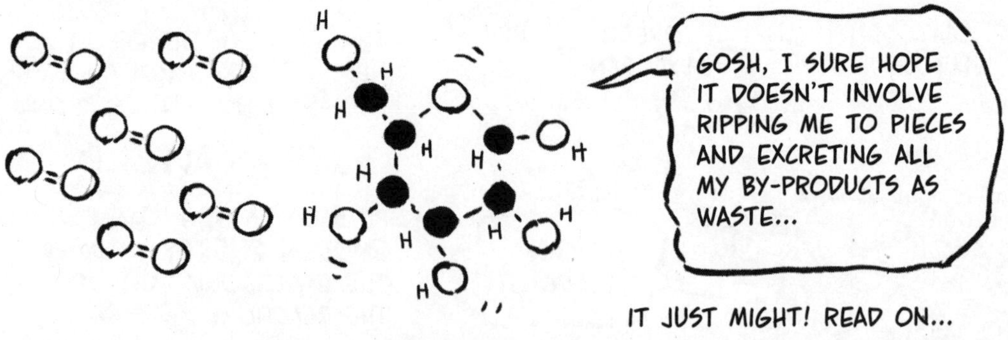

GOSH, I SURE HOPE IT DOESN'T INVOLVE RIPPING ME TO PIECES AND EXCRETING ALL MY BY-PRODUCTS AS WASTE...

IT JUST MIGHT! READ ON...

Chapter 6
CELLULAR RESPIRATION

A TWELVE-STEP PROGRAM

TO ENGLISH-SPEAKING PEOPLE WITH NOSES, "RESPIRATION" MEANS **BREATHING.** TO A BIOLOGIST, IT MEANS SOMETHING ELSE. AT THE CELLULAR LEVEL, RESPIRATION REFERS TO A PARTICULAR WAY OF LIBERATING **CHEMICAL ENERGY** FROM FOOD (TYPICALLY SUGAR) AND APPLYING THAT ENERGY TO **MAKE ATP.**

WE AIR-BREATHERS RELEASE FREE ENERGY BY COMBINING SUGAR WITH **OXYGEN,** BUT MANY ORGANISMS, SUCH AS BACTERIA LIVING DEEP IN ANIMALS' GUTS, CAN DO WITHOUT OXYGEN ENTIRELY. THEY HAVE RESPIRATION WITHOUT AIR!

SIGH
SNURF SNORF
SNIFF
PFFF PFFFF
WHEEZING BEHEMOTHS...
PANT PANT PANT

WE EAT FOR BOTH MATTER AND ENERGY. ON THE ONE HAND, FOOD GIVES BODIES ESSENTIAL STUFF SUCH AS NITROGEN, PHOSPHORUS, AND CARBON; ON THE OTHER HAND, FOOD IS **FUEL** TO POWER ALL SORTS OF BIOCHEMICAL REACTIONS.

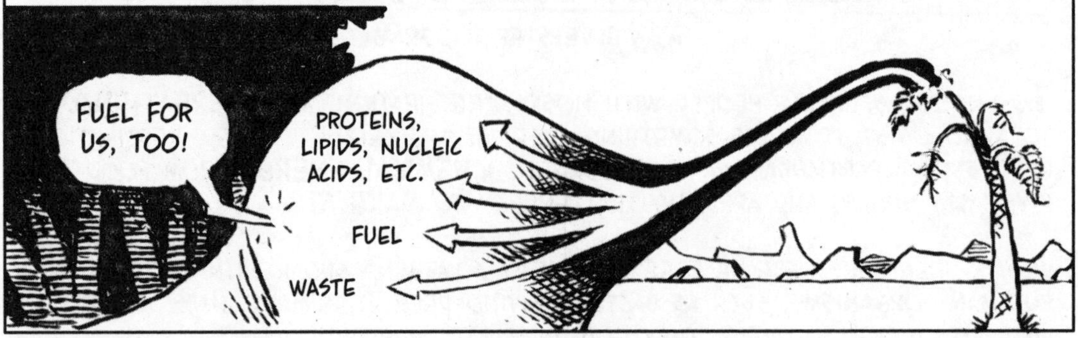

ONE FUEL IS NEARLY EVERYONE'S FAVORITE: **GLUCOSE,** $C_6H_{12}O_6$.

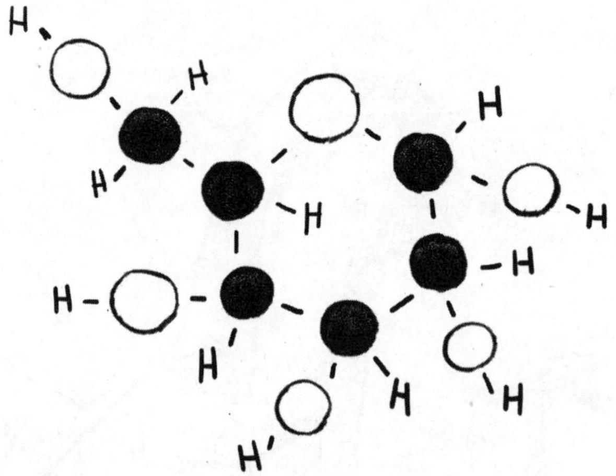

OXYGEN FROM THE AIR **OXIDIZES GLUCOSE** IN THIS EXERGONIC REACTION:

$C_6H_{12}O_6 + 6O_2 \longrightarrow 6CO_2 + 6H_2O$.

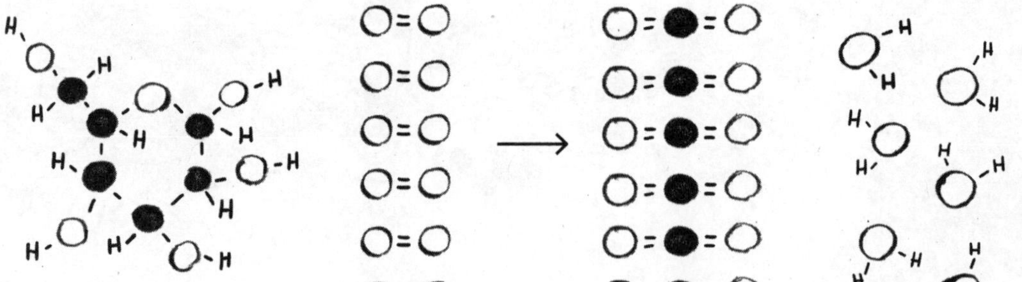

CELLS USE THE FREE ENERGY RELEASED BY THIS REACTION TO DRIVE THE SYNTHESIS OF ATP, WHICH MAKES BODIES GO. THAT'S WHY WE **COUNT CALORIES:** A CORN CHIP'S CALORIES ARE THE ENERGY RELEASED WHEN THE MORSEL IS OXIDIZED.

I EAT TO GET THE ENERGY I NEED TO FIND SOMETHING TO EAT.
AND THEY SAY LIFE HAS NO MEANING...

AIR-BREATHERS GET ENOUGH ENERGY FROM A SINGLE GLUCOSE MOLECULE TO MAKE ABOUT **36** MOLECULES OF ATP. THE EXERGONIC OXIDATION OF GLUCOSE IS **COUPLED** TO THE ENDERGONIC SYNTHESIS OF ATP. WE SUMMARIZE THE COUPLING LIKE SO:

$$C_6H_{12}O_6 + 6O_2 \longrightarrow 6CO_2 + 6H_2O$$

ENERGY-RELEASING REACTION

$36ADP$
$36\ P_i$ $\longrightarrow$ $36ATP$

ENERGETIC PAYOFF

(REMINDER: P_i STANDS FOR INORGANIC PHOSPHATE, DISSOLVED PO_4^{3-} IONS NOT ATTACHED TO ANY ORGANIC MOLECULE.)

H^+
P
THEY'RE PYRAMID-SHAPED!

FOR THE REACTION'S ENERGY SOURCE, WE HAVE TO LOOK AT THE "FINE PRINT," THE ATOMS' **ELECTRONS.**

OXYGEN PULLS ELECTRONS MORE STRONGLY THAN DO CARBON AND HYDROGEN. IN WATER, ELECTRONS HOVER CLOSER TO OXYGEN THAN TO HYDROGEN.
SO LONELY OUT HERE...

THIS MEANS THAT ELECTRONS USUALLY **RELEASE ENERGY** WHEN "FALLING" TO OXYGEN. FOR EXAMPLE, IN THE OXIDATION OF **METHANE,** CH_4, (SEE P. 63) OXYGEN GAINS EIGHT ELECTRONS PER METHANE MOLECULE.

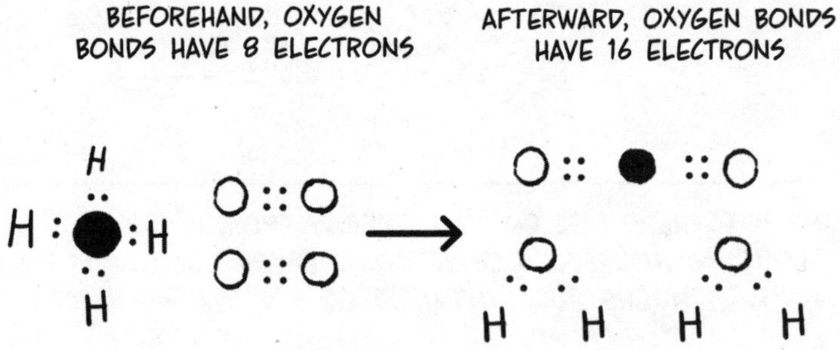
BEFOREHAND, OXYGEN BONDS HAVE 8 ELECTRONS
AFTERWARD, OXYGEN BONDS HAVE 16 ELECTRONS
H
H
H
H
H H H H

$$CH_4 + 2O_2 \longrightarrow CO_2 + 2H_2O$$

GLUCOSE OXIDATION MOVES **24** ELECTRONS FROM A GLUCOSE MOLECULE TO OXYGEN. BEFORE OXIDATION, 48 ELECTRONS ARE IN BONDS WITH OXYGEN; AFTERWARD, 72. THE DIFFERENCE, 24, IS THE NUMBER OF ELECTRONS ACQUIRED BY OXYGEN.

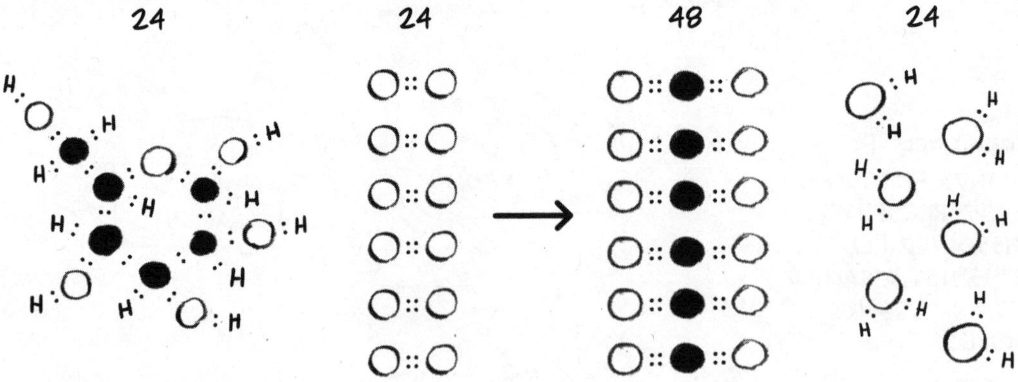
24
24
48
24

THAT'S A LOT OF ELECTRONS RELEASING A LOT OF ENERGY. IF GLUCOSE WERE TO OXIDIZE FULLY ALL AT ONCE, THE BLAST WOULD CRIPPLE THE CELL.

LIKE TRYING TO RUN A TRUCK BY BLOWING UP ITS GAS TANK!
BOOP

SO, IN A BRILLIANT METABOLIC MANEUVER, THE CELL TAKES GLUCOSE APART BIT BY BIT IN A SERIES OF TWELVE **MINI-OXIDATIONS.** AS GLUCOSE BREAKS DOWN, ITS FRAGMENTS GIVE UP TWO ELECTRONS AT A TIME—BUT **NOT TO OXYGEN.**

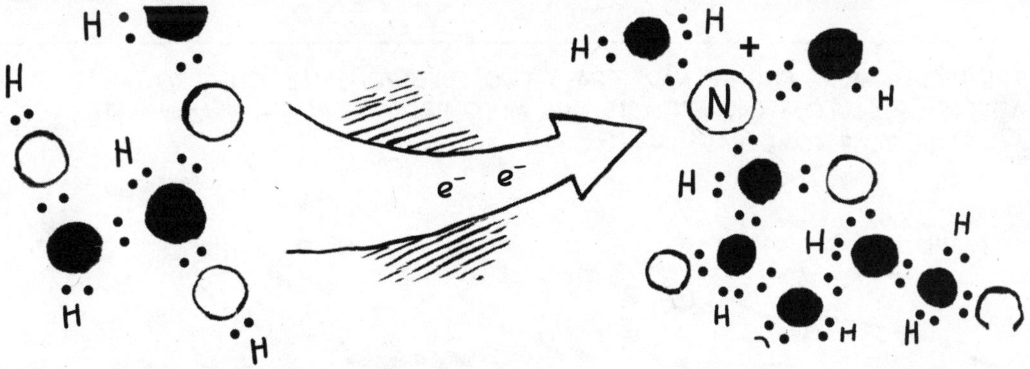

e^- e^-
+
N

WHAT? OXIDIZED WITHOUT OXYGEN? YES, CHEMISTS HAVE DECIDED THAT "OXIDATION" MEANS **GIVING AWAY ELECTRONS,** NO MATTER TO WHOM OR TO WHAT. THE ELECTRON **DONOR** IS OXIDIZED. THE ELECTRON RECEIVER IS SAID TO BE **REDUCED,** AND THE ELECTRON TRANSFER IS CALLED A REDUCTION-OXIDATION, OR **REDOX** REACTION.

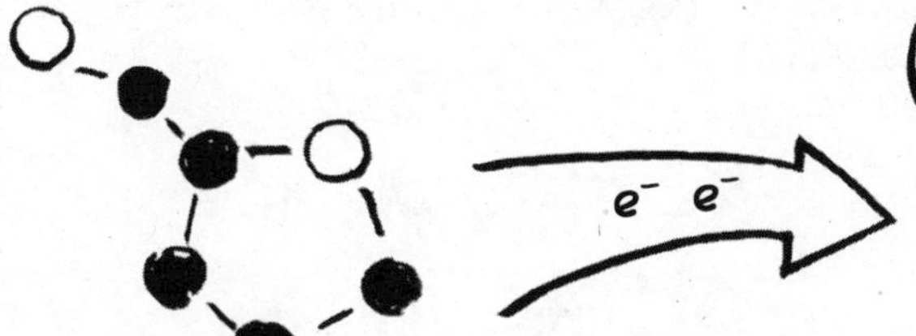

e^- e^-

NOW I'M EVEN MORE NEGATIVE THAN USUAL!

ORGANIC MOLECULE OXIDIZED BY TYRANNOSAURUS REX. T. REX IS REDUCED—IN CHARGE IF NOT IN SIZE OR FEROCITY.

THE MAIN OXIDIZING AGENT (ELECTRON RECEIVER) OF GLUCOSE IS A CHARGED, ORGANIC MOLECULE: NICOTINAMIDE ADENINE DINUCLEOTIDE, OR **NAD$^+$** FOR SHORT.

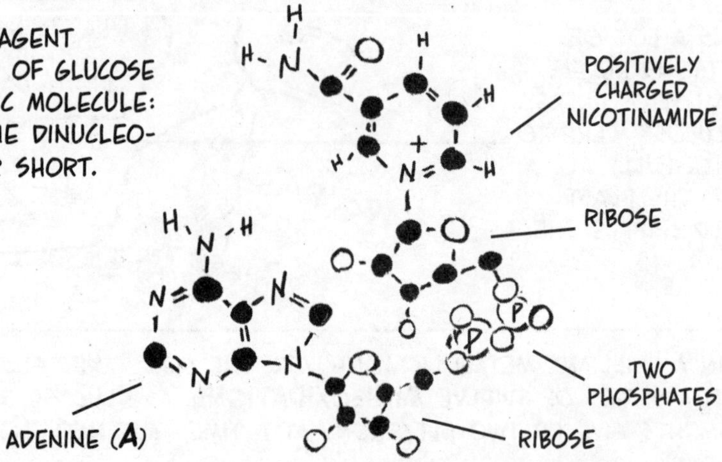

THE NICOTINAMIDE END OF NAD$^+$ GRABS TWO ELECTRONS AND ONE PROTON FROM A PIECE OF GLUCOSE AND ATTRACTS ONE MORE PROTON FROM A NEARBY WATER MOLECULE TO BECOME **NADH + H$^+$**.

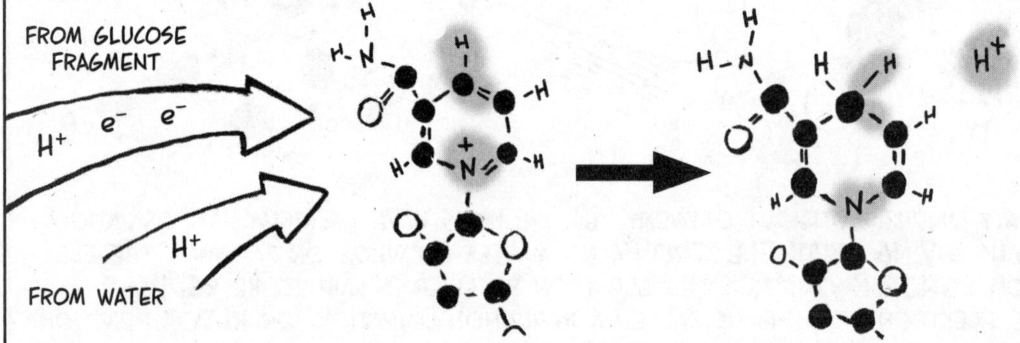

NOW FORGET ABOUT THE DETAILED ATOMIC STRUCTURE AND THINK OF NAD$^+$ AS A **PIGGY BANK** THAT CAN CARRY TWO ELECTRONS. THE MINI-REDOX REACTIONS LOOK LIKE THIS (LEAVING OUT THE WATER MOLECULES):

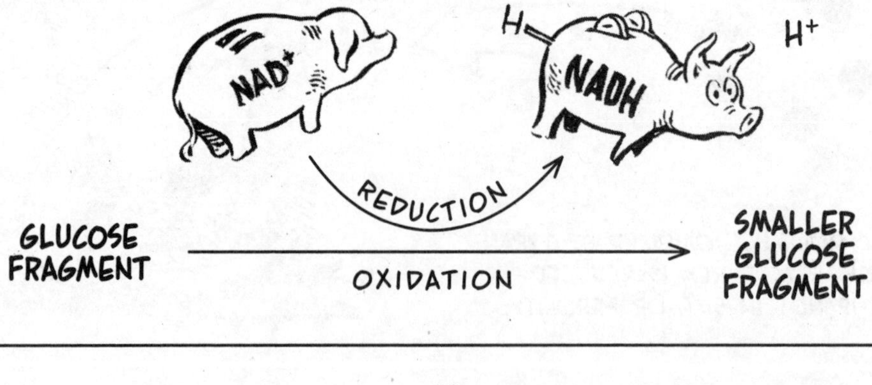

WHY DRAW NAD+ AND NADH AS PIGGY BANKS?
BECAUSE NADH IS THE CELL'S WAY OF **STORING HIGH-ENERGY ELECTRONS.**
THE TWO ELECTRONS ACQUIRED IN THE REDUCTION NAD+ ⟶ NADH + H+ KEEP MOST OF THEIR ENERGY. THE ΔG HERE IS MODEST.*
ΔG
NAD+
NADH + H+
DURING RESPIRATION, THE CELL "BANKS" 24 OF GLUCOSE'S HIGH-ENERGY ELECTRONS, TWO AT A TIME, FOR LATER USE.
NADH (AND A SIMILAR CARRIER, FADH2) WILL SPILL THESE ELECTRONS IN A STEADY STREAM OR CASCADE THAT DRIVES THE SYNTHESIS OF ATP.
BUT FIRST, SOME PREP WORK HAS TO HAPPEN. IT'S NOT SO EASY TO PULL A PAIR OF ELECTRONS OUT OF THE RING-SHAPED GLUCOSE MOLECULE.
SO THE CELL HAS TO BREAK THE SUGAR IN HALF...
*STRICTLY SPEAKING, ΔG DEPENDS ON WHICH GLUCOSE FRAGMENT IS BEING OXIDIZED, BUT ITS MAGNITUDE IS RARELY SIGNIFICANT.

GLYCOLYSIS

(GLY-**COL**-I-SISS, MEANING GLUCOSE-BREAKING)

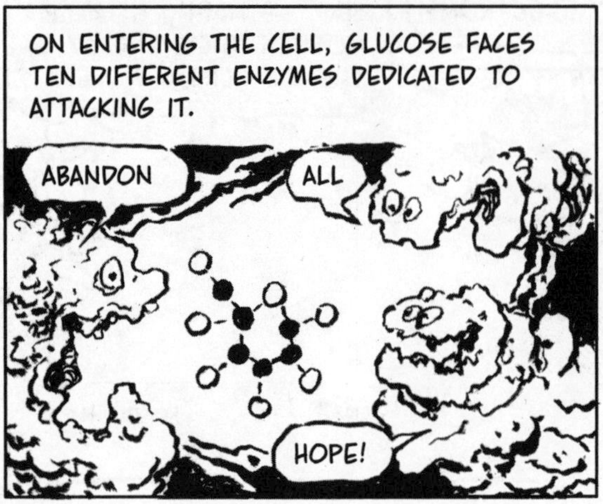
ON ENTERING THE CELL, GLUCOSE FACES TEN DIFFERENT ENZYMES DEDICATED TO ATTACKING IT.
ABANDON
ALL
HOPE!

THE FIRST THREE ADD P_i, REARRANGE THE SUGAR, AND ADD ANOTHER P_i WITH TWO KICKS FROM ATP.

THAT IS, GLYCOLYSIS FIRST **SPENDS** (OR "INVESTS") TWO ATP MOLECULES BEFORE MAKING ANY NEW ONES. THE OUTCOME IS AN UNSTABLE MOLECULE, FRUCTOSE 1,6-BISPHOSPHATE. (WE OMIT HYDROGEN ATOMS.)

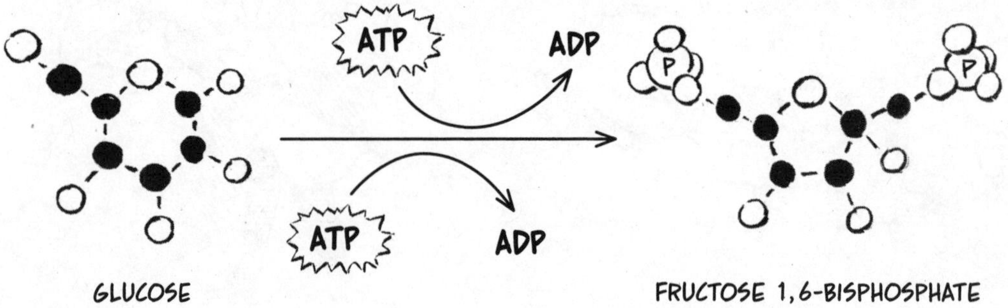

A FOURTH ENZYME EASILY CRACKS FRUCTOSE 1,6-BISPHOSPHATE IN HALF...
SNAP
SNAP
FLEX

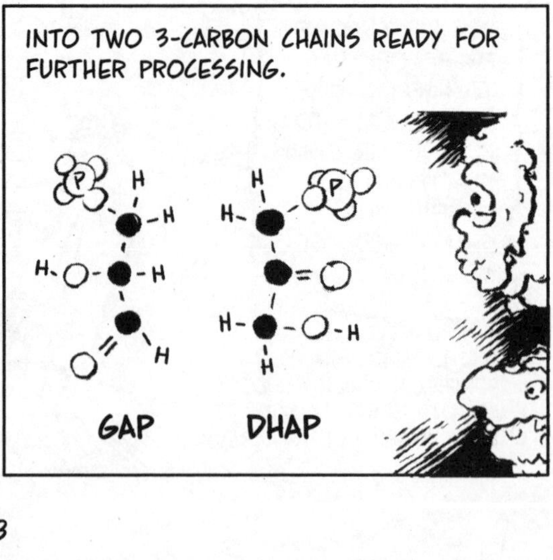
INTO TWO 3-CARBON CHAINS READY FOR FURTHER PROCESSING.
P
P
GAP
DHAP

NOW COMES THE FIRST ATP PAYOFF. ***GAP*** IS OXIDIZED BY NAD^+, FREEING ENOUGH ENERGY TO ADD A SECOND ***P_i***, WHICH IS TRANSFERRED TO ADP TO MAKE ATP.

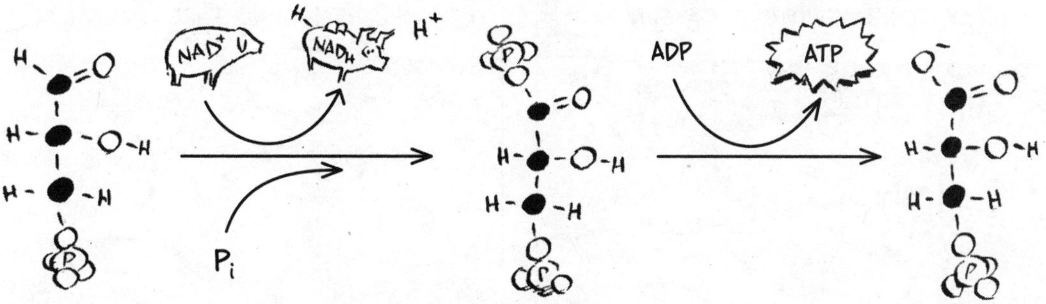

THE TRANSFER TO ATP, ENABLED BY THE ENZYME **PHOSPHOGLYCERATE KINASE,** MOVES THE PHOSPHATE ION FROM THE ORGANIC BISPHOSPHATE (THE ENZYME'S SUBSTRATE) TO ADP. THIS **SUBSTRATE-LEVEL ATP SYNTHESIS** IS THE FIRST OF FOUR IN GLYCOLYSIS.

TWO PHOSPHATES HERE

ONE PHOSPHATE HERE

ANOTHER ENZYME, **PYRUVATE KINASE,** TWEAKS THE PRODUCT AND MOVES ITS LONE PHOSPHATE ONTO ADP, MAKING YET ANOTHER ATP.

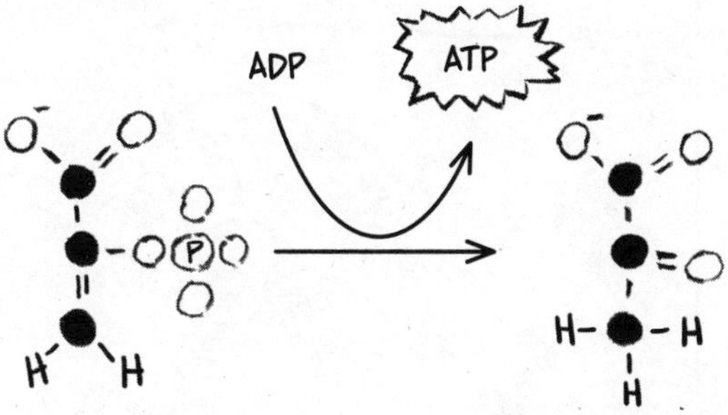

PYRUVATE

THEN EVERY STEP ON THIS PAGE HAPPENS **AGAIN** TO THE SECOND 3-CARBON CHAIN. RESULT:

$$\text{GLUCOSE} + 2\text{ATP} + 2\text{NAD}^+ \longrightarrow 2\text{PYRUVATE} + 4\text{ATP} + 2\text{NADH} + 2\text{H}^+$$

GLYCOLYSIS MAKES TWO NEW ATP MOLECULES PER BROKEN GLUCOSE MOLECULE. THE PROCESS GIVES THE CELL USABLE ENERGY, TOTALLY WITHOUT OXYGEN.
BIOLOGISTS SUSPECT THAT GLYCOLYSIS IS AN ANCIENT PROCESS DATING FROM BILLIONS OF YEARS AGO, WHEN THE ATMOSPHERE HAD NO FREE OXYGEN.
WAKE UP AND SMELL THE AMMONIA!
GLYCOLYSIS IS **INEFFICIENT.** ONLY FOUR ELECTRONS MOVE FROM GLUCOSE TO NAD^+; GLUCOSE IS ONLY SLIGHTLY OXIDIZED, AND THOSE FOUR ELECTRONS STILL CARRY A LOT OF ENERGY.
OXYGEN, WHICH BEGAN TO ACCUMULATE IN THE ATMOSPHERE AFTER SOME EONS, CAN PULL **24** ELECTRONS FROM GLUCOSE AND LIBERATE FAR MORE ENERGY...
SO NOW THAT THERE'S LOTS OF OXYGEN IN THE AIR, TAKE A DEEP BREATH... FILL YOUR LUNGS WITH IT...
AH!
YOU'LL NEED IT TO UNDERSTAND THE FINAL, EFFICIENT STAGES OF AEROBIC CELLULAR RESPIRATION!

Into the Mitochondrion

WHEN OXYGEN IS AVAILABLE, ATP SYNTHESIS GOES INTO OVERDRIVE.
INSTEAD OF **TWO** PALTRY ATP FROM GLYCOLYSIS, THE CELL CAN MAKE NEARLY **TWENTY TIMES** THAT NUMBER AEROBICALLY.
IN EUKARYOTES, ALL REMAINING GLUCOSE OXIDATION AND ATP SYNTHESIS HAPPEN INSIDE A SPECIALIZED ORGANELLE CALLED A **MITOCHONDRION.**
INSIDE ITS OUTER MEMBRANE, THE MITOCHONDRION HAS A HIGHLY CONVOLUTED INNER MEMBRANE ENCLOSING A REGION CALLED THE **MATRIX.**
O=O

WHEN OXYGEN IS PRESENT, EUKARYOTES MOVE ALL KEY INGREDIENTS INTO A MITOCHONDRION. HERE IS WHAT GOES INTO THE MATRIX:

MOLECULAR OXYGEN, O_2

2 PYRUVATE FROM EACH GLYCOLYSIS

4 ELECTRONS FROM 2NADH (PUMPED AT A COST OF 2 ATP. NADH ITSELF DOES NOT PASS THROUGH THE MEMBRANE.)

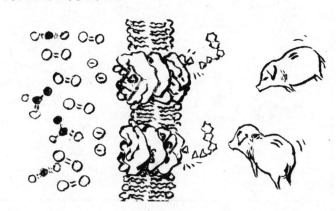

ON ENTERING THE MITOCHONDRIAL MATRIX, PYRUVATE IS OXIDIZED BY NAD^+, AND A CO_2 MOLECULE FALLS OFF.

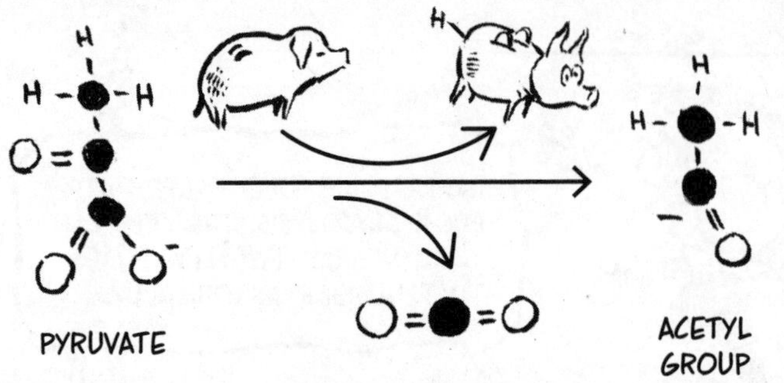

THE ACETYL GROUP IS THEN **FULLY** OXIDIZED BY THE

KREBS CYCLE,

A SERIES OF REACTIONS THAT CHASES ITS OWN TAIL.

THE CYCLE BEGINS WHEN THE ACETYL GROUP JOINS A 4-CARBON CHAIN, OXALOACETATE, TO MAKE A 6-CARBON CITRATE. AS REACTIONS PROCEED, $8e^-$ GO TO NAD^+ AND ITS COUSIN FAD; TWO CO_2 FALL AWAY; AND... THIS MAKES OXALOACETATE AGAIN! IT PICKS UP A NEW ACETYL GROUP, AND THE CYCLE RESUMES. (WE SUPPRESS SOME DETAILS HERE.)

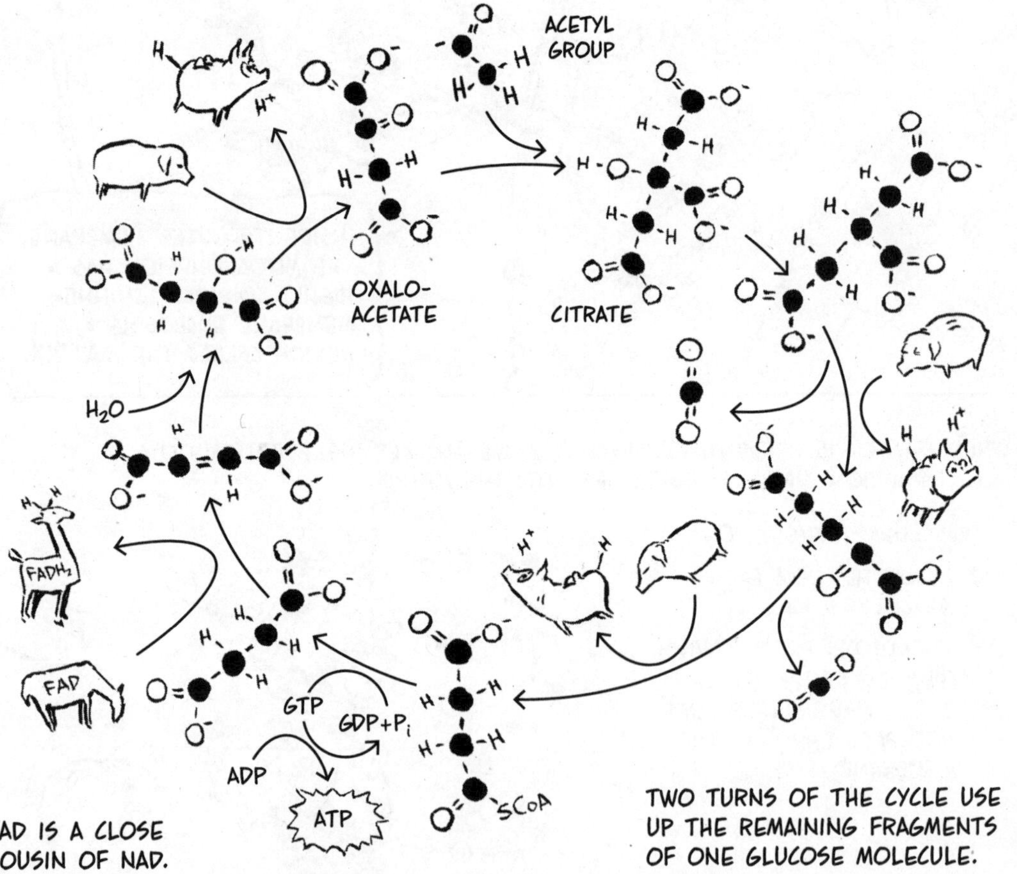

FAD IS A CLOSE COUSIN OF NAD.

TWO TURNS OF THE CYCLE USE UP THE REMAINING FRAGMENTS OF ONE GLUCOSE MOLECULE.

MY HEAD IS CYCLING...

GLUCOSE IS NOW FULLY OXIDIZED. THE OUTPUT OF ALL THE MINI-REDOX REACTIONS, GOING ALL THE WAY BACK TO GLYCOLYSIS, IS THIS:

10 OXIDATIONS BY NAD^+ AND 2 BY FAD HAVE PULLED **24** ELECTRONS OFF GLUCOSE FRAGMENTS. TOTAL OXIDATION!

THESE ALSO BRING **24** PROTONS TO THE PARTY.

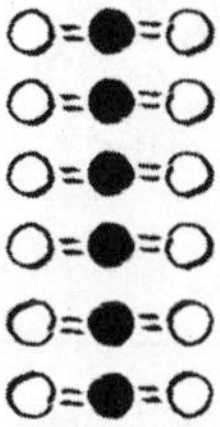

6 CO_2 MOLECULES HAVE BUBBLED AWAY AS WASTE.

NET ATP PRODUCTION STANDS AT A MERE **2 MOLECULES,** AND MOLECULAR OXYGEN HAS YET TO APPEAR. (THE OXYGEN ATOMS IN $6CO_2$ ALL CAME FROM GLUCOSE AND WATER.)

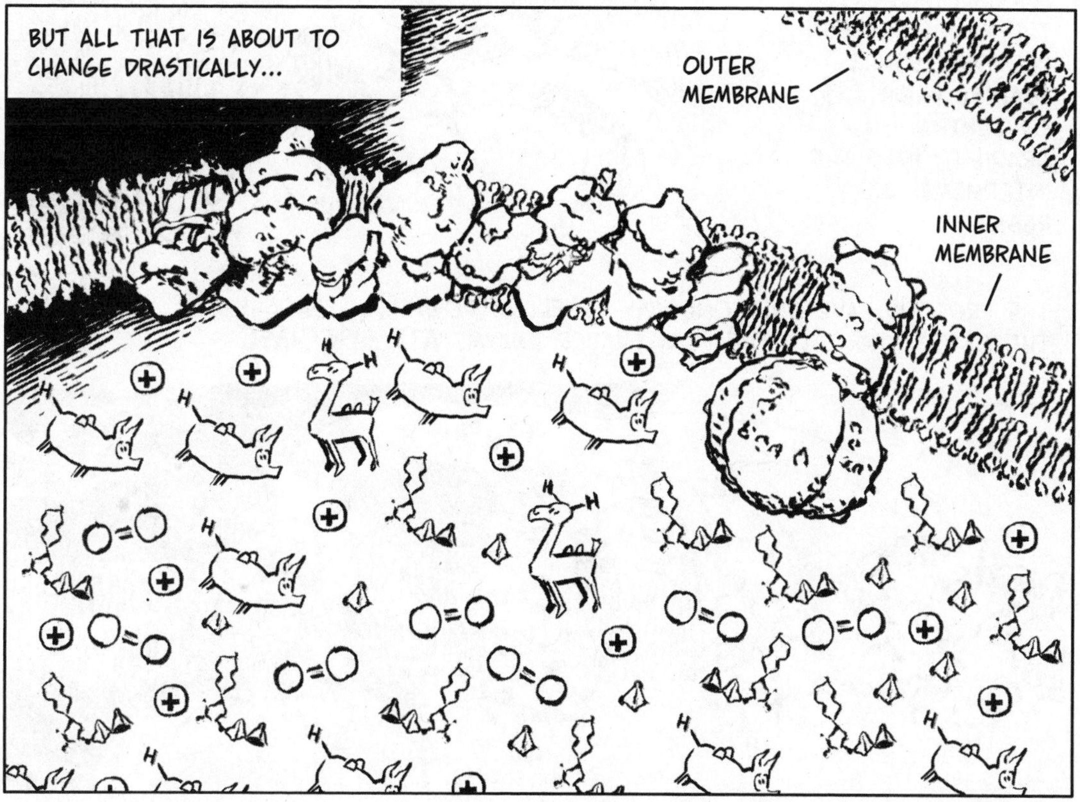
BUT ALL THAT IS ABOUT TO CHANGE DRASTICALLY...
OUTER MEMBRANE
INNER MEMBRANE

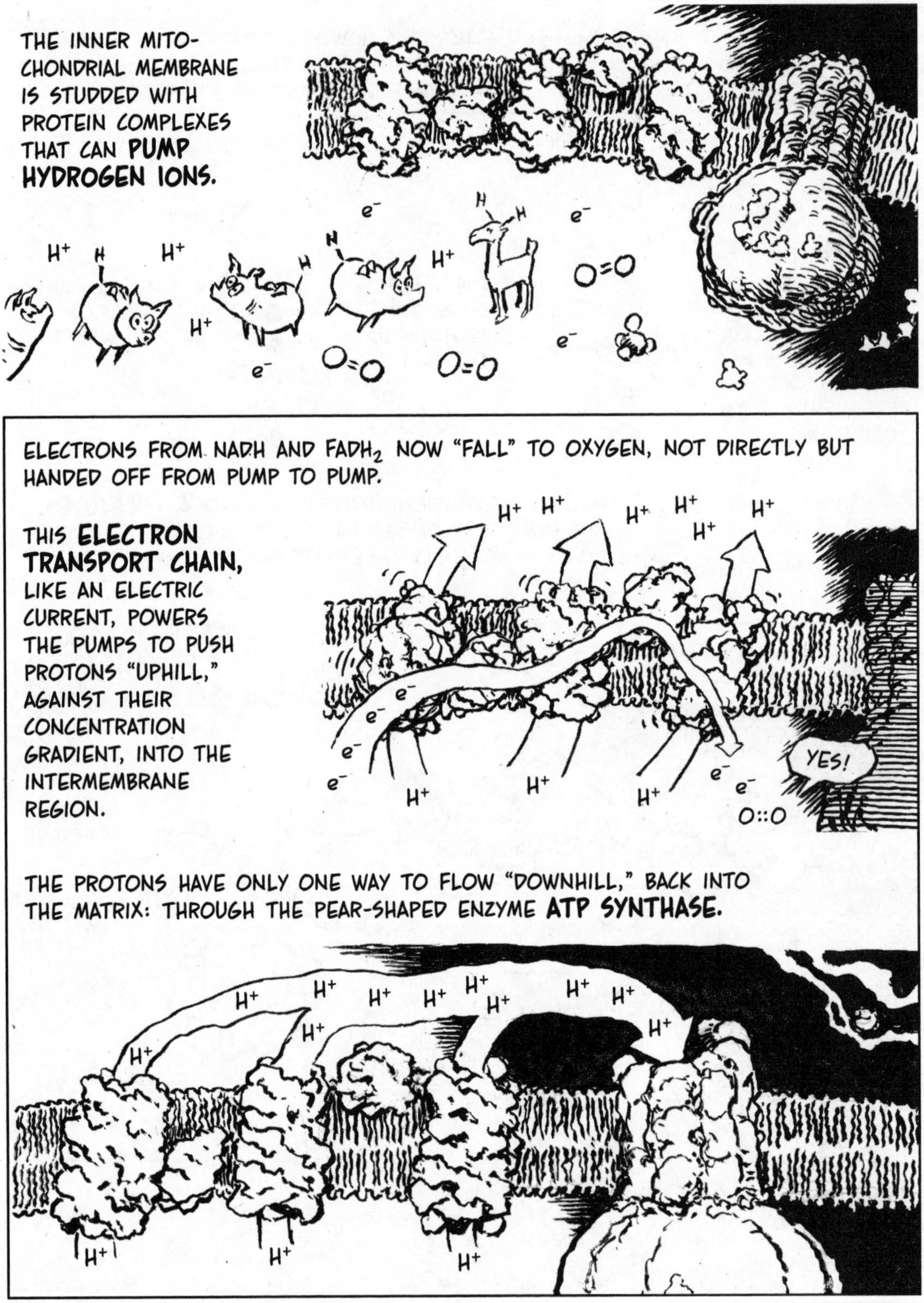
THE INNER MITOCHONDRIAL MEMBRANE IS STUDDED WITH PROTEIN COMPLEXES THAT CAN **PUMP HYDROGEN IONS.**
H⁺
e⁻
O=O
ELECTRONS FROM NADH AND FADH$_2$ NOW "FALL" TO OXYGEN, NOT DIRECTLY BUT HANDED OFF FROM PUMP TO PUMP.
THIS **ELECTRON TRANSPORT CHAIN,** LIKE AN ELECTRIC CURRENT, POWERS THE PUMPS TO PUSH PROTONS "UPHILL," AGAINST THEIR CONCENTRATION GRADIENT, INTO THE INTERMEMBRANE REGION.
YES!
O::O
THE PROTONS HAVE ONLY ONE WAY TO FLOW "DOWNHILL," BACK INTO THE MATRIX: THROUGH THE PEAR-SHAPED ENZYME **ATP SYNTHASE.**

ATP SYNTHASE IS AN ATP-MAKING MACHINE.
FOR A JOB WELL DONE, CALL A SPECIALIST!

ITS BULBOUS BASE HAS THREE BINDING SITES FOR BOTH ADP AND P_i.

THE ENERGY OF PROTON FLOW DRIVES A ROTOR, WHICH SLAPS P_i ONTO ADP AND KNOCKS NEW ATP OFF THE ENZYME.
SLAP

THE PROTON FLOW IS CALLED **CHEMIOSMOSIS** AND THIS ATP PRODUCTION IS CALLED **OXIDATIVE SYNTHESIS** OR **OXIDATIVE PHOSPHORYLATION.**

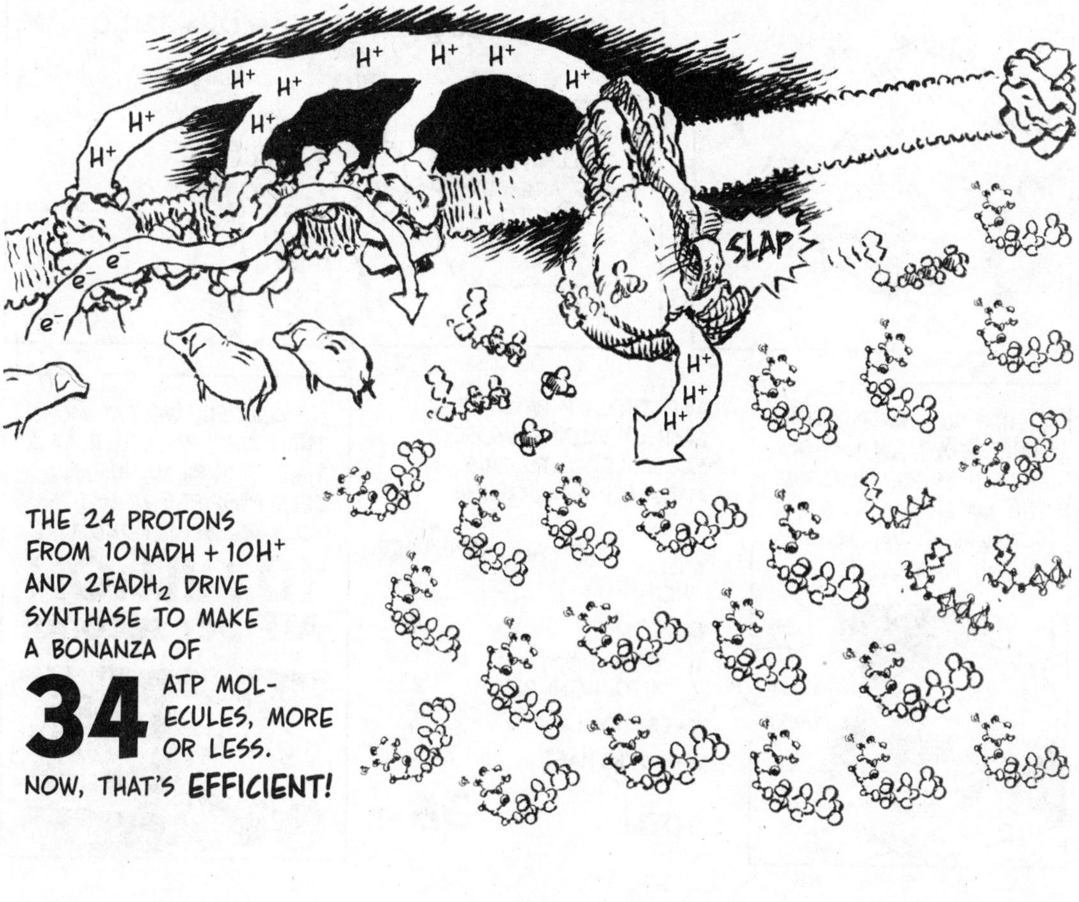

THE 24 PROTONS FROM 10NADH + $10H^+$ AND $2FADH_2$ DRIVE SYNTHASE TO MAKE A BONANZA OF

34 ATP MOLECULES, MORE OR LESS.

NOW, THAT'S **EFFICIENT!**

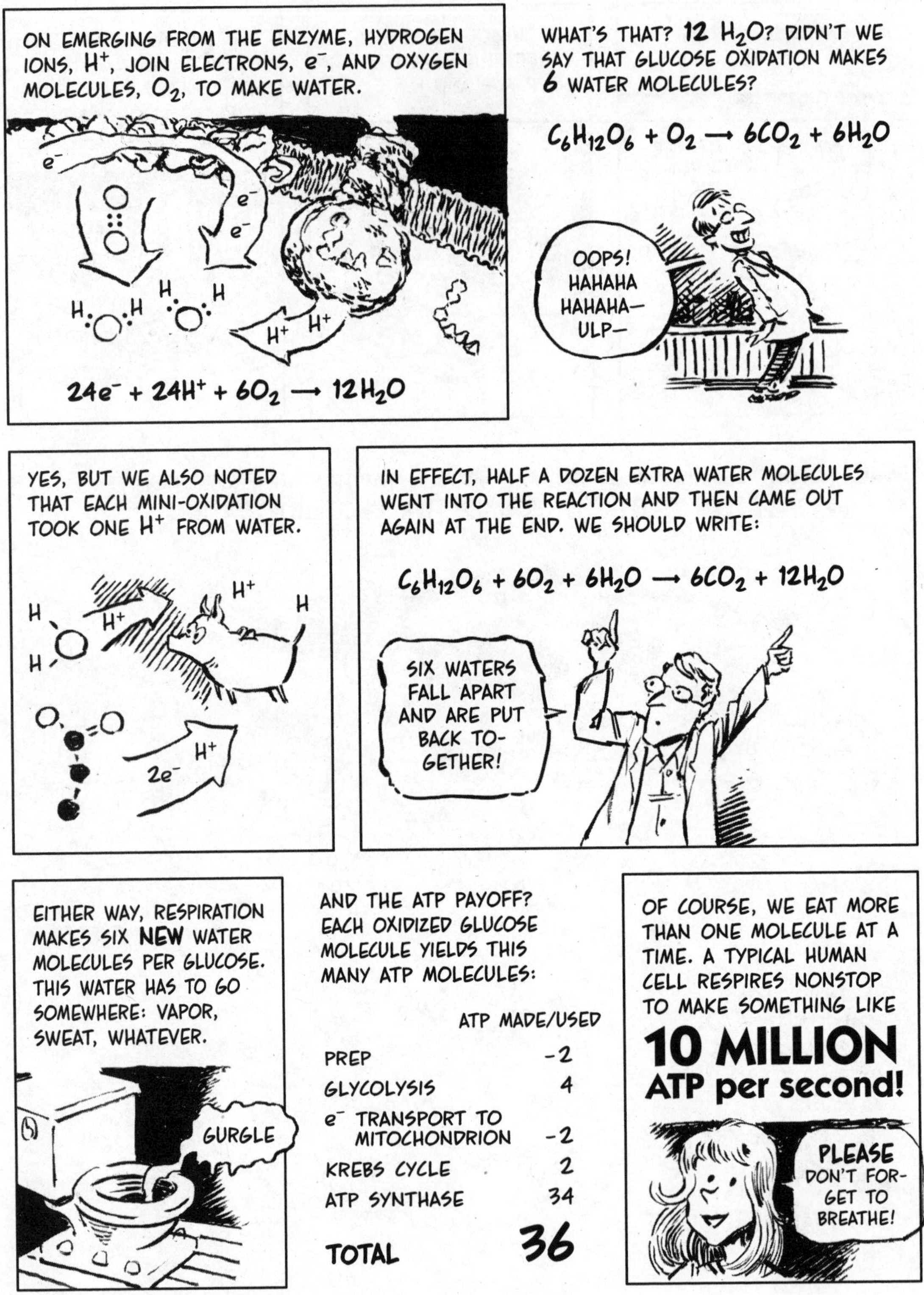

ON EMERGING FROM THE ENZYME, HYDROGEN IONS, H^+, JOIN ELECTRONS, e^-, AND OXYGEN MOLECULES, O_2, TO MAKE WATER.
e^-
e^-
e^-
H H
H H
H^+ H^+
$24e^- + 24H^+ + 6O_2 \longrightarrow 12H_2O$
WHAT'S THAT? 12 H_2O? DIDN'T WE SAY THAT GLUCOSE OXIDATION MAKES 6 WATER MOLECULES?
$C_6H_{12}O_6 + O_2 \longrightarrow 6CO_2 + 6H_2O$
OOPS! HAHAHA HAHAHA— ULP—
YES, BUT WE ALSO NOTED THAT EACH MINI-OXIDATION TOOK ONE H^+ FROM WATER.
H
H
H^+
H^+
H
H^+
$2e^-$
IN EFFECT, HALF A DOZEN EXTRA WATER MOLECULES WENT INTO THE REACTION AND THEN CAME OUT AGAIN AT THE END. WE SHOULD WRITE:
$C_6H_{12}O_6 + 6O_2 + 6H_2O \longrightarrow 6CO_2 + 12H_2O$
SIX WATERS FALL APART AND ARE PUT BACK TOGETHER!
EITHER WAY, RESPIRATION MAKES SIX **NEW** WATER MOLECULES PER GLUCOSE. THIS WATER HAS TO GO SOMEWHERE: VAPOR, SWEAT, WHATEVER.
GURGLE
AND THE ATP PAYOFF? EACH OXIDIZED GLUCOSE MOLECULE YIELDS THIS MANY ATP MOLECULES:
ATP MADE/USED
PREP -2
GLYCOLYSIS 4
e^- TRANSPORT TO MITOCHONDRION -2
KREBS CYCLE 2
ATP SYNTHASE 34
TOTAL 36
OF COURSE, WE EAT MORE THAN ONE MOLECULE AT A TIME. A TYPICAL HUMAN CELL RESPIRES NONSTOP TO MAKE SOMETHING LIKE
10 MILLION ATP per second!
PLEASE DON'T FORGET TO BREATHE!

TO SUMMARIZE, WE MAY DIVIDE AEROBIC RESPIRATION INTO TWO MAIN STEPS. STEP ONE FULLY OXIDIZES GLUCOSE AND EMITS CO_2. THIS STEP INCLUDES GLYCOLYSIS, ENTRY INTO THE MITOCHONDRION, AND THE KREBS CYCLE.

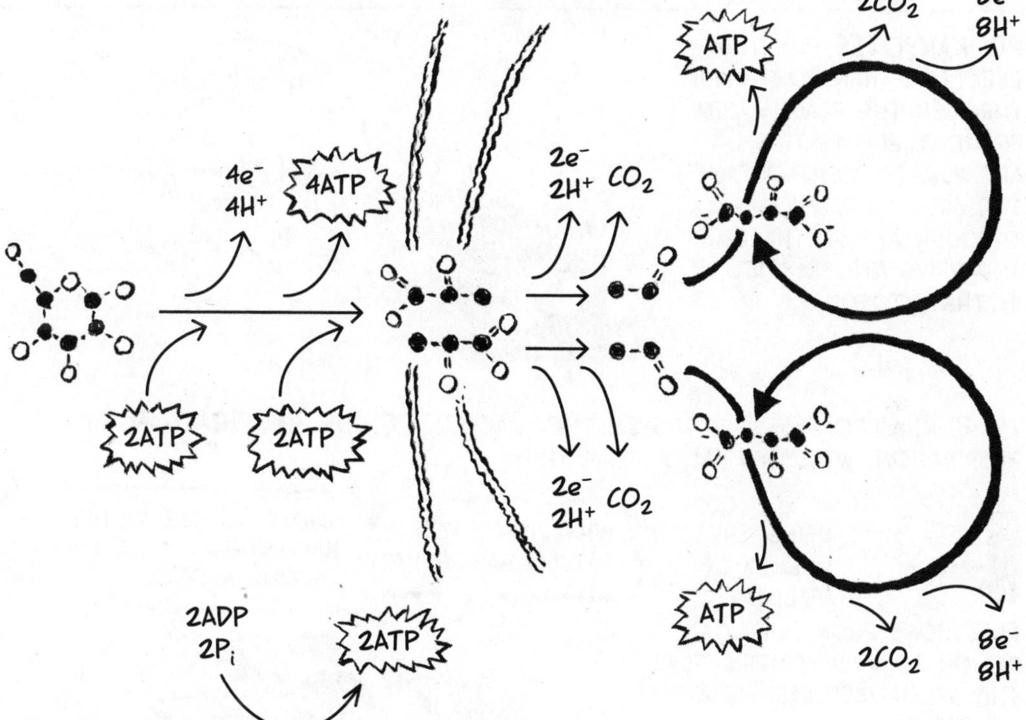

$$C_6H_{12}O_6 + 6H_2O \longrightarrow 6CO_2 + 24e^- + 24H^+$$

IN STEP TWO, ELECTRONS "FALL" TO OXYGEN AND RELEASE ENERGY THAT DRIVES WHOLESALE ATP SYNTHESIS. PROTONS AND ELECTRONS JOIN OXYGEN TO FORM WATER.

$$24e^- + 24H^+ + 6O_2 \longrightarrow 12H_2O$$

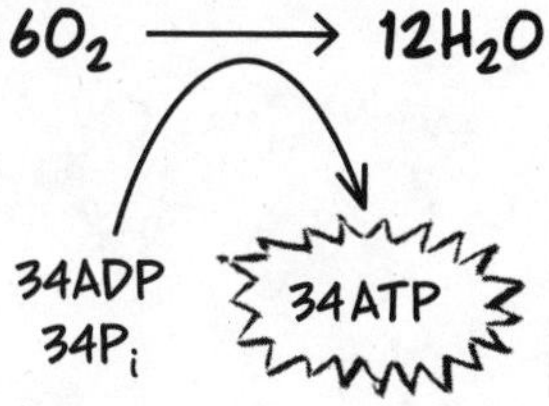

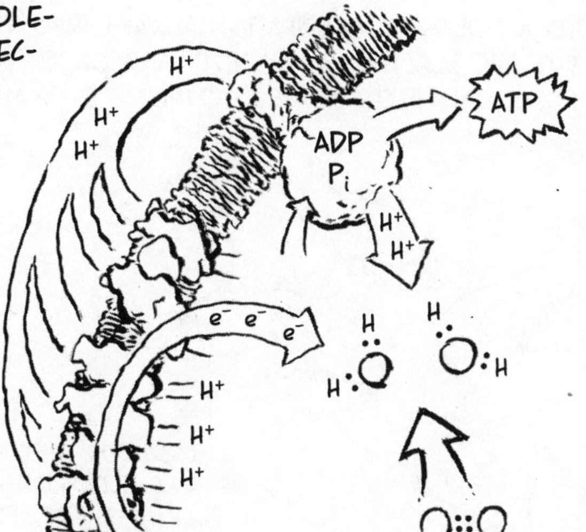

Respiration Defined (Finally!)

EUKARYOTES LIKE OURSELVES MAKE NEARLY ALL OUR ATP INSIDE MITOCHONDRIA.

PROKARYOTES RUN THEIR ELECTRON TRANSPORT CHAIN THROUGH THE PLASMA MEMBRANE ITSELF. PROTONS ARE PUMPED RIGHT OUT OF THE CELL; THEY FLOW BACK THROUGH ATP SYNTHASE AND DRIVE ATP PRODUCTION IN THE CYTOSOL.

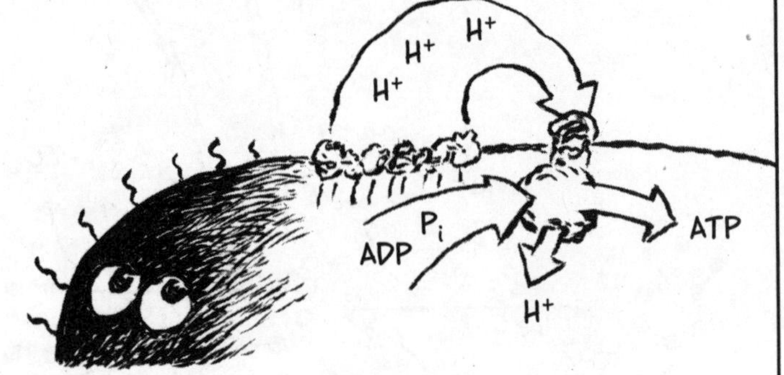

THESE KEY PROCESSES, WHEREVER THEY HAPPEN, **DEFINE RESPIRATION.** BY RESPIRATION, WE SPECIFICALLY MEAN THIS:

USING THE ENERGY OF "FALLING" ELECTRONS FROM FUEL OXIDATION TO DRIVE (INDIRECTLY) THE SPECIALIZED ENZYME ATP SYNTHASE.

WHY DIDN'T YOU SAY SO EARLIER??
'CAUSE NO ONE WOULD HAVE KNOWN WHAT I WAS TALKING ABOUT?

TO A BIOLOGIST, RESPIRATION ALWAYS INVOLVES AN ELECTRON TRANSPORT CHAIN, BUT THE ELECTRONS MAY NOT LAND ON **OXYGEN.** IN **ANAEROBIC** RESPIRATION, ELECTRONS FLOW TO SOME OTHER "TERMINAL ELECTRON ACCEPTOR."

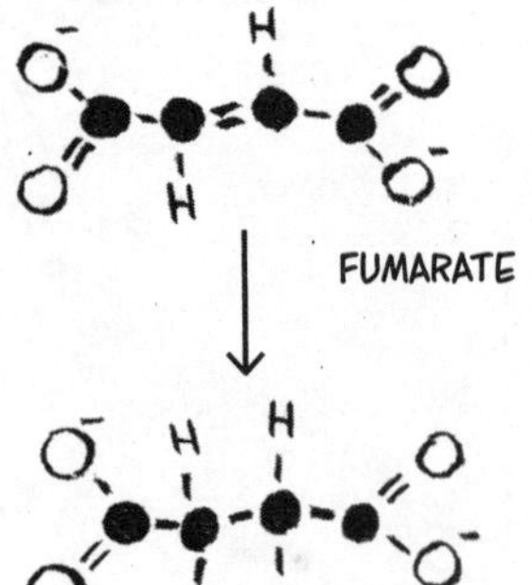

E. COLI, FOR INSTANCE, USES ORGANIC ELECTRON ACCEPTORS, SUCH AS (HERE) FUMARATE.

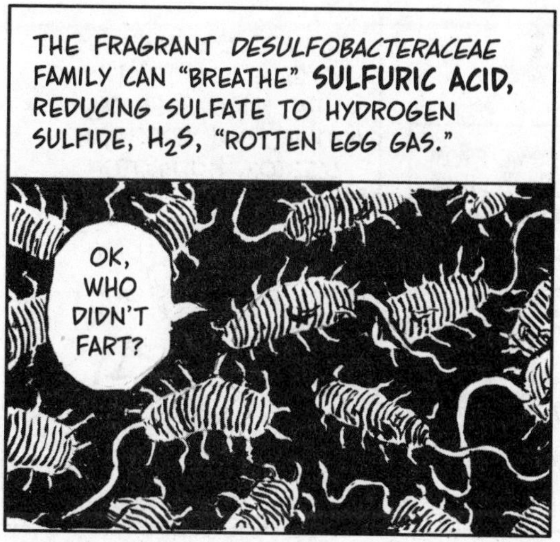
THE FRAGRANT *DESULFOBACTERACEAE* FAMILY CAN "BREATHE" **SULFURIC ACID,** REDUCING SULFATE TO HYDROGEN SULFIDE, H_2S, "ROTTEN EGG GAS."
OK, WHO DIDN'T FART?

SHEWANELLA ONEIDENSIS REDUCES URANIUM(!) AND OTHER HEAVY METALS, A TALENT BEING PUT TO USE IN CLEANING UP RADIOACTIVE WASTEWATER.
NOW **THAT'S** WEIRD...

VERSATILITY ALSO PAYS IN THE CHOICE OF **FUEL**. MOST CELLS FAVOR SUGARS, BUT ALL SORTS OF ORGANIC MOLECULES CAN BE OXIDIZED.

I'M OFF CARBS!

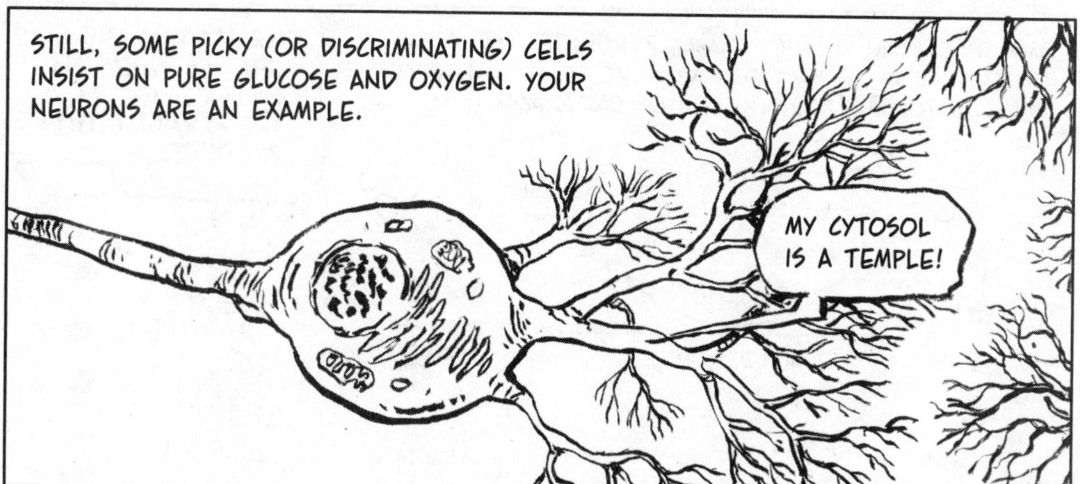
STILL, SOME PICKY (OR DISCRIMINATING) CELLS INSIST ON PURE GLUCOSE AND OXYGEN. YOUR NEURONS ARE AN EXAMPLE.
MY CYTOSOL IS A TEMPLE!

FERMENTATION

IN RESPIRATORS, NEARLY ALL ATP IS MADE BY ATP SYNTHASE...

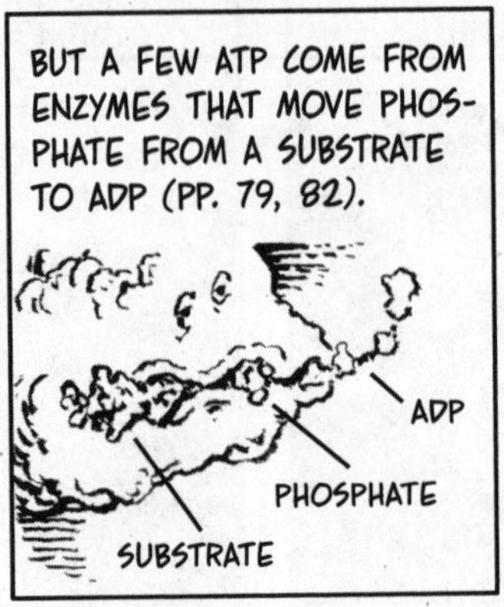
BUT A FEW ATP COME FROM ENZYMES THAT MOVE PHOSPHATE FROM A SUBSTRATE TO ADP (PP. 79, 82).
ADP
PHOSPHATE
SUBSTRATE

THIS PROCESS INVOLVES NO ELECTRON TRANSPORT CHAIN, NO OXYGEN OR OTHER FINAL ELECTRON ACCEPTOR, NO PROTON FLOW, NO ATP SYNTHASE.

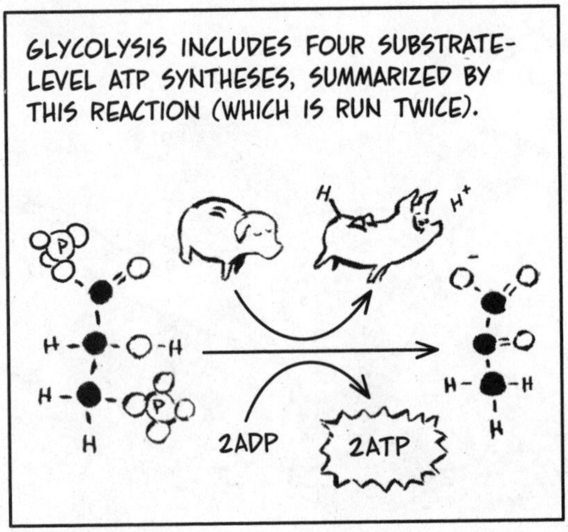
GLYCOLYSIS INCLUDES FOUR SUBSTRATE-LEVEL ATP SYNTHESES, SUMMARIZED BY THIS REACTION (WHICH IS RUN TWICE).
2ADP
2ATP

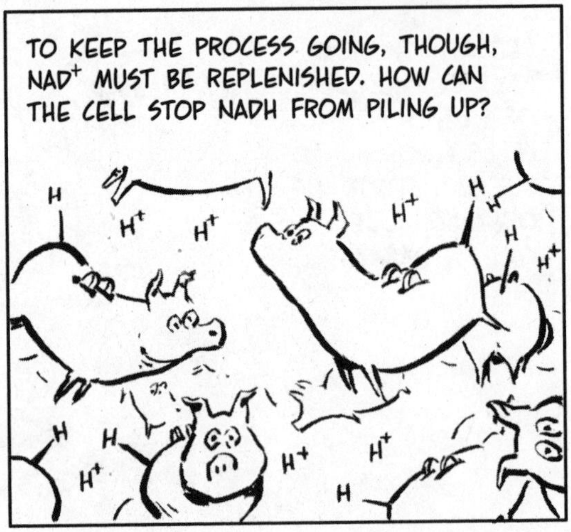
TO KEEP THE PROCESS GOING, THOUGH, NAD+ MUST BE REPLENISHED. HOW CAN THE CELL STOP NADH FROM PILING UP?

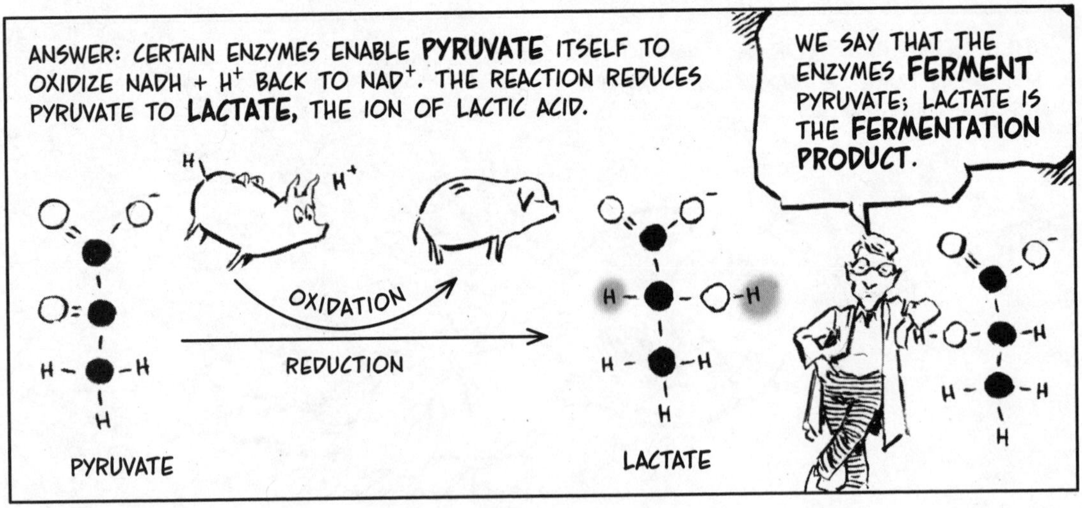
ANSWER: CERTAIN ENZYMES ENABLE PYRUVATE ITSELF TO OXIDIZE NADH + H+ BACK TO NAD+. THE REACTION REDUCES PYRUVATE TO LACTATE, THE ION OF LACTIC ACID.
OXIDATION
REDUCTION
PYRUVATE
LACTATE
WE SAY THAT THE ENZYMES FERMENT PYRUVATE; LACTATE IS THE FERMENTATION PRODUCT.

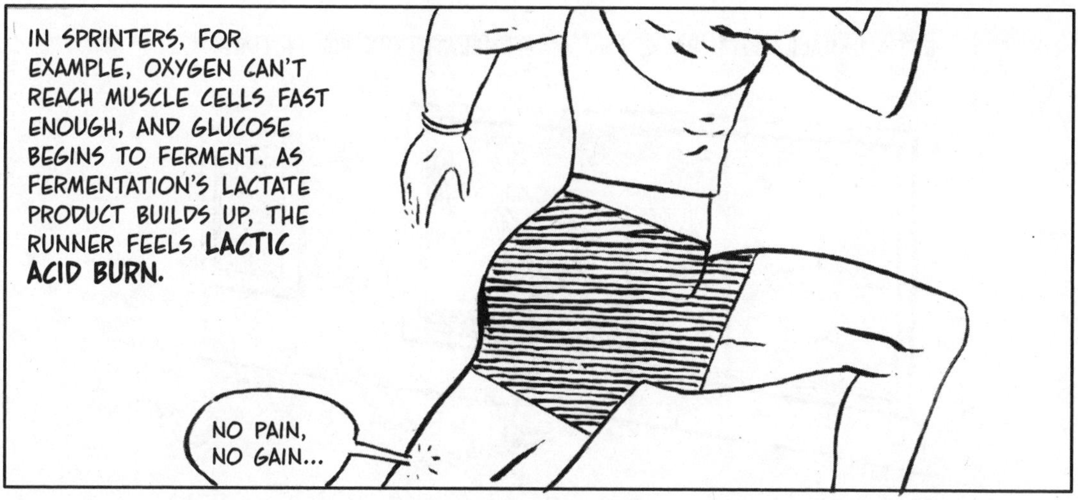
IN SPRINTERS, FOR EXAMPLE, OXYGEN CAN'T REACH MUSCLE CELLS FAST ENOUGH, AND GLUCOSE BEGINS TO FERMENT. AS FERMENTATION'S LACTATE PRODUCT BUILDS UP, THE RUNNER FEELS **LACTIC ACID BURN.**
NO PAIN, NO GAIN...

A DIFFERENT FERMENTATION PROCESS BREAKS DOWN PYRUVATE, RELEASES CO_2, AND OXIDIZES NADH + H^+ TO YIELD **ETHYL ALCOHOL.**

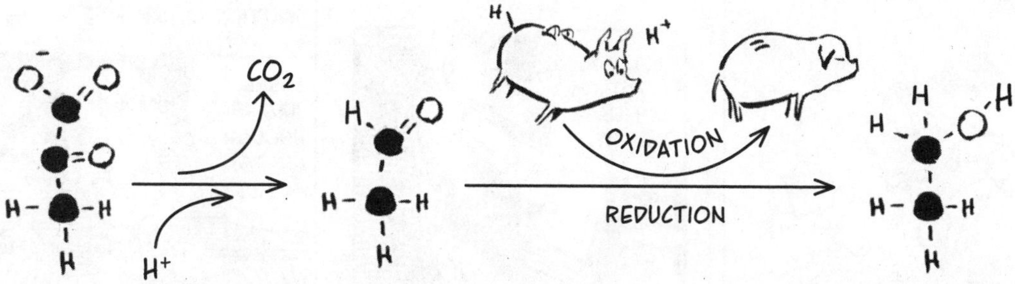
CO_2
H^+
OXIDATION
REDUCTION

PYRUVATE ACETALDEHYDE ETHYL ALCOHOL

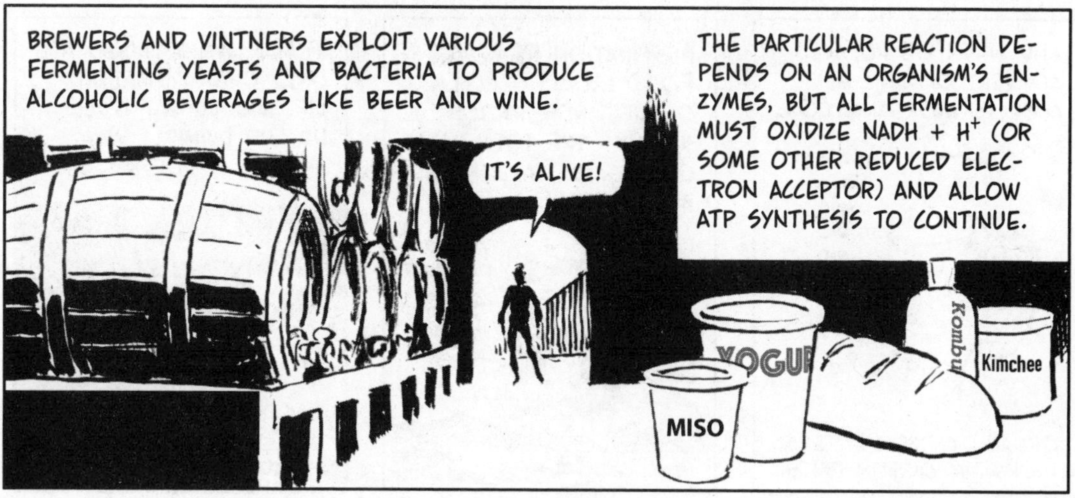
BREWERS AND VINTNERS EXPLOIT VARIOUS FERMENTING YEASTS AND BACTERIA TO PRODUCE ALCOHOLIC BEVERAGES LIKE BEER AND WINE.
IT'S ALIVE!
THE PARTICULAR REACTION DEPENDS ON AN ORGANISM'S ENZYMES, BUT ALL FERMENTATION MUST OXIDIZE NADH + H^+ (OR SOME OTHER REDUCED ELECTRON ACCEPTOR) AND ALLOW ATP SYNTHESIS TO CONTINUE.
MISO
Kimchee

OKAYYYY... LET'S EXHALE, STEP BACK, AND CONTEMPLATE THE BIG PICTURE.

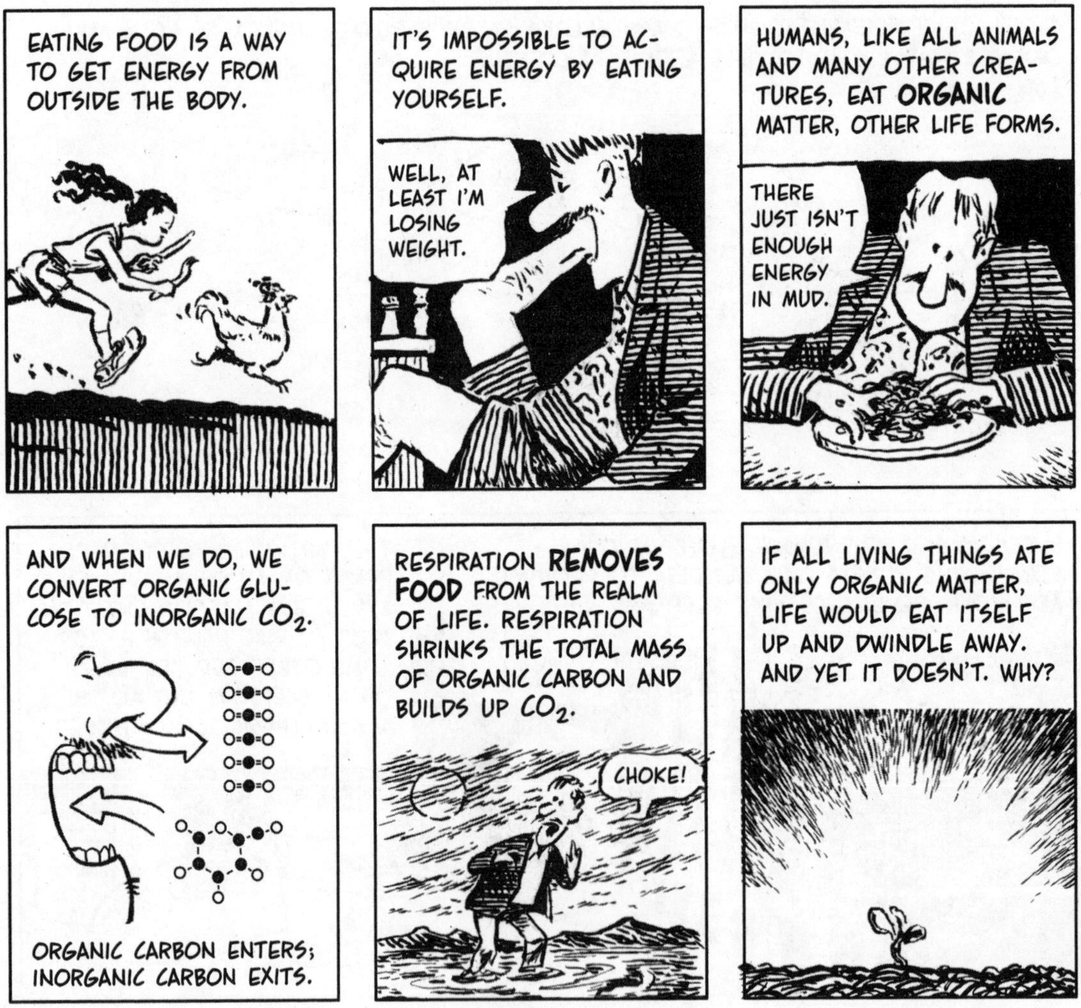
EATING FOOD IS A WAY TO GET ENERGY FROM OUTSIDE THE BODY.
IT'S IMPOSSIBLE TO ACQUIRE ENERGY BY EATING YOURSELF.
WELL, AT LEAST I'M LOSING WEIGHT.
HUMANS, LIKE ALL ANIMALS AND MANY OTHER CREATURES, EAT **ORGANIC** MATTER, OTHER LIFE FORMS.
THERE JUST ISN'T ENOUGH ENERGY IN MUD.
AND WHEN WE DO, WE CONVERT ORGANIC GLUCOSE TO INORGANIC CO_2.
ORGANIC CARBON ENTERS; INORGANIC CARBON EXITS.
RESPIRATION **REMOVES FOOD** FROM THE REALM OF LIFE. RESPIRATION SHRINKS THE TOTAL MASS OF ORGANIC CARBON AND BUILDS UP CO_2.
CHOKE!
IF ALL LIVING THINGS ATE ONLY ORGANIC MATTER, LIFE WOULD EAT ITSELF UP AND DWINDLE AWAY. AND YET IT DOESN'T. WHY?

Chapter 7
PHOTOSYNTHESIS

RADIANT ENERGY DOES SOME LIGHT WORK

ALL AROUND US, ANIMALS ARE TAKING FOOD OUT OF THE WORLD. WITH EVERY BREATH, THEY MOVE CARBON FROM ORGANIC GLUCOSE TO INORGANIC CO_2. UNLESS SOMETHING, SOMEHOW, PUSHES INORGANIC CARBON BACK INTO ORGANIC MOLECULES, FOOD WOULD RUN OUT. THAT, READERS, IS WHAT **GREEN PLANTS** DO. THEY TAKE CO_2 FROM THE ATMOSPHERE AND MAKE IT ORGANIC. **PLANTS EAT AIR.**

THIS TRICK IS CALLED **CARBON FIXING.** "LOOSE" CARBON, AS AIRBORNE CO_2, ENTERS PLANT CELLS, WHICH **AFFIX** IT TO AN ORGANIC MOLECULE. BY DOING SO, PLANTS MAKE FOOD, BOTH FOR THEMSELVES AND FOR EVERYBODY ELSE.

THANKS FROM THE BOTTOM OF MY HEART AND OTHER ORGANS...

THE REACTION JOINS CO_2 AND WATER TO A DOUBLY PHOSPHORYLATED 5-CARBON SUGAR, **RIBULOSE 1,5-BISPHOSPHATE (RuBP),** WHICH THEN SPLITS INTO TWO PHOSPHORYLATED 3-CARBON CHAINS.

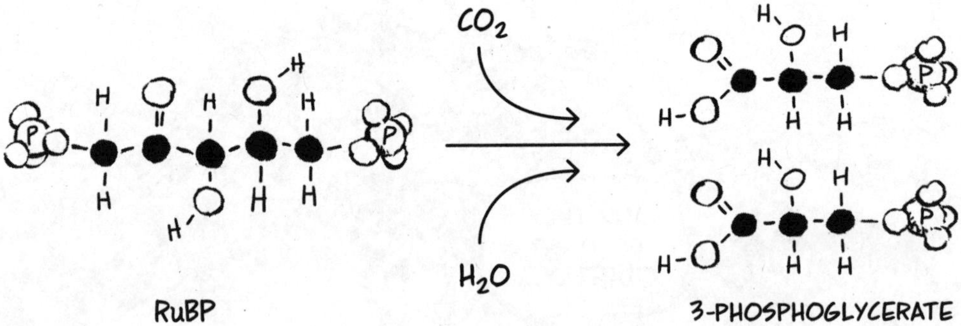

NATURALLY, AN ENZYME ENABLES THE REACTION: A GIGANTIC PROTEIN CALLED **RUBISCO,** MOLECULAR WEIGHT AROUND 500,000, AND THE MOST ABUNDANT AND IMPORTANT PROTEIN ON EARTH.

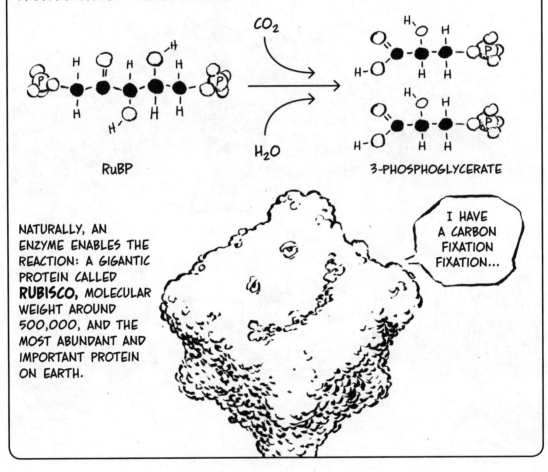

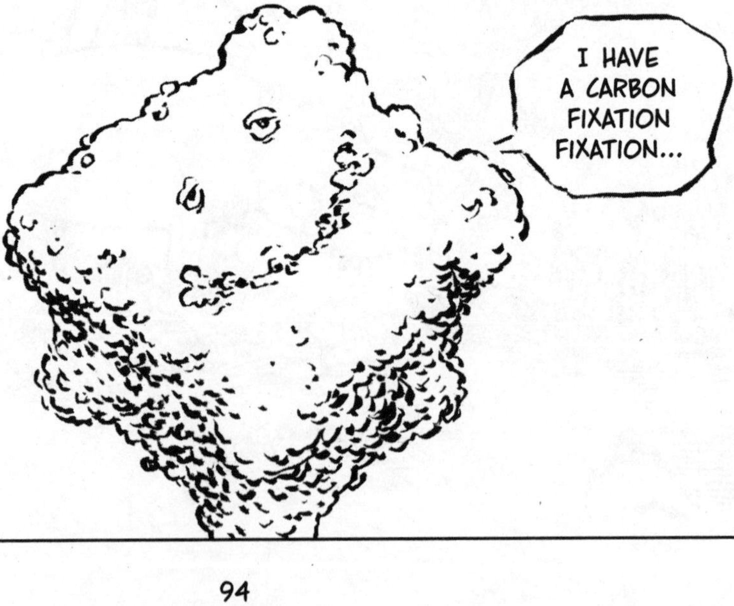

I HAVE A CARBON FIXATION FIXATION...

AS ENZYMES GO, RUBISCO IS DEAD SLOW, CATALYZING A MERE THREE REACTIONS PER SECOND. (MOST ENZYMES DO HUNDREDS OR MORE IN THAT TIME.)
COME ON!!!
IT IS ALSO ERROR-PRONE, OCCASIONALLY PUTTING O_2 WHERE CO_2 IS SUPPOSED TO GO. (OTHER ENZYMES STEP IN TO REPAIR THE BAD OUTPUT.)
OH, MANNN...
DESPITE (OR BECAUSE OF?) THIS INEFFICIENCY, PLANTS MAKE LOADS OF RUBISCO, ENOUGH TO DRAW TENS OF BILLIONS OF TONS OF CARBON OUT OF THE ATMOSPHERE EVERY DAY WORLDWIDE.
ALWAYS READY FOR A JOB DONE BADLY!
EVEN SO, RUBISCO IS NOT ENOUGH. FOR PROPER METABOLISM, PLANTS NEED TO REDUCE RUBISCO'S PHOSPHOGLYCERATE OUTPUT, AS WELL AS TO CONTINUALLY MAKE MORE SUBSTRATE, RuBP. RUBISCO IS MERELY A CRITICAL INTERMEDIARY.

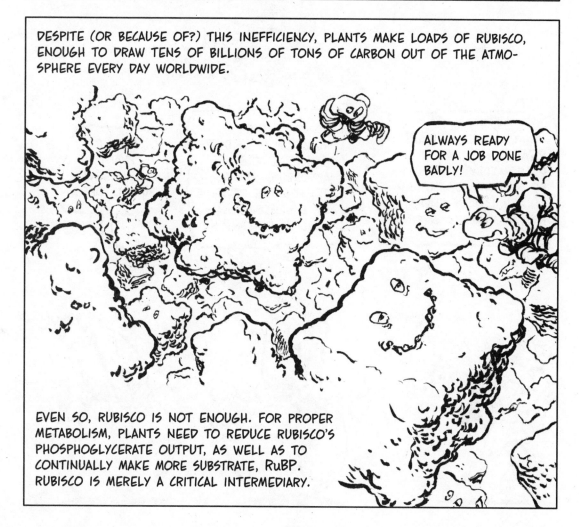

ULTIMATELY, IN FACT, THE PLANT MAKES **GLUCOSE** FROM CARBON DIOXIDE AND WATER.

$$6CO_2 + 6H_2O \longrightarrow C_6H_{12}O_6 + 6O_2$$

LOOK FAMILIAR? IT OUGHT TO! THIS IS NOTHING BUT GLUCOSE OXIDATION IN REVERSE!
OR, TO PUT IT ANOTHER WAY, IT INVOLVES THE SPLITTING OF WATER.

THAT'S A REALLY HARD REACTION! OXYGEN DESPERATELY "WANTS" TO BE WITH HYDROGEN, AND IT TAKES A LOT OF ENERGY TO PRY THEM APART. GRRR...
THAT IS, THE REACTION IS ENDER-GONIC, ENERGY-ABSORBING. IT HAS TO BE: IT'S THE REVERSE OF EXERGONIC GLU-COSE OXIDATION!
AND YET PLANTS SOMEHOW PULL IT OFF...

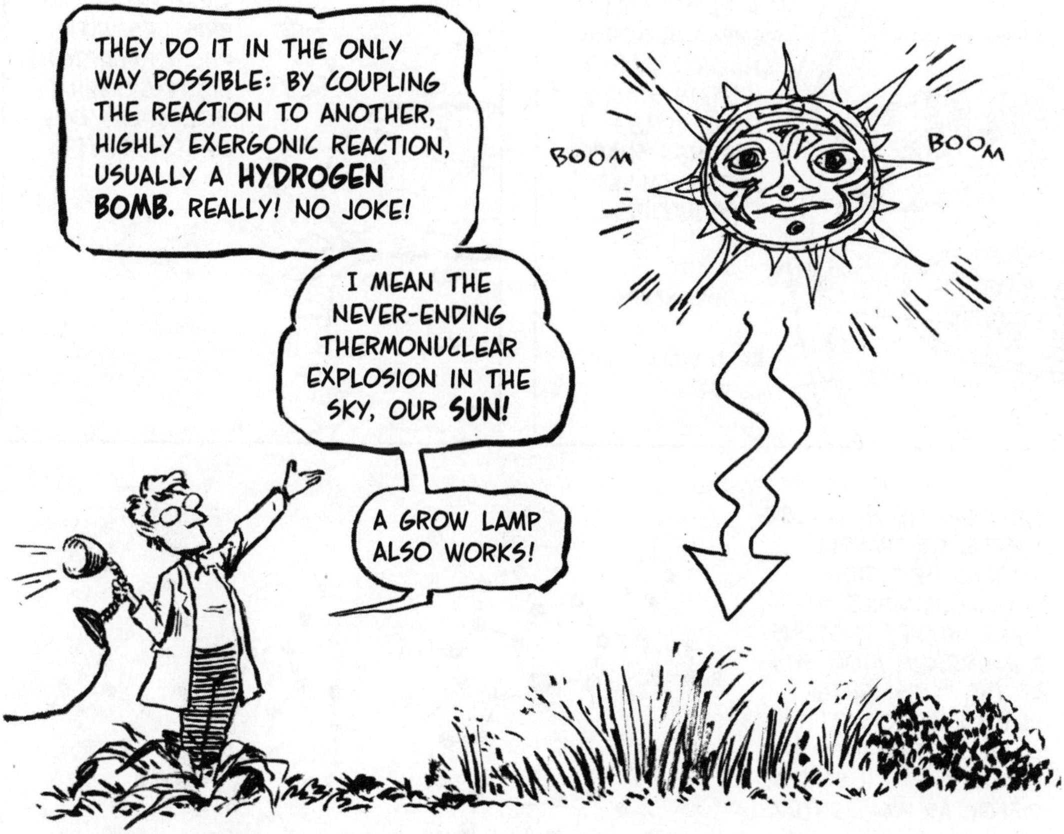
THEY DO IT IN THE ONLY WAY POSSIBLE: BY COUPLING THE REACTION TO ANOTHER, HIGHLY EXERGONIC REACTION, USUALLY A **HYDROGEN BOMB.** REALLY! NO JOKE!
I MEAN THE NEVER-ENDING THERMONUCLEAR EXPLOSION IN THE SKY, OUR **SUN!**
A GROW LAMP ALSO WORKS!
BOOM
BOOM

THE SYNTHESIS OF GLUCOSE RUNS ON **PHOTONS,** ENERGETIC PARTICLES OF LIGHT, WHICH GIVE THE PROCESS ITS NAME:

PHOTO-SYNTHESIS!

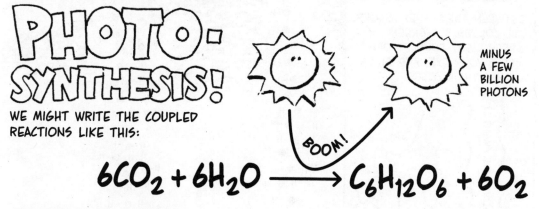

MINUS A FEW BILLION PHOTONS
BOOM!

WE MIGHT WRITE THE COUPLED REACTIONS LIKE THIS:

$$6CO_2 + 6H_2O \longrightarrow C_6H_{12}O_6 + 6O_2$$

PLANT CELLS HAVE SPECIAL MACHINERY FOR CAPTURING SOLAR ENERGY.

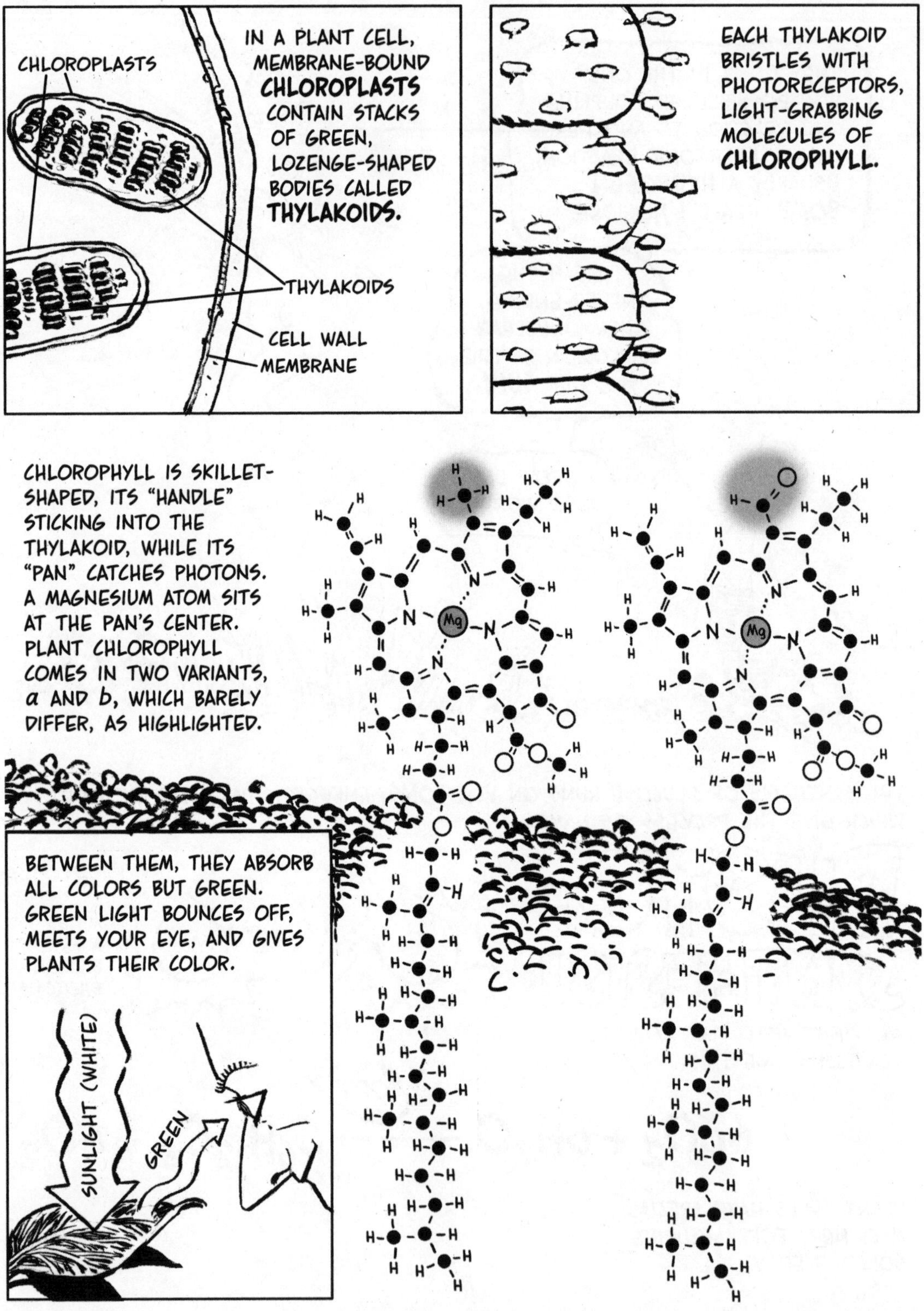
IN A PLANT CELL, MEMBRANE-BOUND **CHLOROPLASTS** CONTAIN STACKS OF GREEN, LOZENGE-SHAPED BODIES CALLED **THYLAKOIDS.**
CHLOROPLASTS
THYLAKOIDS
CELL WALL
MEMBRANE
EACH THYLAKOID BRISTLES WITH PHOTORECEPTORS, LIGHT-GRABBING MOLECULES OF **CHLOROPHYLL.**
CHLOROPHYLL IS SKILLET-SHAPED, ITS "HANDLE" STICKING INTO THE THYLAKOID, WHILE ITS "PAN" CATCHES PHOTONS. A MAGNESIUM ATOM SITS AT THE PAN'S CENTER. PLANT CHLOROPHYLL COMES IN TWO VARIANTS, *a* AND *b*, WHICH BARELY DIFFER, AS HIGHLIGHTED.
BETWEEN THEM, THEY ABSORB ALL COLORS BUT GREEN. GREEN LIGHT BOUNCES OFF, MEETS YOUR EYE, AND GIVES PLANTS THEIR COLOR.
SUNLIGHT (WHITE)
GREEN

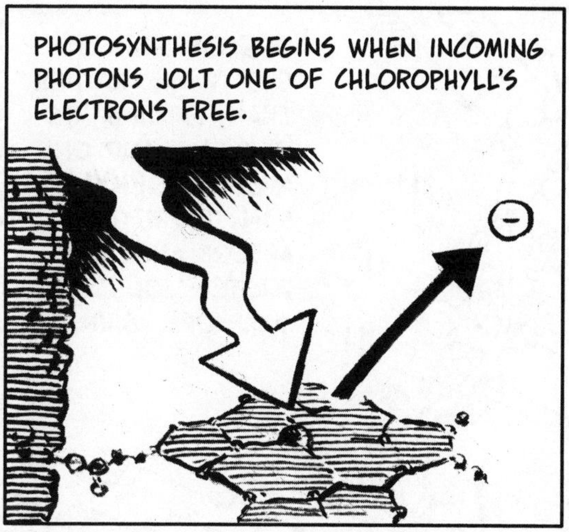
PHOTOSYNTHESIS BEGINS WHEN INCOMING PHOTONS JOLT ONE OF CHLOROPHYLL'S ELECTRONS FREE.

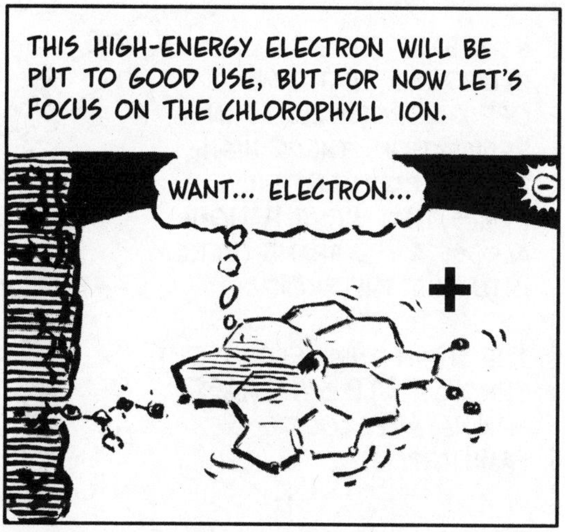
THIS HIGH-ENERGY ELECTRON WILL BE PUT TO GOOD USE, BUT FOR NOW LET'S FOCUS ON THE CHLOROPHYLL ION.
WANT... ELECTRON...

CHLOROPHYLL$^+$ CRAVES A NEW ELECTRON SO INTENSELY THAT IT CAN OUTBID THE CHAMPION ELECTRON-HUGGER ITSELF, **OXYGEN.** CHLOROPHYLL$^+$ **OXIDIZES** (TAKES AN ELECTRON FROM) **WATER,** NO MEAN FEAT! A PROTON ALSO LEAVES.

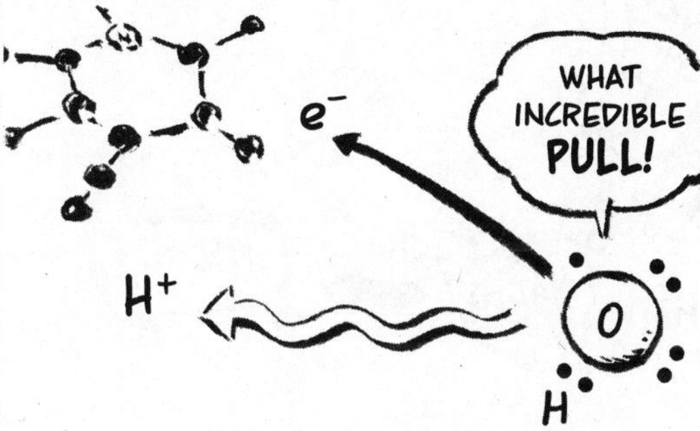
WHAT INCREDIBLE PULL!
e⁻
H⁺
O
H

REPEATED FOUR TIMES, THE REACTION MAKES ONE OXYGEN MOLECULE.

4CHLOROPHYLL$^+$ → 4CHLOROPHYLL

$$2H_2O \longrightarrow 4H^+ + O_2$$

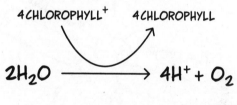

THIS OXYGEN ESCAPES AND ENTERS THE AIR. THE PROTONS REMAIN DISSOLVED.
O₂
AHHHH...

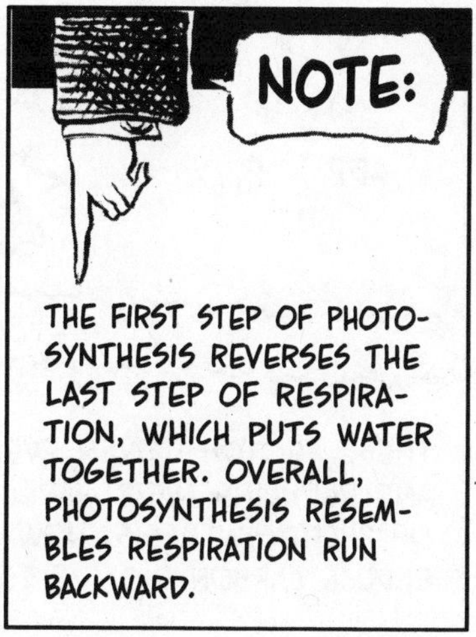
NOTE:
THE FIRST STEP OF PHOTOSYNTHESIS REVERSES THE LAST STEP OF RESPIRATION, WHICH PUTS WATER TOGETHER. OVERALL, PHOTOSYNTHESIS RESEMBLES RESPIRATION RUN BACKWARD.

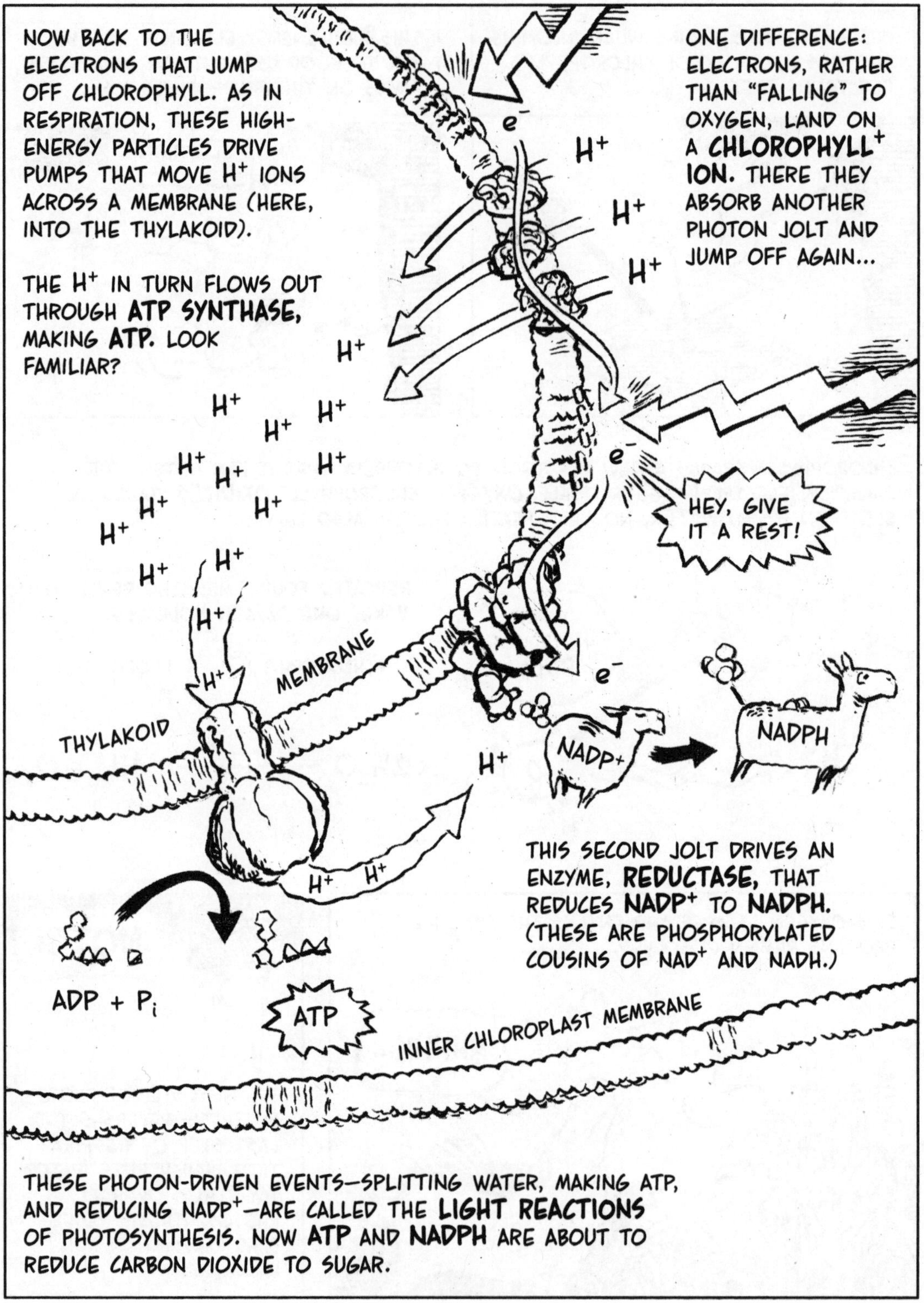
NOW BACK TO THE ELECTRONS THAT JUMP OFF CHLOROPHYLL. AS IN RESPIRATION, THESE HIGH-ENERGY PARTICLES DRIVE PUMPS THAT MOVE H^+ IONS ACROSS A MEMBRANE (HERE, INTO THE THYLAKOID).
THE H^+ IN TURN FLOWS OUT THROUGH **ATP SYNTHASE,** MAKING **ATP.** LOOK FAMILIAR?
ONE DIFFERENCE: ELECTRONS, RATHER THAN "FALLING" TO OXYGEN, LAND ON A **CHLOROPHYLL$^+$ ION.** THERE THEY ABSORB ANOTHER PHOTON JOLT AND JUMP OFF AGAIN...
e^-
H^+
HEY, GIVE IT A REST!
MEMBRANE
THYLAKOID
$NADP^+$
NADPH
THIS SECOND JOLT DRIVES AN ENZYME, **REDUCTASE,** THAT REDUCES **$NADP^+$** TO **NADPH.** (THESE ARE PHOSPHORYLATED COUSINS OF NAD^+ AND NADH.)
ADP + P_i
ATP
INNER CHLOROPLAST MEMBRANE
THESE PHOTON-DRIVEN EVENTS—SPLITTING WATER, MAKING ATP, AND REDUCING $NADP^+$—ARE CALLED THE **LIGHT REACTIONS** OF PHOTOSYNTHESIS. NOW **ATP** AND **NADPH** ARE ABOUT TO REDUCE CARBON DIOXIDE TO SUGAR.

THE **CALVIN CYCLE,** POWERED BY ATP, FIXES CARBON TO RuBP, REDUCES THE PRODUCT TO SUGAR, AND BUILDS MORE RuBP TO START THE PROCESS AGAIN. COUNTING CARBONS, YOU CAN SEE THAT $3CO_2$ AND 3RuBP (18 CARBONS) GENERATE JUST ONE 3-CARBON SUGAR MOLECULE, PLUS 3RuBP.

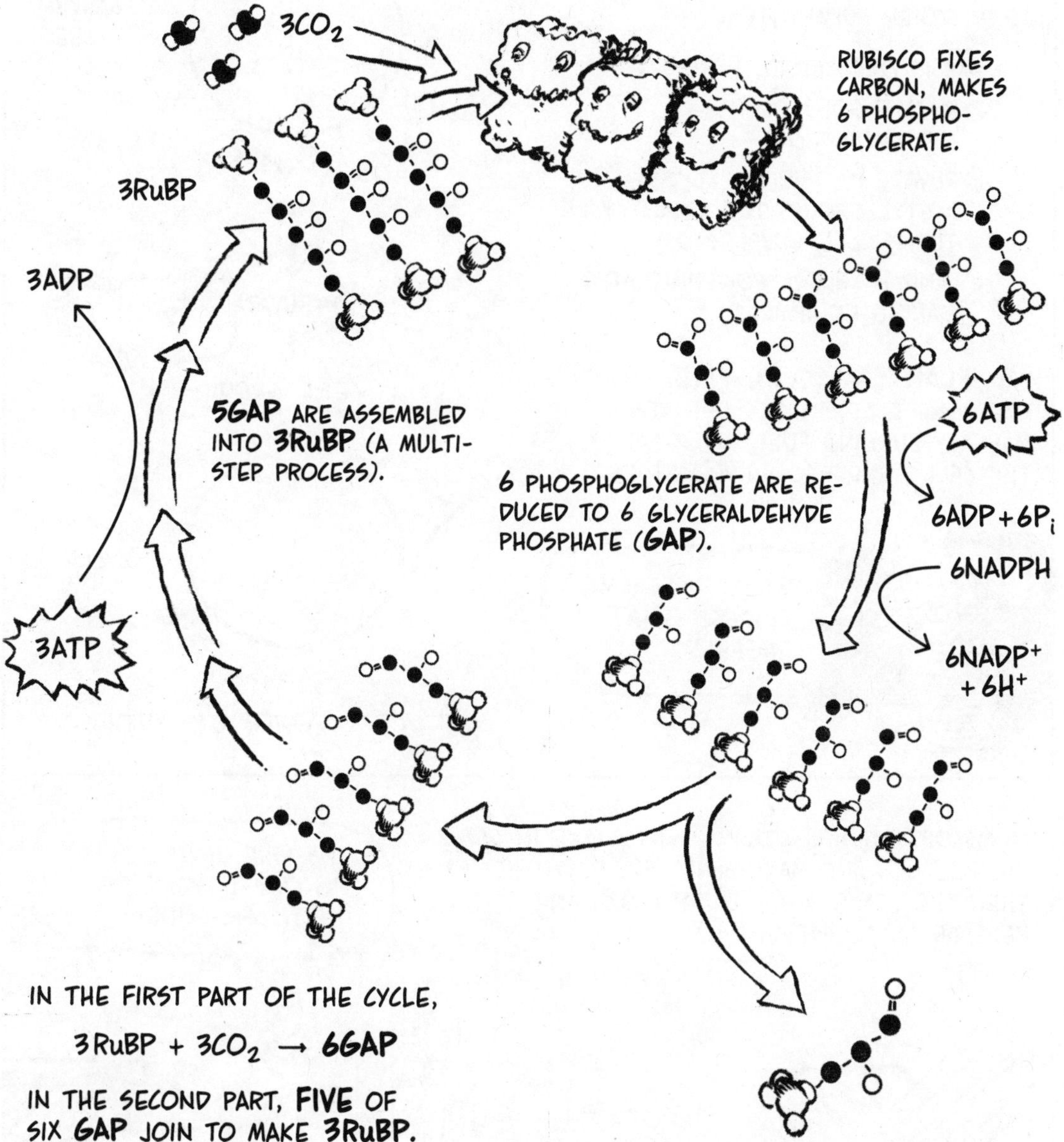

$3CO_2$
RUBISCO FIXES CARBON, MAKES 6 PHOSPHO-GLYCERATE.
3RuBP
3ADP
6ATP
5GAP ARE ASSEMBLED INTO **3RuBP** (A MULTI-STEP PROCESS).
6 PHOSPHOGLYCERATE ARE RE-DUCED TO 6 GLYCERALDEHYDE PHOSPHATE (**GAP**).
$6ADP + 6P_i$
6NADPH
$6NADP^+$
$+ 6H^+$
3ATP

IN THE FIRST PART OF THE CYCLE,

$3RuBP + 3CO_2 \rightarrow$ **6GAP**

IN THE SECOND PART, **FIVE** OF SIX **GAP** JOIN TO MAKE **3RuBP**.

$5GAP \longrightarrow 3RuBP$

THOSE 3 RuBP TAKE UP $3CO_2$, AND THE CYCLE BEGINS AGAIN.

ONE GAP, A 3-CARBON, PHOSPHORYLATED SUGAR, **ESCAPES.** THIS IS THE CYCLE'S **OUTPUT.**

AS THE CALVIN CYCLE TURNS, IT SPITS OUT **GAP**, GLYCERALDEHYDE PHOSPHATE. WHAT DOES A PLANT DO WITH **GAP?**

SOME OF IT GOES TO **RESPIRATION**, AS IN OTHER EUKARYOTES:

- GAP IS OXIDIZED TO PYRUVATE;
- PYRUVATE GOES TO THE MITOCHONDRION;
- PYRUVATE IS OXIDIZED TO AN ACETYL GROUP; THE KREBS CYCLE TURNS; CO_2 BUBBLES OFF;
- e^- AND H^+ FLOW, POWERING ATP-MAKING EQUIPMENT.

YES, **PLANTS ALSO RESPIRE.** LIKE EVERYONE ELSE, PLANTS USE ATP, MADE BY BURNING FUEL, TO DRIVE THE CELL'S ENDERGONIC PROCESSES.

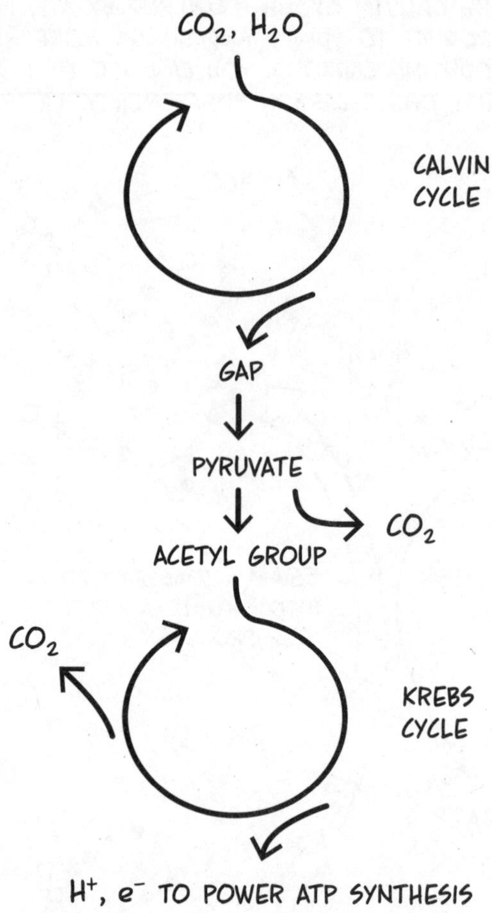

H+, e- TO POWER ATP SYNTHESIS

WE'RE DIFFERENT, BUT NOT THAT DIFFERENT!

IT ALMOST SOUNDS CRAZY: PLANTS TAKE IN CO_2, RELEASE O_2, AND MAKE **GAP** (PHOTOSYNTHESIS). THEN THEY TAKE IN O_2, BURN **GAP**, AND RELEASE CO_2 (RESPIRATION).

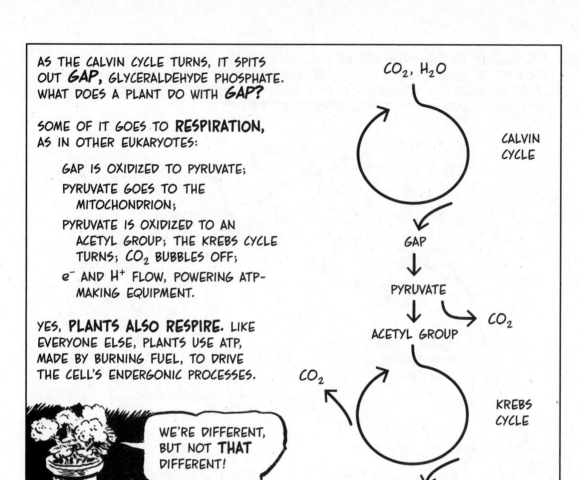

IT'S LIKE USING SOLAR PANELS TO MAKE SYNTHETIC GASOLINE!

BUT—
PLANTS PUT ONLY A FRACTION OF THEIR FRESHLY MADE **GAP** INTO RESPIRATION. THE BULK GOES INTO **STORAGE** AND **STRUCTURE.** THE SUN BLASTS OUT FAR MORE ENERGY THAN A PLANT NEEDS FOR FUEL.
SO WE MAKE HAY WHILE THE SUN SHINES!!!
LITERALLY.

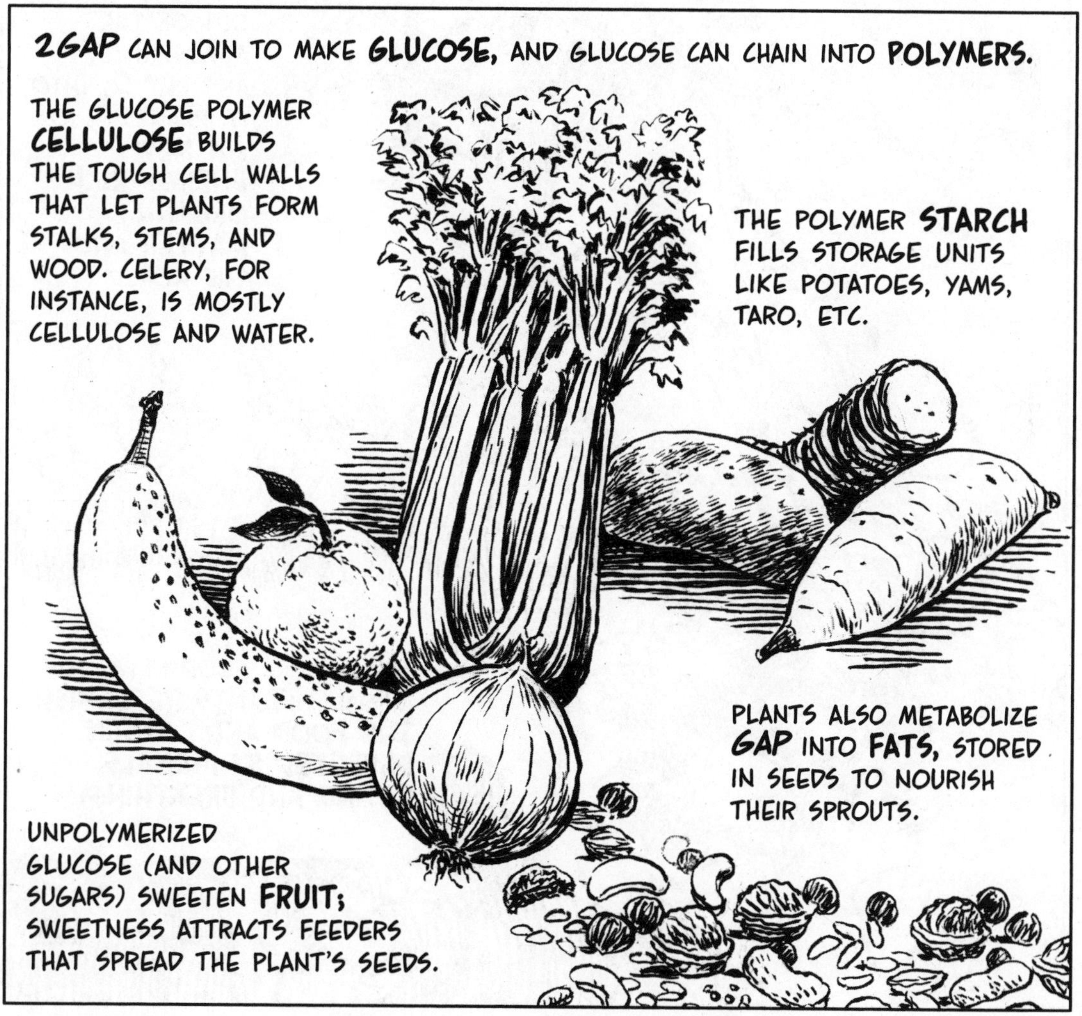
2GAP CAN JOIN TO MAKE **GLUCOSE,** AND GLUCOSE CAN CHAIN INTO **POLYMERS.**
THE GLUCOSE POLYMER **CELLULOSE** BUILDS THE TOUGH CELL WALLS THAT LET PLANTS FORM STALKS, STEMS, AND WOOD. CELERY, FOR INSTANCE, IS MOSTLY CELLULOSE AND WATER.
THE POLYMER **STARCH** FILLS STORAGE UNITS LIKE POTATOES, YAMS, TARO, ETC.
PLANTS ALSO METABOLIZE **GAP** INTO **FATS,** STORED IN SEEDS TO NOURISH THEIR SPROUTS.
UNPOLYMERIZED GLUCOSE (AND OTHER SUGARS) SWEETEN **FRUIT;** SWEETNESS ATTRACTS FEEDERS THAT SPREAD THE PLANT'S SEEDS.

IN SUM, PLANTS FIX FAR MORE CARBON THAN THEY NEED FOR FUEL IN THE MOMENT. THIS IMPLIES THAT **MORE CO_2 COMES OUT OF THE AIR** DURING PHOTOSYNTHESIS THAN GOES BACK DURING RESPIRATION, AND **MORE O_2 GOES INTO THE AIR** THAN COMES OUT.
CO_2
O_2
THE NET EFFECT: PLANTS TAKE CO_2 OUT OF THE ATMOSPHERE, AND PUT O_2 INTO IT. PLANTS ALSO PUT **NEW ENERGY-RICH ORGANIC MATTER** INTO THE WORLD.
THIS IS VERY GOOD NEWS FOR ANIMALS! **PLANTS REPLENISH THE FOOD AND OXYGEN DEPLETED BY ANIMALS' EATING AND BREATHING.**

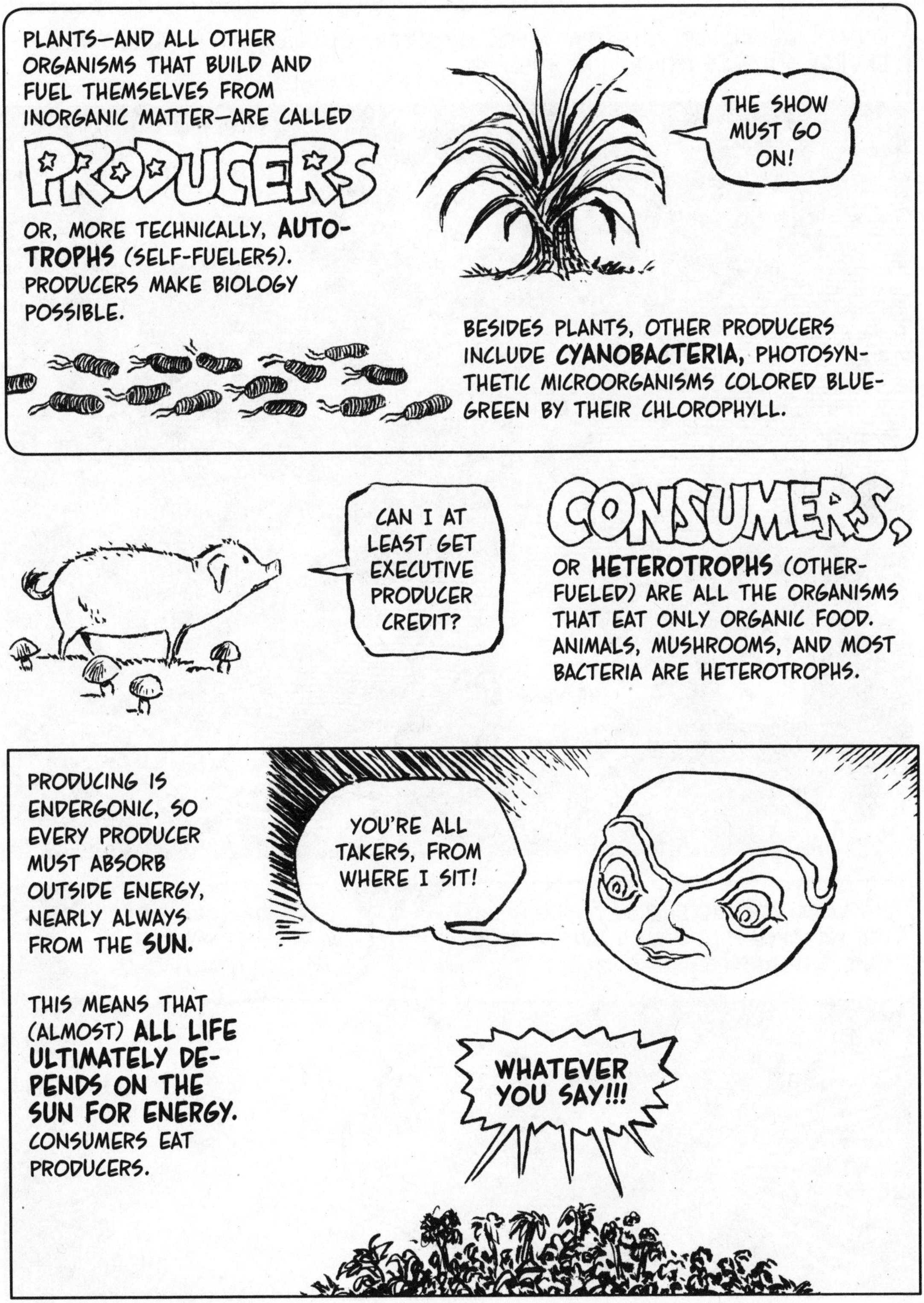
PLANTS—AND ALL OTHER ORGANISMS THAT BUILD AND FUEL THEMSELVES FROM INORGANIC MATTER—ARE CALLED
PRODUCERS
OR, MORE TECHNICALLY, AUTOTROPHS (SELF-FUELERS). PRODUCERS MAKE BIOLOGY POSSIBLE.
THE SHOW MUST GO ON!
BESIDES PLANTS, OTHER PRODUCERS INCLUDE CYANOBACTERIA, PHOTOSYNTHETIC MICROORGANISMS COLORED BLUE-GREEN BY THEIR CHLOROPHYLL.
CAN I AT LEAST GET EXECUTIVE PRODUCER CREDIT?
CONSUMERS,
OR HETEROTROPHS (OTHER-FUELED) ARE ALL THE ORGANISMS THAT EAT ONLY ORGANIC FOOD. ANIMALS, MUSHROOMS, AND MOST BACTERIA ARE HETEROTROPHS.
PRODUCING IS ENDERGONIC, SO EVERY PRODUCER MUST ABSORB OUTSIDE ENERGY, NEARLY ALWAYS FROM THE SUN.
YOU'RE ALL TAKERS, FROM WHERE I SIT!
THIS MEANS THAT (ALMOST) ALL LIFE ULTIMATELY DEPENDS ON THE SUN FOR ENERGY. CONSUMERS EAT PRODUCERS.
WHATEVER YOU SAY!!!

FINALLY, WE SHOULD NOTE THAT A FEW PRODUCERS CAN LIVE OFF **ALTERNATIVE ENERGY SOURCES** OTHER THAN SUNLIGHT.
TOTALLY IN THE DARK—AND ALIVE!
SOME LIVE DEEP IN THE OCEAN, WHERE OOZING MOLTEN ROCK CREATES HELLISH SURROUNDINGS OF SULFUROUS GAS AND SUPERHEATED WATER.
VARIOUS STRANGE PROKARYOTES THERE CAN HARNESS THIS HEAT, RATHER THAN SUNLIGHT, FOR MAKING FUEL.
AND WE'RE IN HEAVEN!

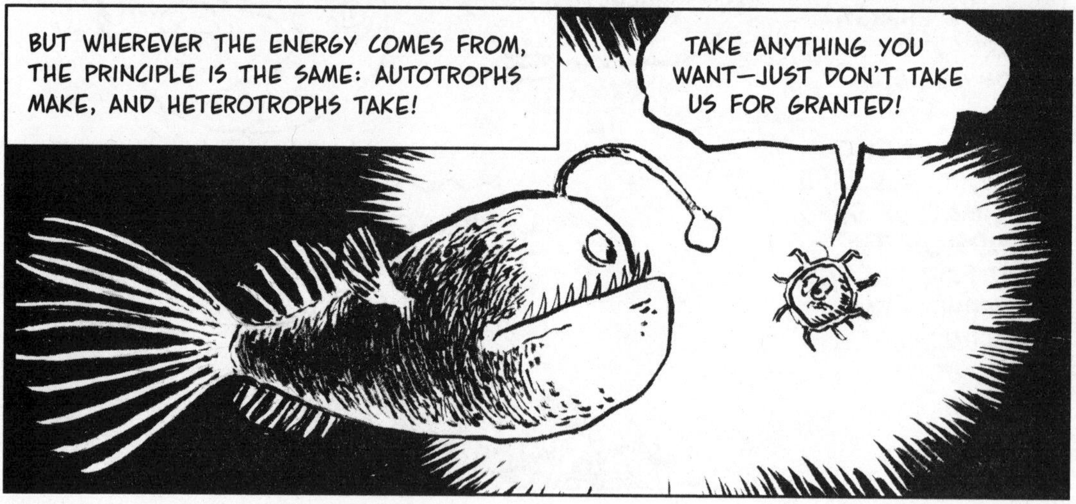
BUT WHEREVER THE ENERGY COMES FROM, THE PRINCIPLE IS THE SAME: AUTOTROPHS MAKE, AND HETEROTROPHS TAKE!
TAKE ANYTHING YOU WANT—JUST DON'T TAKE US FOR GRANTED!

Chapter 8
COMMUNICATION

BY NOW, IT'S CLEAR THAT PROTEINS DO THE CELL'S WORK. THEY CATALYZE, OR COAX ALONG, ALL METABOLIC REACTIONS; THEY HELP MAINTAIN CELL CHEMISTRY BY MOVING MATERIAL ACROSS THE MEMBRANE; THEY PULL VESICLES FULL OF MATERIAL AROUND; THEY FORM FIBERS AND OTHER STRUCTURES; AND MORE.

GO!
MOVE IT!
DANGER!
A TYPICAL CELL DEPLOYS MILLIONS OF PROTEIN MOLECULES ALL AT ONCE, EACH WITH ITS OWN SPECIFIC JOB. WHY DON'T THEY GET IN EACH OTHER'S WAY? HOW CAN SO MANY MOLECULAR MACHINES **WORK TOGETHER?** THE ANSWER IS: THEY **COMMUNICATE.**
IN A MINUTE!
NO, **THAT** WAY...
EXCUSE ME.
SORRY.
GET OFF!

PROTEINS RESPOND TO ONLY THE SIMPLEST MESSAGES, LIKE "START," "STOP," "MORE," "LESS," "OPEN," "CLOSE," OR JUST "HERE I AM!" BUT WHAT DO YOU EXPECT? A PROTEIN IS ONLY A DUMB MOLECULE!

CRITIQUE OF PURINE REASON
?

THESE SIGNALS COME IN THE "LANGUAGE" OF **CHEMISTRY.** SUPPOSE AN ENZYME IS DUTIFULLY MUNCHING AWAY ON ITS SUBSTRATE, WHEN A SMALL SIGNAL MOLECULE ARRIVES. THE MOLECULE FITS A SITE ON THE ENZYME, SEPARATE FROM THE "ACTIVE SITE" THAT BINDS THE SUBSTRATE.

BY UPSETTING THE BIG MOLECULE'S DELICATE BALANCE OF CHARGES, THE BOUND MESSENGER MAKES THE ENZYME **CHANGE SHAPE...**

THE ENZYME STOPS FUNCTIONING AND NO LONGER BINDS THE SUBSTRATE. THAT LITTLE MOLECULE SAID **"STOP."**

AK! AK!
AK! AK!
AK! AK!

CHEMICAL SHAPE-SHIFTING IS CALLED

ALLOSTERY

(ALLO = OTHER; STERY = SHAPE). WE'VE ALREADY MET AT LEAST ONE ALLOSTERIC PROTEIN: A **LIGAND-GATED CHANNEL** (SEE P. 46). THESE PROTEIN PIPES ARE CLOSED UNTIL LIGAND MOLECULES BIND AND OPEN THE GATE.*

EACH OF THESE CHANGES IS SMALL IN ITSELF, BUT TOGETHER THEY COMBINE TO PRODUCE COMPLEXITY, JUST AS COMPUTER SOFTWARE ARISES FROM NOTHING BUT ONES AND ZEROES. CONSIDER THE ALLOSTERIC MEMBRANE PROTEINS CALLED **RECEPTORS.** THEY GIVE THE CELL INFORMATION ABOUT THE OUTSIDE WORLD.

A GLUCOSE RECEPTOR, FOR INSTANCE, BINDS GLUCOSE OUTSIDE THE CELL. THE MESSAGE IS **"FOOD!"**

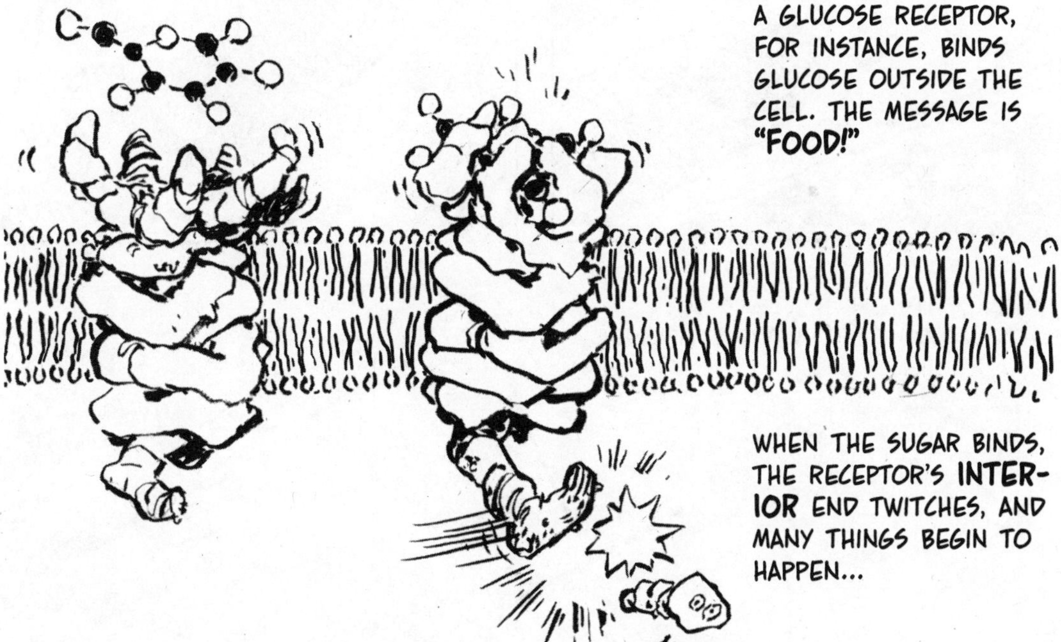

WHEN THE SUGAR BINDS, THE RECEPTOR'S **INTERIOR** END TWITCHES, AND MANY THINGS BEGIN TO HAPPEN...

*THERE ARE ALSO GATED CHANNELS THAT REMAIN OPEN UNTIL A LIGAND CLOSES THEM.

LET'S SEE WHAT GLUCOSE RECEPTORS CAN DO FOR BACTERIA THAT SWIM BY THE ACTION OF A WHIPLIKE **FLAGELLUM.**

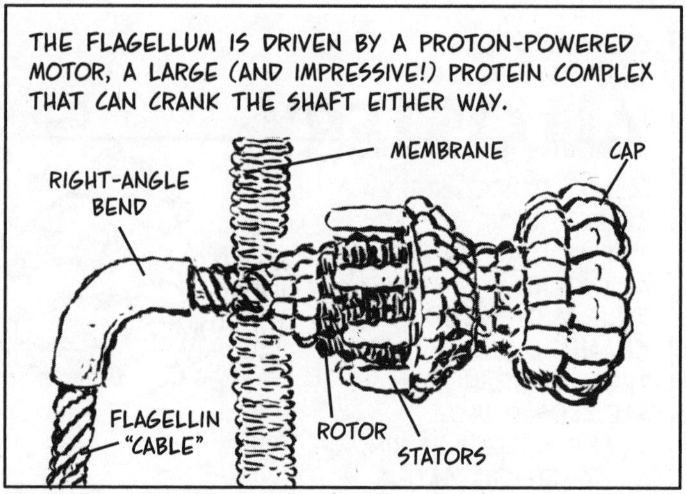
THE FLAGELLUM IS DRIVEN BY A PROTON-POWERED MOTOR, A LARGE (AND IMPRESSIVE!) PROTEIN COMPLEX THAT CAN CRANK THE SHAFT EITHER WAY.
MEMBRANE
CAP
RIGHT-ANGLE BEND
FLAGELLIN "CABLE"
ROTOR
STATORS

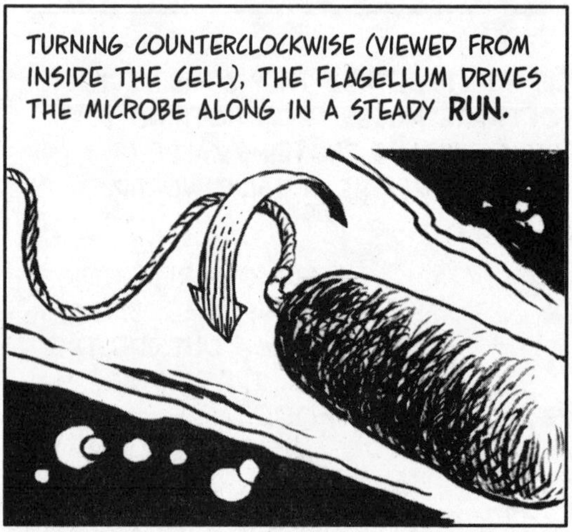
TURNING COUNTERCLOCKWISE (VIEWED FROM INSIDE THE CELL), THE FLAGELLUM DRIVES THE MICROBE ALONG IN A STEADY **RUN.**

WHEN THE FLAGELLUM SPINS THE OTHER WAY, THE BACTERIUM TUMBLES AROUND AIMLESSLY.

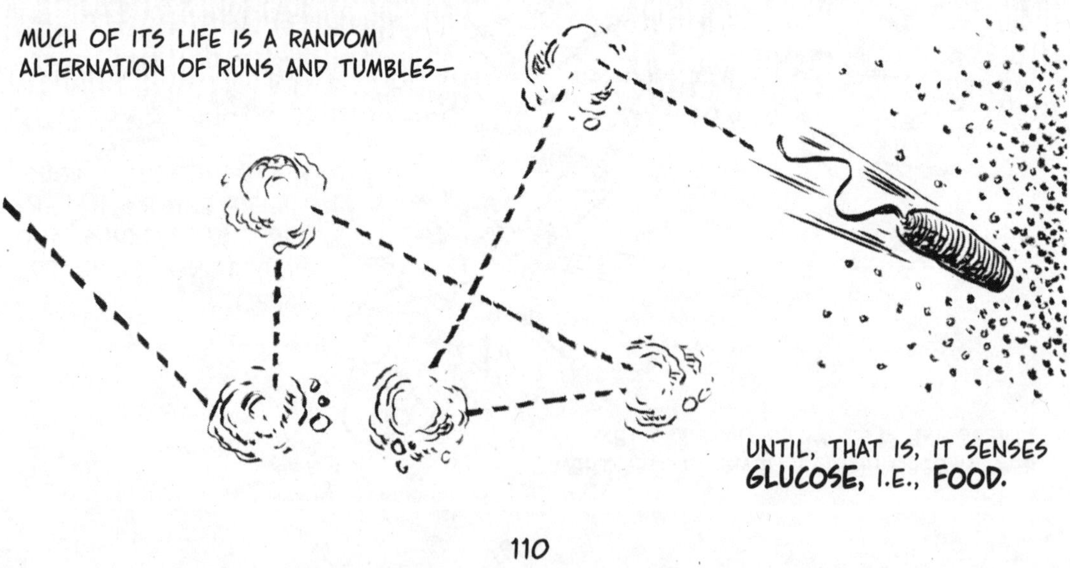
MUCH OF ITS LIFE IS A RANDOM ALTERNATION OF RUNS AND TUMBLES—
UNTIL, THAT IS, IT SENSES **GLUCOSE,** I.E., **FOOD.**

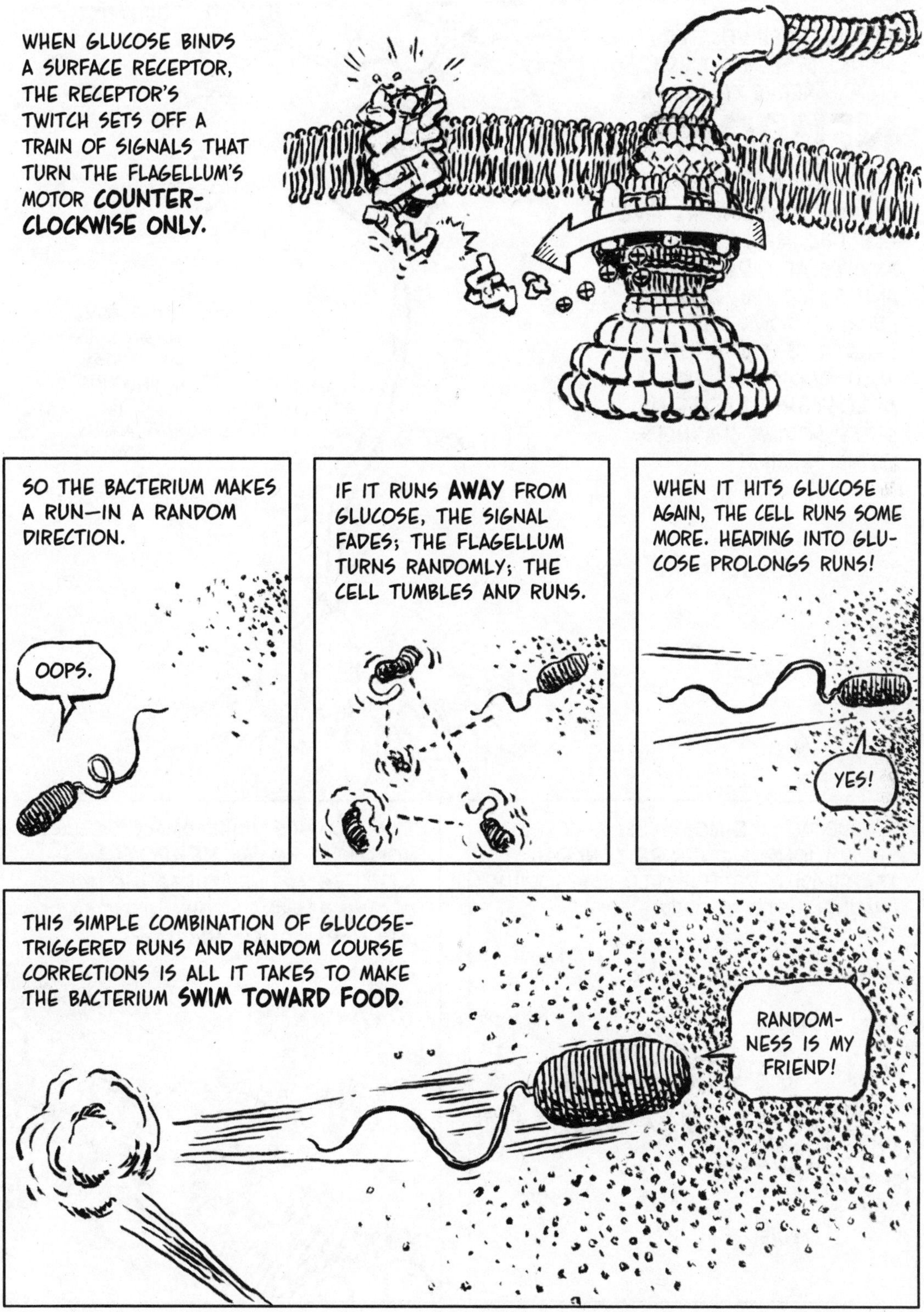
WHEN GLUCOSE BINDS A SURFACE RECEPTOR, THE RECEPTOR'S TWITCH SETS OFF A TRAIN OF SIGNALS THAT TURN THE FLAGELLUM'S MOTOR **COUNTER-CLOCKWISE ONLY.**
SO THE BACTERIUM MAKES A RUN—IN A RANDOM DIRECTION.
OOPS.
IF IT RUNS **AWAY** FROM GLUCOSE, THE SIGNAL FADES; THE FLAGELLUM TURNS RANDOMLY; THE CELL TUMBLES AND RUNS.
WHEN IT HITS GLUCOSE AGAIN, THE CELL RUNS SOME MORE. HEADING INTO GLUCOSE PROLONGS RUNS!
YES!
THIS SIMPLE COMBINATION OF GLUCOSE-TRIGGERED RUNS AND RANDOM COURSE CORRECTIONS IS ALL IT TAKES TO MAKE THE BACTERIUM **SWIM TOWARD FOOD.**
RANDOMNESS IS MY FRIEND!

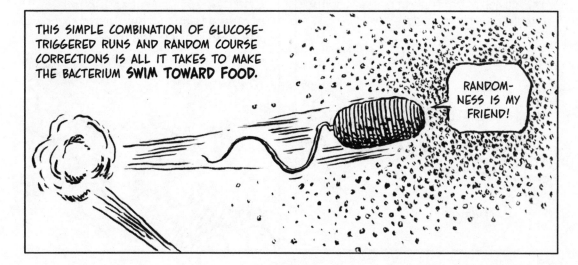

A MORE SOPHISTICATED SIGNAL SYSTEM GOVERNS ANIMAL NERVE CELLS, OR

NEURONS.

NEURONS, THE BODY'S WIRING, SEND AND RECEIVE ELECTRIC IMPULSES. INPUT ARRIVES AT A **DENDRITE** AND SHOOTS AN ELECTRIC CURRENT DOWN THE LENGTH OF THE **AXON**. WHAT MAKES THIS WORK? **ALLOSTERIC PROTEINS:** GATED SODIUM CHANNELS, TO BE PRECISE!

DENDRITE

AXON

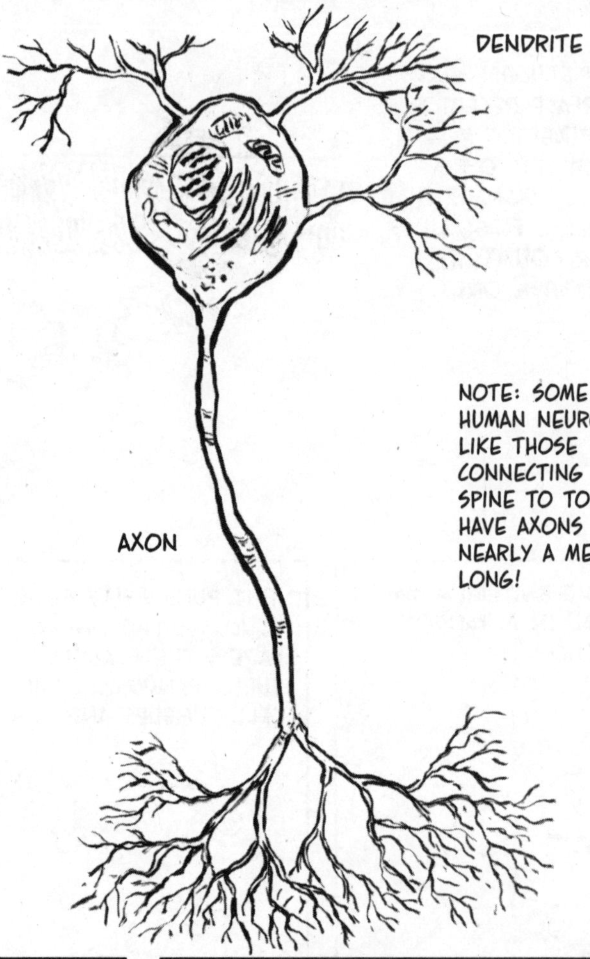

NOTE: SOME HUMAN NEURONS, LIKE THOSE CONNECTING SPINE TO TOES, HAVE AXONS NEARLY A METER LONG!

A NEURON, LIKE MOST CELLS, KEEPS SODIUM IONS, Na^+, MORE CONCENTRATED OUTSIDE THE CELL. ITS SODIUM CHANNELS STAY CLOSED.

Na^+ Na^+ Na^+ Na^+

Na^+ Na^+

Na^+

OUTSIDE

INSIDE

Na^+

Na^+

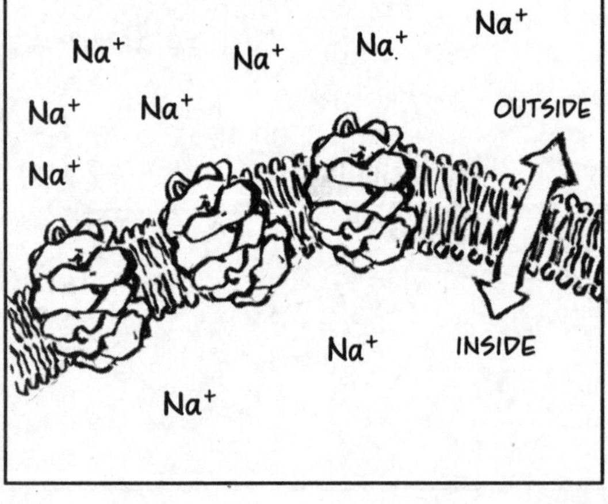

THIS CHANGES WHEN SMALL SIGNAL MOLECULES CALLED **NEUROTRANSMITTERS** ARRIVE. THESE INCLUDE DOPAMINE, EPINEPHRINE, NITRIC OXIDE, AND MANY OTHER SUBSTANCES.

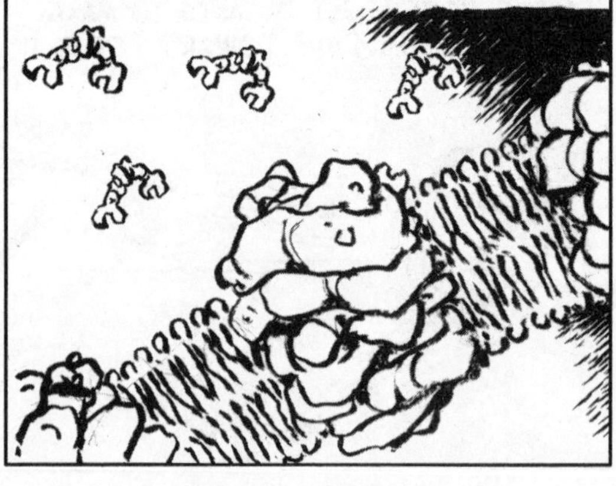

SOME NEUROTRANSMITTERS ACT AS LIGANDS THAT OPEN THE GATES OF SOME SODIUM CHANNELS.
Na+ Na+ Na+ Na+ Na+ Na+
Na+ Na+
AS Na+ DIFFUSES INTO THE CELL, THE CHARGE DIFFERENCE, OR VOLTAGE, ACROSS THE MEMBRANE EASES A BIT.
Na+ Na+ Na+ Na+ Na+ Na+
Na+ Na+ Na+
IF THE SIGNAL IS STRONG, AND THE CROSS-MEMBRANE VOLTAGE FALLS FAR ENOUGH, THE ACTION HEATS UP.
Na+ Na+ Na+ Na+ Na+ Na+ Na+
Na+ Na+ Na+ Na+

THAT'S BECAUSE MOST OF THE NEURON'S SODIUM CHANNELS ARE **VOLTAGE-GATED**: THEY OPEN ONLY WHEN **VOLTAGE** DROPS BELOW A CRITICAL THRESHOLD. AT THAT POINT, SODIUM IONS RUSH INTO THE NEURON IN A WAVE THAT SHOOTS DOWN THE AXON LIKE LIGHTNING.

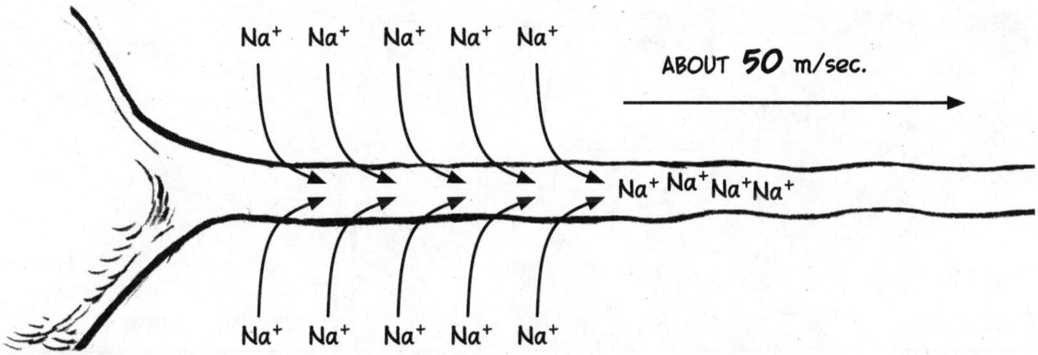
Na+ Na+ Na+ Na+ Na+
ABOUT 50 m/sec.
Na+ Na+ Na+ Na+
Na+ Na+ Na+ Na+ Na+

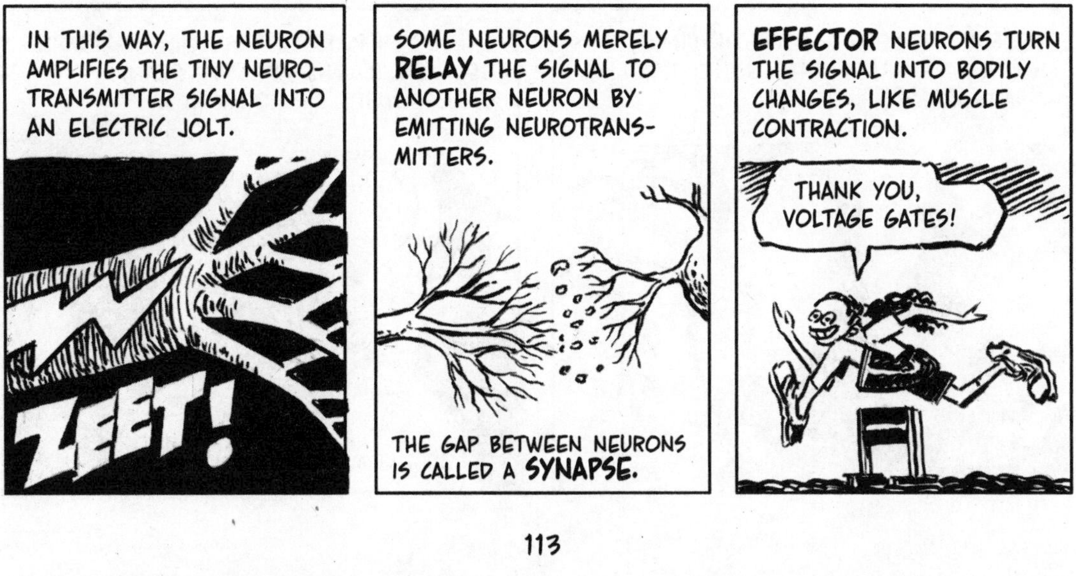
IN THIS WAY, THE NEURON AMPLIFIES THE TINY NEUROTRANSMITTER SIGNAL INTO AN ELECTRIC JOLT.
ZEET!
SOME NEURONS MERELY RELAY THE SIGNAL TO ANOTHER NEURON BY EMITTING NEUROTRANSMITTERS.
THE GAP BETWEEN NEURONS IS CALLED A SYNAPSE.
EFFECTOR NEURONS TURN THE SIGNAL INTO BODILY CHANGES, LIKE MUSCLE CONTRACTION.
THANK YOU, VOLTAGE GATES!

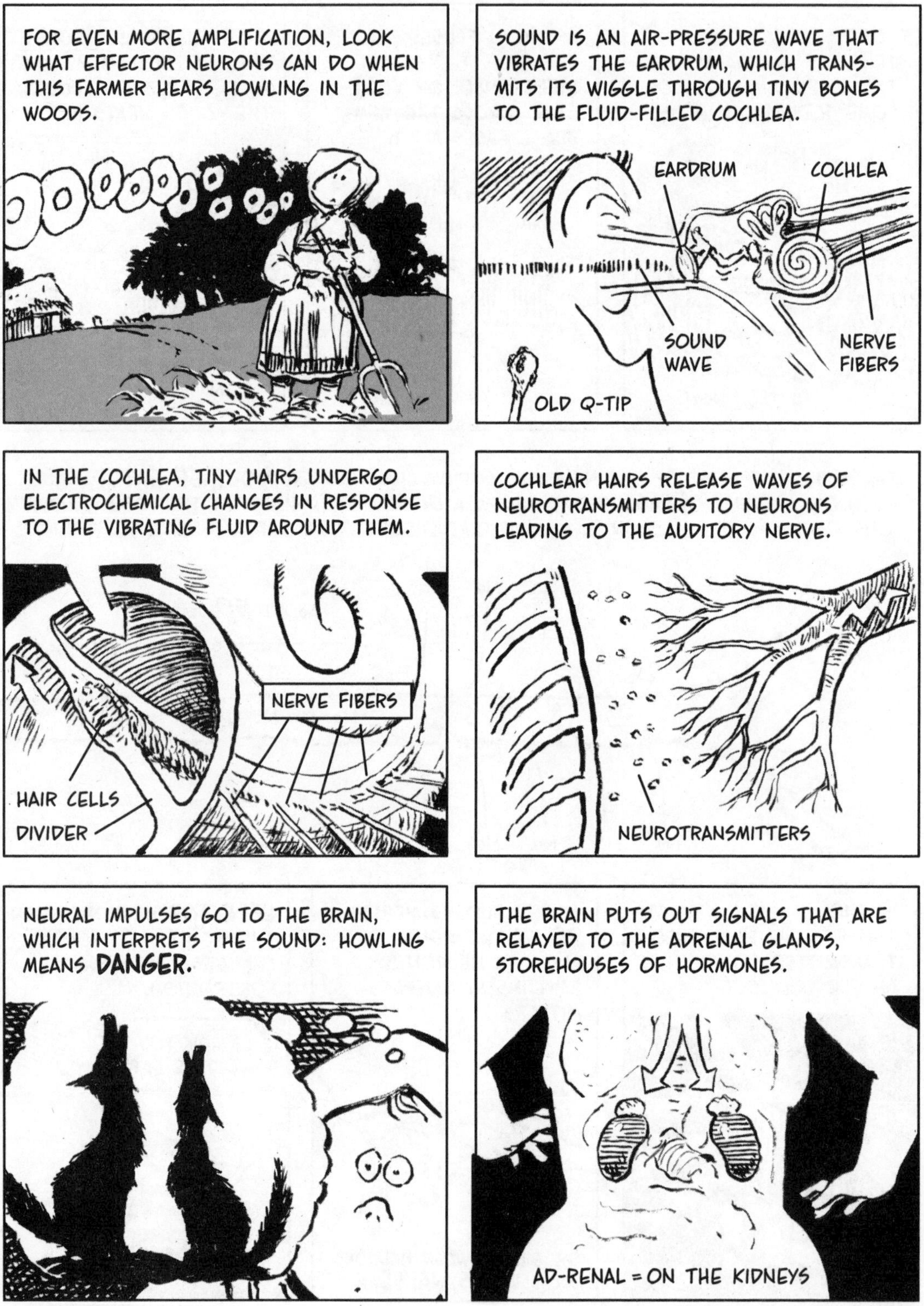
FOR EVEN MORE AMPLIFICATION, LOOK WHAT EFFECTOR NEURONS CAN DO WHEN THIS FARMER HEARS HOWLING IN THE WOODS.
SOUND IS AN AIR-PRESSURE WAVE THAT VIBRATES THE EARDRUM, WHICH TRANSMITS ITS WIGGLE THROUGH TINY BONES TO THE FLUID-FILLED COCHLEA.
EARDRUM
COCHLEA
SOUND WAVE
NERVE FIBERS
OLD Q-TIP
IN THE COCHLEA, TINY HAIRS UNDERGO ELECTROCHEMICAL CHANGES IN RESPONSE TO THE VIBRATING FLUID AROUND THEM.
NERVE FIBERS
HAIR CELLS
DIVIDER
COCHLEAR HAIRS RELEASE WAVES OF NEUROTRANSMITTERS TO NEURONS LEADING TO THE AUDITORY NERVE.
NEUROTRANSMITTERS
NEURAL IMPULSES GO TO THE BRAIN, WHICH INTERPRETS THE SOUND: HOWLING MEANS DANGER.
THE BRAIN PUTS OUT SIGNALS THAT ARE RELAYED TO THE ADRENAL GLANDS, STOREHOUSES OF HORMONES.
AD-RENAL = ON THE KIDNEYS

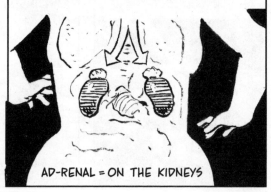

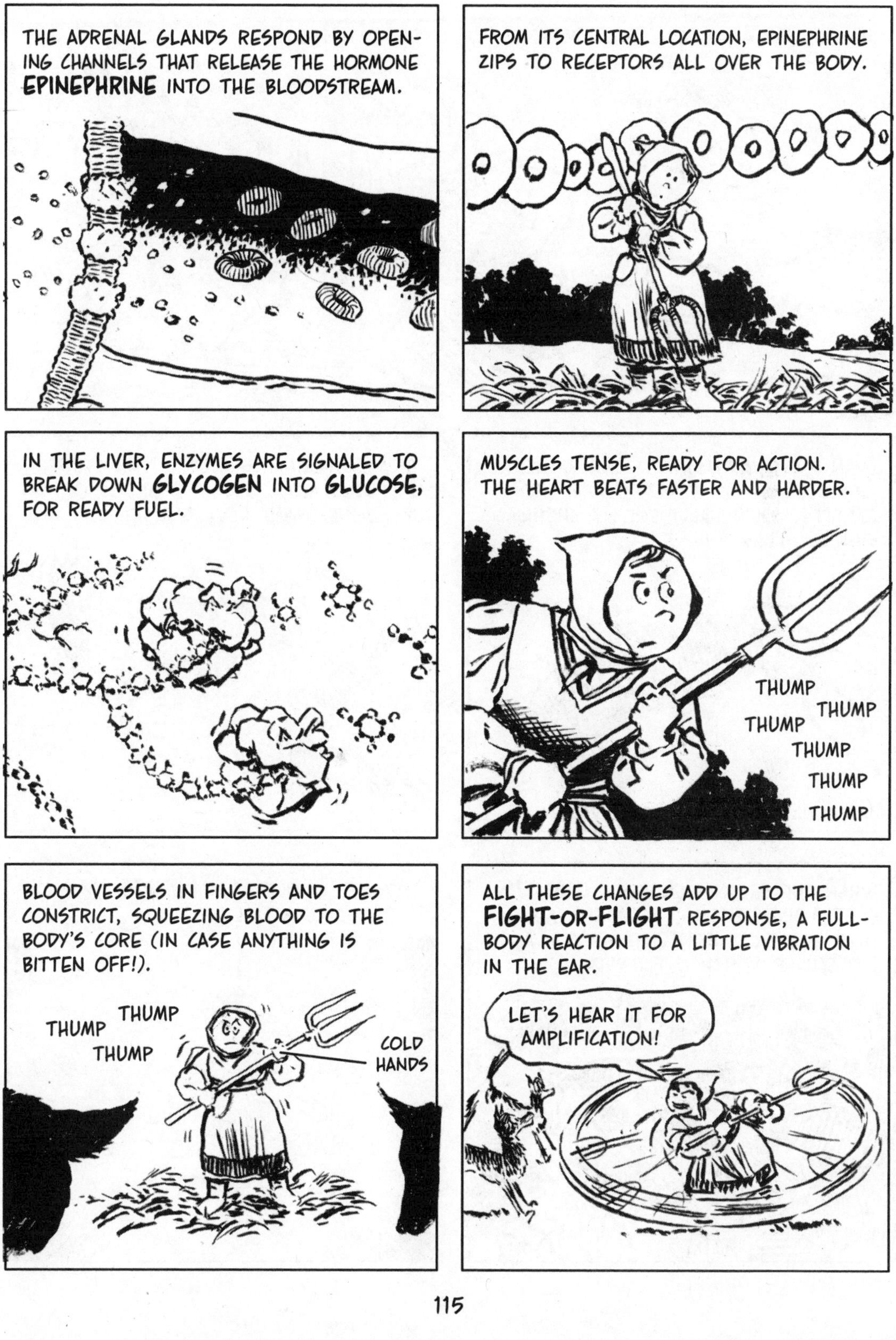

THE ADRENAL GLANDS RESPOND BY OPENING CHANNELS THAT RELEASE THE HORMONE **EPINEPHRINE** INTO THE BLOODSTREAM.

FROM ITS CENTRAL LOCATION, EPINEPHRINE ZIPS TO RECEPTORS ALL OVER THE BODY.

IN THE LIVER, ENZYMES ARE SIGNALED TO BREAK DOWN **GLYCOGEN** INTO **GLUCOSE,** FOR READY FUEL.

MUSCLES TENSE, READY FOR ACTION. THE HEART BEATS FASTER AND HARDER.

BLOOD VESSELS IN FINGERS AND TOES CONSTRICT, SQUEEZING BLOOD TO THE BODY'S CORE (IN CASE ANYTHING IS BITTEN OFF!).

ALL THESE CHANGES ADD UP TO THE **FIGHT-OR-FLIGHT** RESPONSE, A FULL-BODY REACTION TO A LITTLE VIBRATION IN THE EAR.

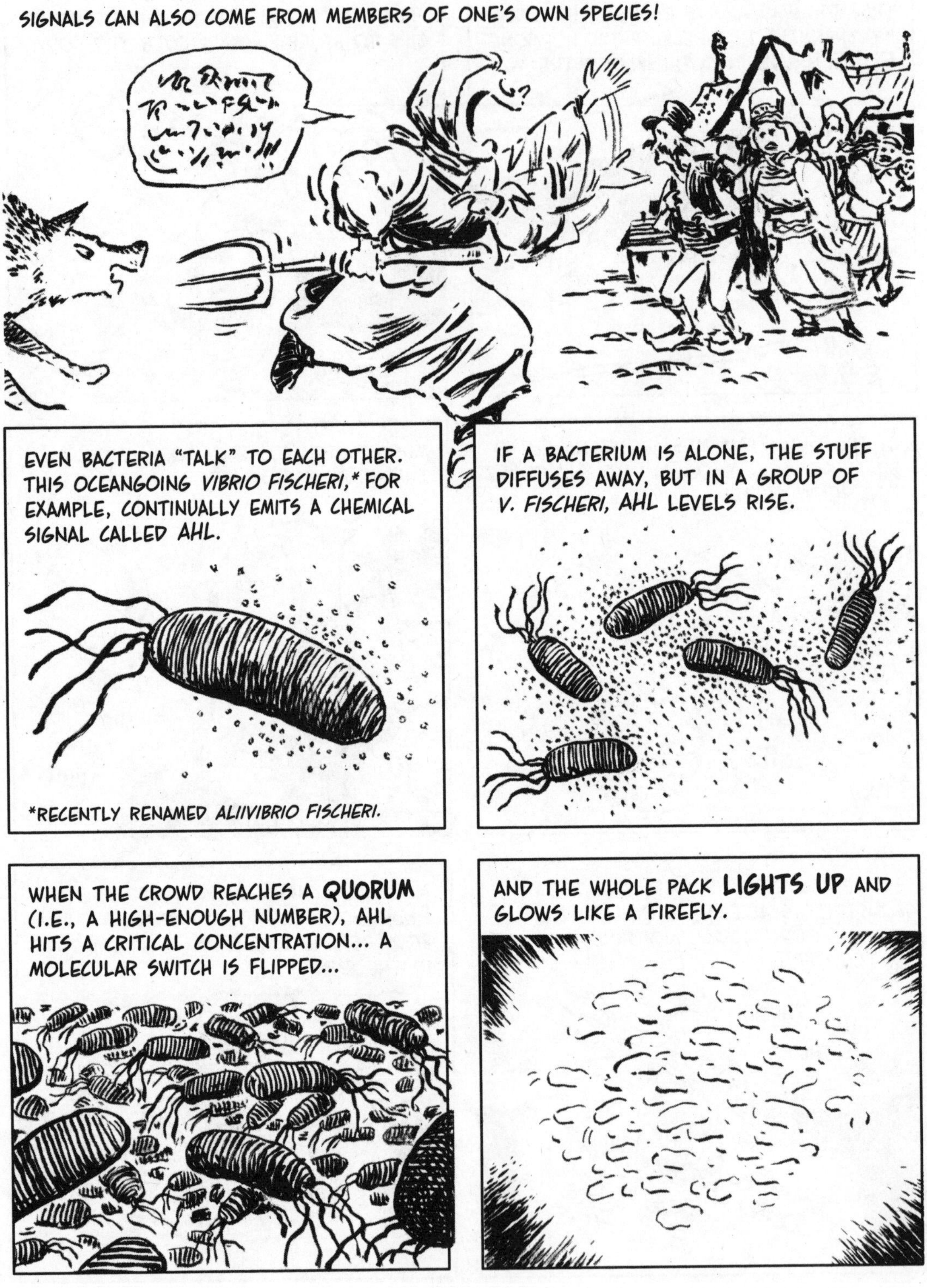
SIGNALS CAN ALSO COME FROM MEMBERS OF ONE'S OWN SPECIES!
EVEN BACTERIA "TALK" TO EACH OTHER. THIS OCEANGOING *VIBRIO FISCHERI*,* FOR EXAMPLE, CONTINUALLY EMITS A CHEMICAL SIGNAL CALLED AHL.
*RECENTLY RENAMED *ALIIVIBRIO FISCHERI*.
IF A BACTERIUM IS ALONE, THE STUFF DIFFUSES AWAY, BUT IN A GROUP OF *V. FISCHERI*, AHL LEVELS RISE.
WHEN THE CROWD REACHES A **QUORUM** (I.E., A HIGH-ENOUGH NUMBER), AHL HITS A CRITICAL CONCENTRATION... A MOLECULAR SWITCH IS FLIPPED...
AND THE WHOLE PACK **LIGHTS UP** AND GLOWS LIKE A FIREFLY.

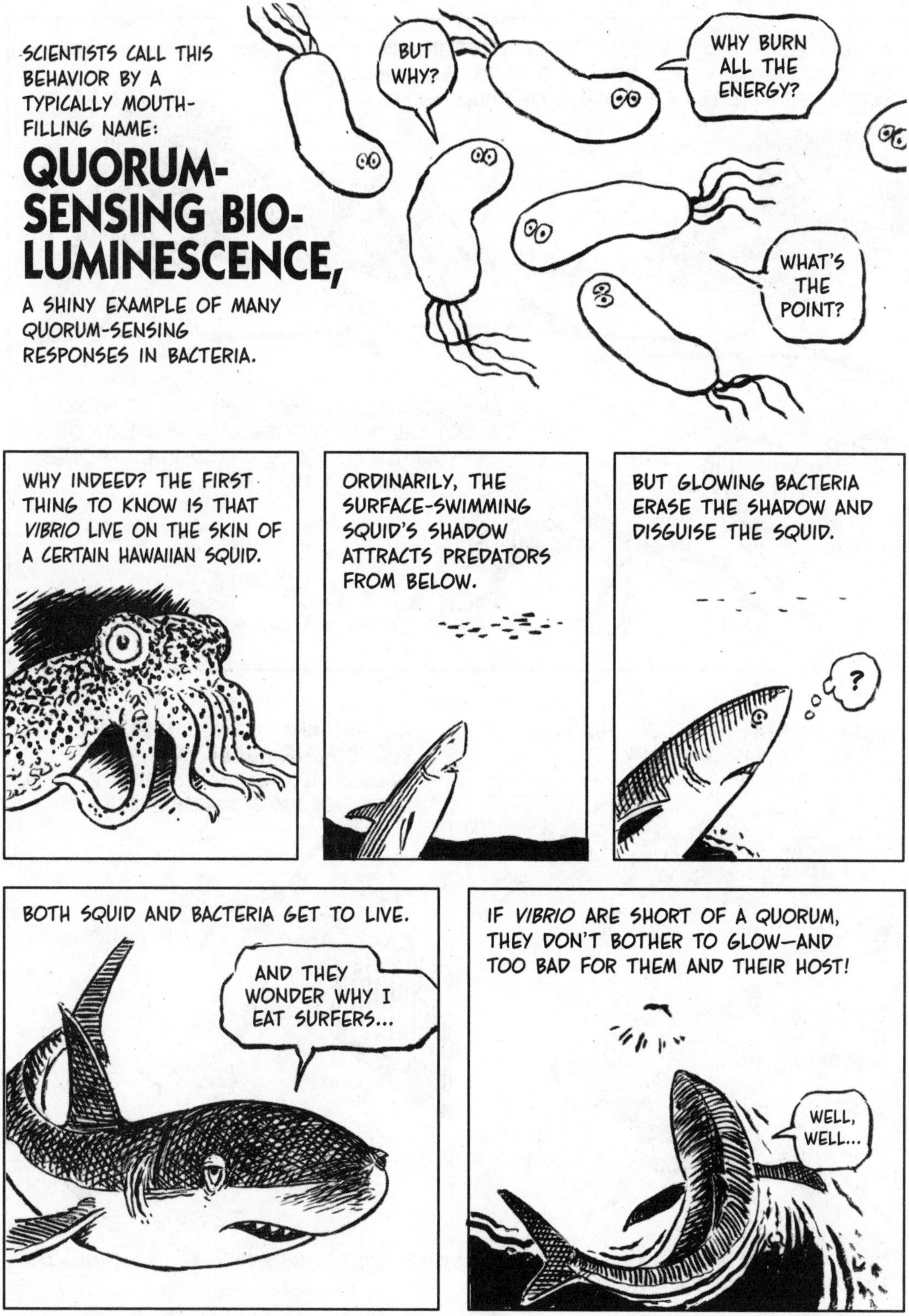
SCIENTISTS CALL THIS BEHAVIOR BY A TYPICALLY MOUTH-FILLING NAME:
QUORUM-SENSING BIO-LUMINESCENCE,
A SHINY EXAMPLE OF MANY QUORUM-SENSING RESPONSES IN BACTERIA.
BUT WHY?
WHY BURN ALL THE ENERGY?
WHAT'S THE POINT?
WHY INDEED? THE FIRST THING TO KNOW IS THAT *VIBRIO* LIVE ON THE SKIN OF A CERTAIN HAWAIIAN SQUID.
ORDINARILY, THE SURFACE-SWIMMING SQUID'S SHADOW ATTRACTS PREDATORS FROM BELOW.
BUT GLOWING BACTERIA ERASE THE SHADOW AND DISGUISE THE SQUID.
?
BOTH SQUID AND BACTERIA GET TO LIVE.
AND THEY WONDER WHY I EAT SURFERS...
IF *VIBRIO* ARE SHORT OF A QUORUM, THEY DON'T BOTHER TO GLOW—AND TOO BAD FOR THEM AND THEIR HOST!
WELL, WELL...

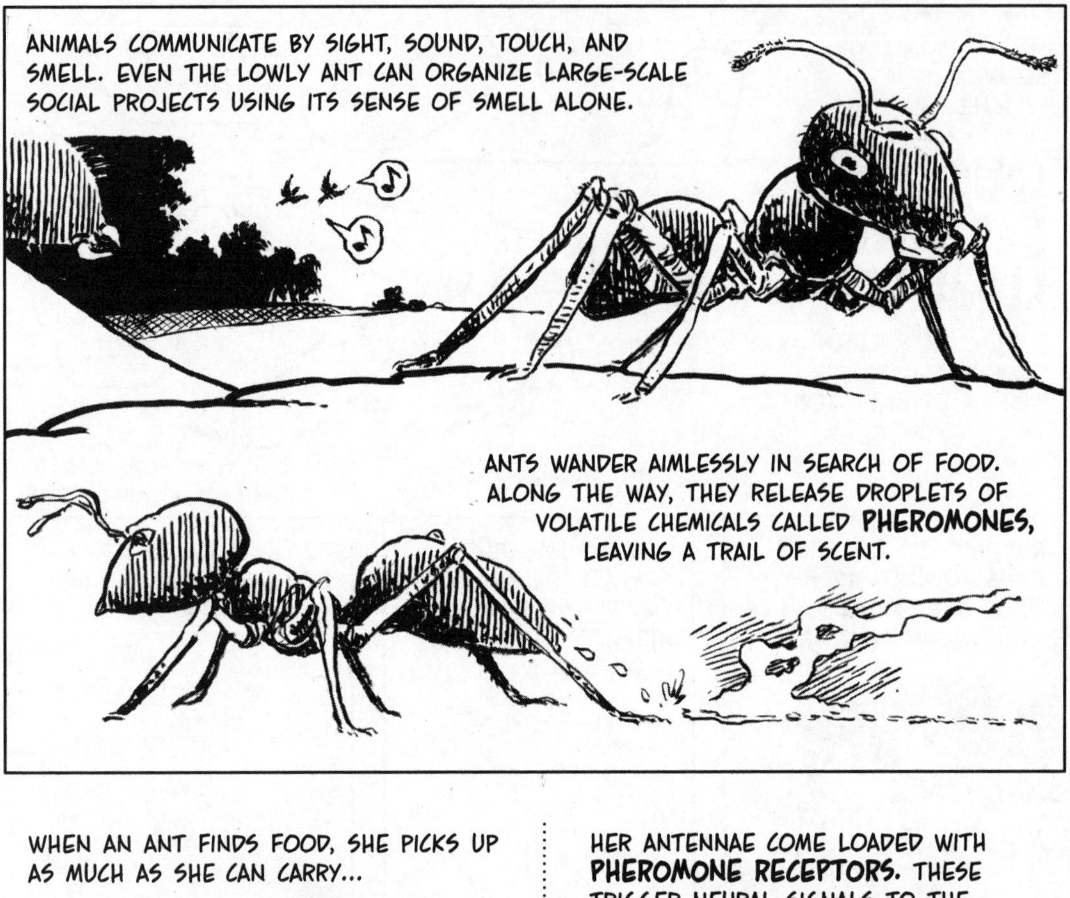
ANIMALS COMMUNICATE BY SIGHT, SOUND, TOUCH, AND SMELL. EVEN THE LOWLY ANT CAN ORGANIZE LARGE-SCALE SOCIAL PROJECTS USING ITS SENSE OF SMELL ALONE.
ANTS WANDER AIMLESSLY IN SEARCH OF FOOD. ALONG THE WAY, THEY RELEASE DROPLETS OF VOLATILE CHEMICALS CALLED **PHEROMONES,** LEAVING A TRAIL OF SCENT.
WHEN AN ANT FINDS FOOD, SHE PICKS UP AS MUCH AS SHE CAN CARRY...
THEN TURNS AROUND TO FIND HER OWN PHEROMONE TRAIL.
HER ANTENNAE COME LOADED WITH **PHEROMONE RECEPTORS.** THESE TRIGGER NEURAL SIGNALS TO THE ANT'S NOT INCONSIDERABLE BRAIN.*

*WITH SOME 250,000 NEURONS, AN ANT'S BRAIN IS AROUND 20 TIMES THE SIZE OF A SNAIL'S AND 1/12,000 THE SIZE OF A MONKEY'S.

THE ANT'S FOOTSTEPS FOLLOW THE SCENT IN REVERSE, GIVING THE TRAIL A SECOND JUICING OF PHEROMONES.
ANY OTHER ANT MEETING THE TRAIL WILL FOLLOW THE SCENT AND ADD MORE PHEROMONES.
IF THE NEW ANT CHOOSES THE WRONG DIRECTION AND ENDS UP AT THE NEST, SHE REVERSES COURSE.
OOPS!

THE MORE ANTS ON THE TRAIL, THE STRONGER THE AROMA, AND SO IT GOES, UNTIL THEY BECOME A MARCHING ARMY. THE PARADE GOES ON UNTIL THE FOOD RUNS OUT, AT WHICH POINTS ANTS WANDER OFF AND THE TRAIL EVAPORATES.

EVIDENTLY, THOSE 250,000 NEURONS IN THE ANT'S BRAIN MANAGE TO STORE A SET OF SIX **RULES:**

IF THIS:	**THEN** DO THIS:
NOT ON TRAIL, NOT CARRYING	WALK RANDOMLY, LAY PHEROMONE
ON TRAIL, NOT CARRYING	FOLLOW TRAIL, LAY PHEROMONE
ARRIVE HOME ON TRAIL, NOT CARRYING	TURN AROUND, FOLLOW TRAIL THE OTHER WAY, LAY PHEROMONE
REACH FOOD	PICK UP FOOD, TURN AROUND, FIND TRAIL
CARRYING FOOD	FOLLOW TRAIL, LAY PHEROMONE
REACH HOME WITH FOOD	DEPOSIT FOOD, TURN AROUND, FOLLOW TRAIL

SOUNDS LOGICAL...

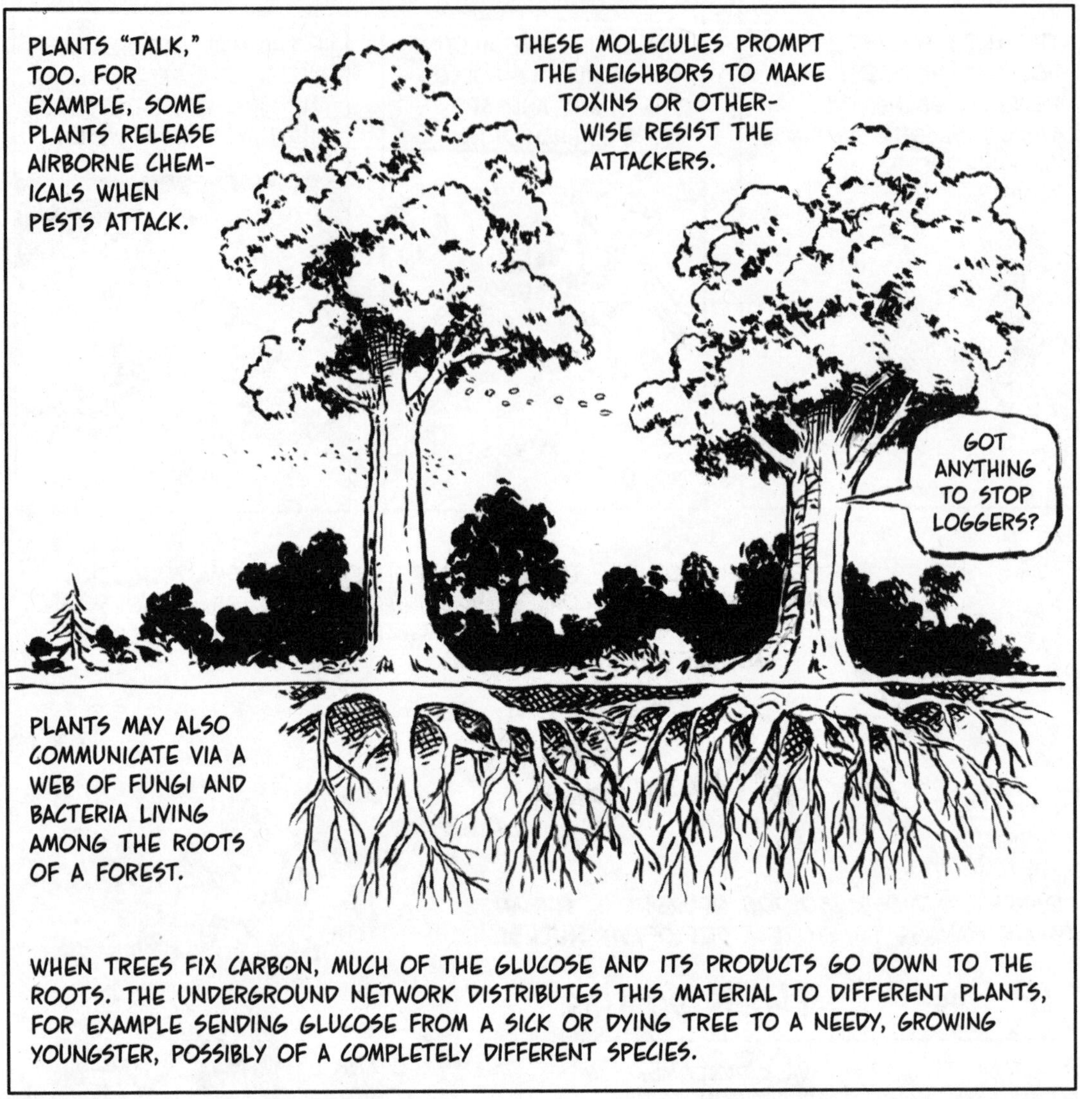
PLANTS "TALK," TOO. FOR EXAMPLE, SOME PLANTS RELEASE AIRBORNE CHEMICALS WHEN PESTS ATTACK.
THESE MOLECULES PROMPT THE NEIGHBORS TO MAKE TOXINS OR OTHERWISE RESIST THE ATTACKERS.
GOT ANYTHING TO STOP LOGGERS?
PLANTS MAY ALSO COMMUNICATE VIA A WEB OF FUNGI AND BACTERIA LIVING AMONG THE ROOTS OF A FOREST.
WHEN TREES FIX CARBON, MUCH OF THE GLUCOSE AND ITS PRODUCTS GO DOWN TO THE ROOTS. THE UNDERGROUND NETWORK DISTRIBUTES THIS MATERIAL TO DIFFERENT PLANTS, FOR EXAMPLE SENDING GLUCOSE FROM A SICK OR DYING TREE TO A NEEDY, GROWING YOUNGSTER, POSSIBLY OF A COMPLETELY DIFFERENT SPECIES.

IN THIS WAY, AN ENTIRE FOREST MAY SEEK TO MAINTAIN HOMEOSTASIS AS IF IT WERE A SINGLE LIVING THING. (NOTE: THIS IDEA IS SOMEWHAT CONTROVERSIAL.)

DO YOU EVER FEEL LIKE PART OF SOMETHING INFINITELY GREATER?
GREATER THAN ME?

IN THIS CHAPTER, WE'VE SEEN LIFE COMMUNICATING AT MANY LEVELS:
WITHIN A COMPLEX SYSTEM OF INDIVIDUALS
WITHIN AN INDIVIDUAL
BETWEEN INDIVIDUALS
BETWEEN THE OUT-SIDE WORLD AND AN INDIVIDUAL.

ULTIMATELY, EVERY EXAMPLE HAS THE SAME MOLECULAR BASIS: CHEMICAL (AND ELECTROCHEMICAL) SIGNALS MAKE PROTEINS CHANGE SHAPE. ORGANIC SIGNALING BOILS DOWN TO A SERIES OF ALLOSTERIC TWITCHES.

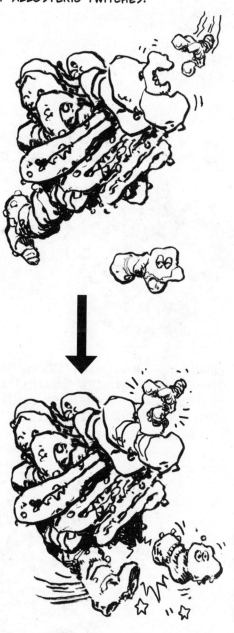

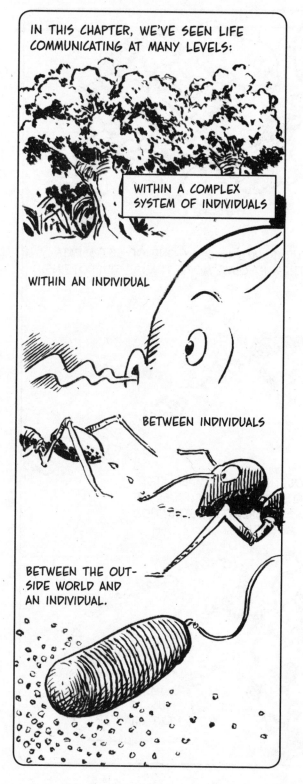

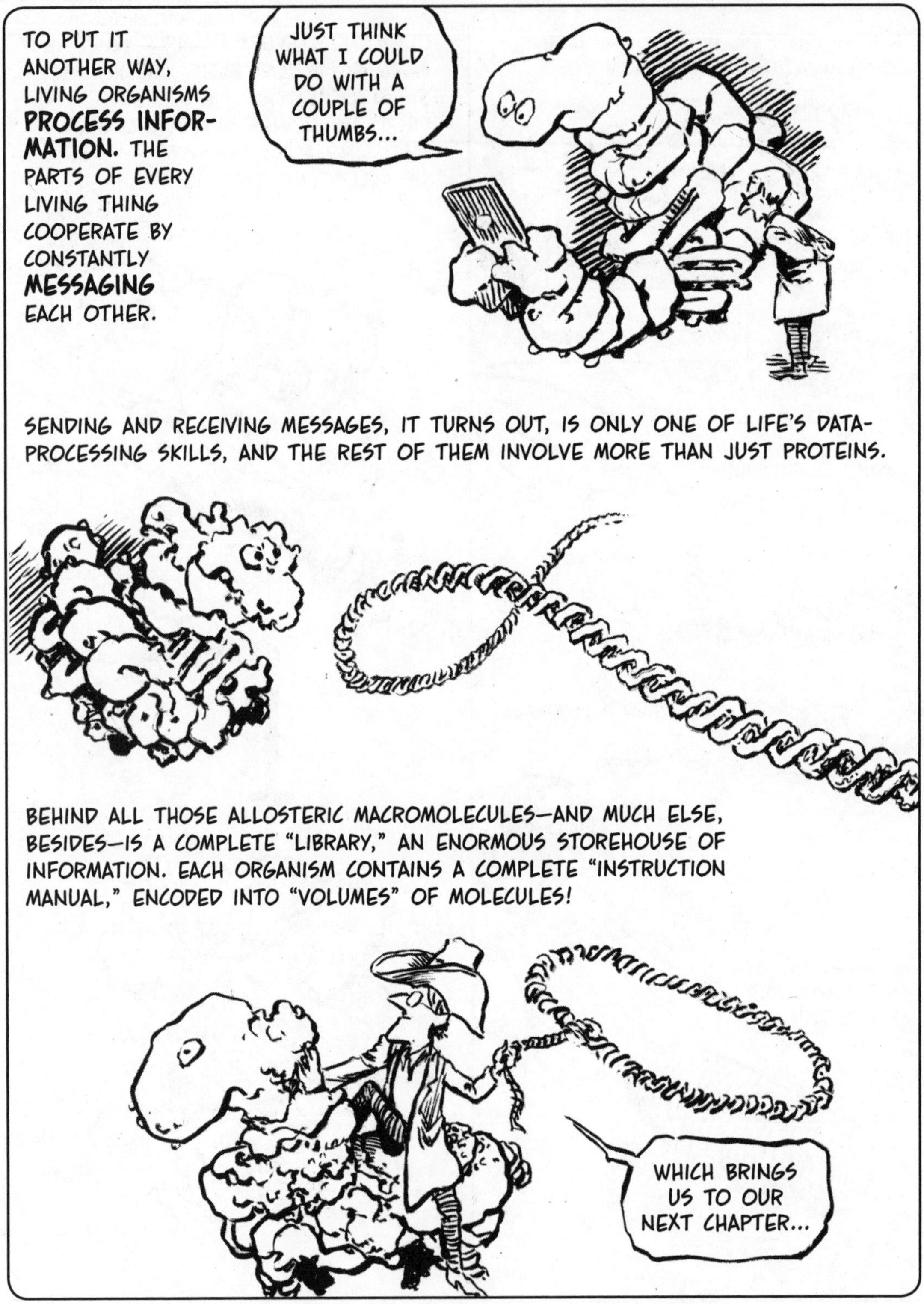
TO PUT IT ANOTHER WAY, LIVING ORGANISMS **PROCESS INFORMATION.** THE PARTS OF EVERY LIVING THING COOPERATE BY CONSTANTLY **MESSAGING** EACH OTHER.
JUST THINK WHAT I COULD DO WITH A COUPLE OF THUMBS...
SENDING AND RECEIVING MESSAGES, IT TURNS OUT, IS ONLY ONE OF LIFE'S DATA-PROCESSING SKILLS, AND THE REST OF THEM INVOLVE MORE THAN JUST PROTEINS.
BEHIND ALL THOSE ALLOSTERIC MACROMOLECULES—AND MUCH ELSE, BESIDES—IS A COMPLETE "LIBRARY," AN ENORMOUS STOREHOUSE OF INFORMATION. EACH ORGANISM CONTAINS A COMPLETE "INSTRUCTION MANUAL," ENCODED INTO "VOLUMES" OF MOLECULES!
WHICH BRINGS US TO OUR NEXT CHAPTER...

Chapter 9
MEET THE GENOME
HOW PROTEINS ARE MADE

AS WE SAID IN THE LAST CHAPTER, **PROTEINS MAKE THE CELL!**
AT THE SAME TIME, THE **CELL MAKES PROTEINS.**
THIS SEEMS TOTALLY CIRCULAR!
HOW DOES A CELL MAKE A PROTEIN? THE AMAZING ANSWER, DISCOVERED IN THE SECOND HALF OF THE TWENTIETH CENTURY, **REVOLUTIONIZED** BIOLOGY...

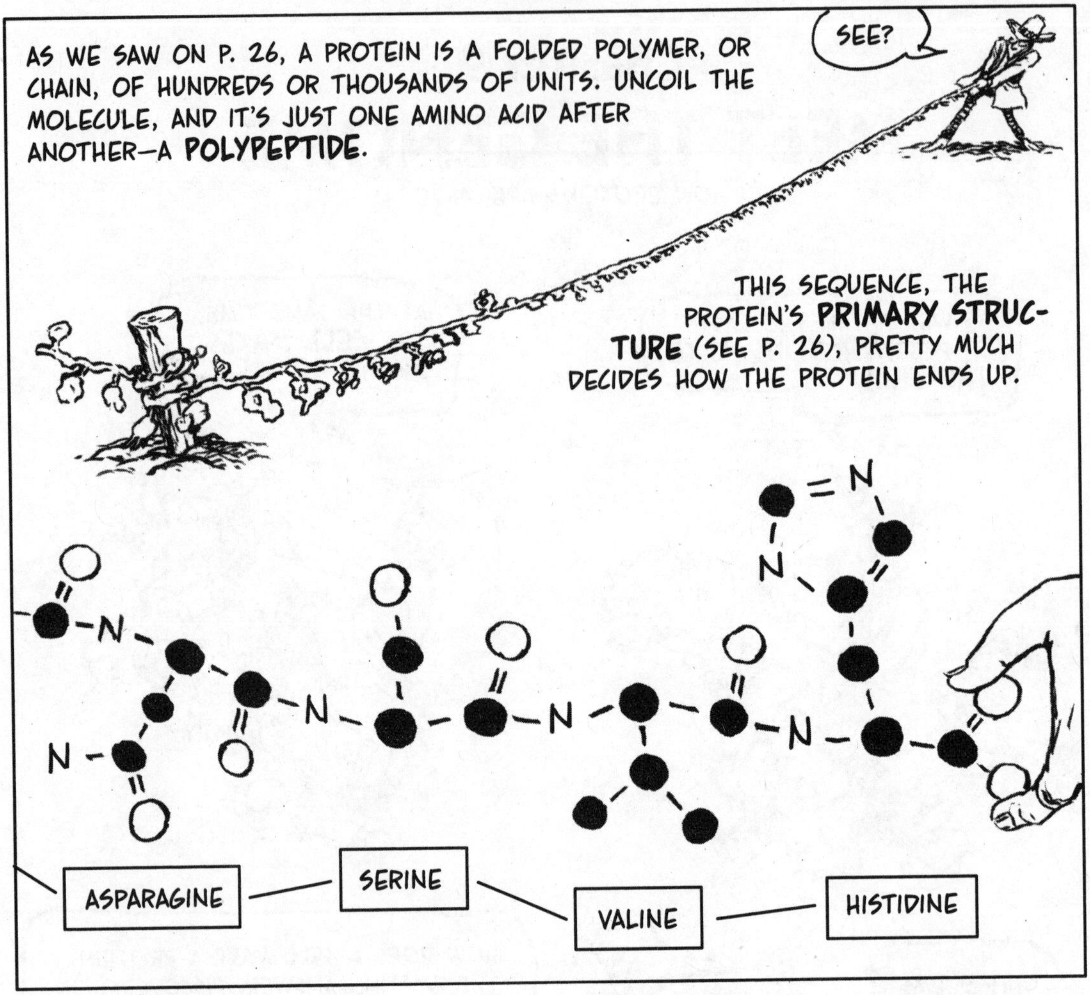
AS WE SAW ON P. 26, A PROTEIN IS A FOLDED POLYMER, OR CHAIN, OF HUNDREDS OR THOUSANDS OF UNITS. UNCOIL THE MOLECULE, AND IT'S JUST ONE AMINO ACID AFTER ANOTHER—A **POLYPEPTIDE**.
SEE?
THIS SEQUENCE, THE PROTEIN'S **PRIMARY STRUCTURE** (SEE P. 26), PRETTY MUCH DECIDES HOW THE PROTEIN ENDS UP.
ASPARAGINE
SERINE
VALINE
HISTIDINE

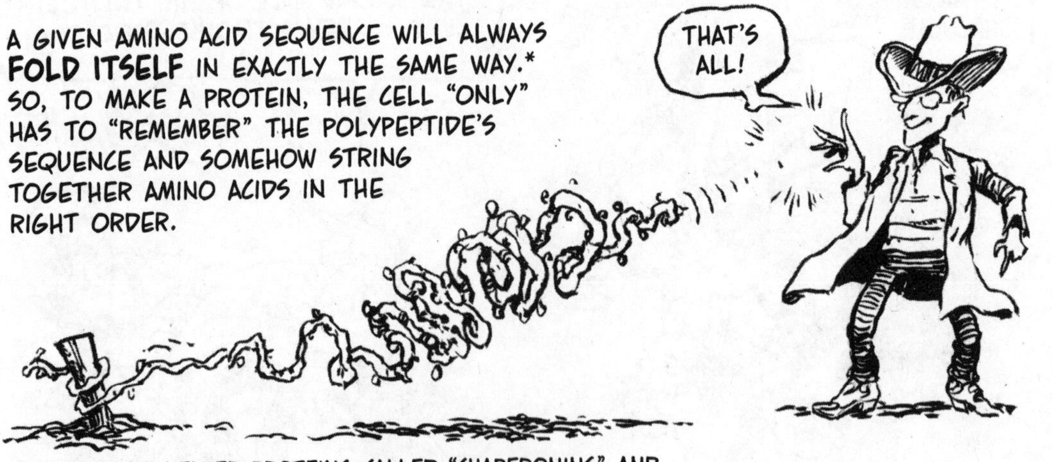
A GIVEN AMINO ACID SEQUENCE WILL ALWAYS **FOLD ITSELF** IN EXACTLY THE SAME WAY.* SO, TO MAKE A PROTEIN, THE CELL "ONLY" HAS TO "REMEMBER" THE POLYPEPTIDE'S SEQUENCE AND SOMEHOW STRING TOGETHER AMINO ACIDS IN THE RIGHT ORDER.
THAT'S ALL!

*OR ALMOST. HELPER PROTEINS CALLED "CHAPERONINS" AND OTHER CHEMICAL FACTORS ALSO MAY HAVE AN EFFECT.

THE CELL WORKS THIS MAGIC BY STORING A **MASTER LIST** OF ALL ITS PROTEIN SEQUENCES. THIS LIST, EMBODIED IN HUGELY LONG MOLECULES OF **DNA** (SEE P. 34), IS CALLED THE CELL'S

GENOME.

!

DNA ACTS AS A SORT OF **RECORDING TAPE** OR DIGITAL STORAGE DEVICE, A **GENE** BEING THAT PART OF THE "TAPE" THAT STORES A SINGLE PROTEIN'S SEQUENCE. EACH GENE, WE SAY, **ENCODES** A PROTEIN. WHEN A PROTEIN IS MADE, WE SAY THE GENE IS **EXPRESSED**.

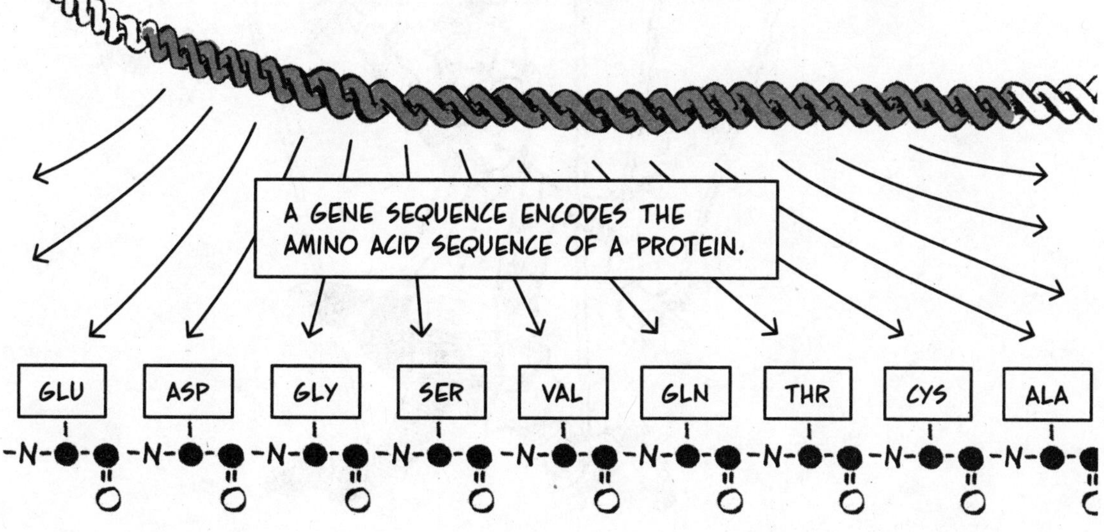
A GENE SEQUENCE ENCODES THE AMINO ACID SEQUENCE OF A PROTEIN.
GLU
ASP
GLY
SER
VAL
GLN
THR
CYS
ALA

GENES PLAY A UNIQUE BIOLOGICAL ROLE: THEY **STORE INFORMATION.** THE GENOME SOMEHOW PRESERVES THE PATTERN OF THE ESSENTIAL MOLECULES OF LIFE—IN A **READABLE FORMAT.** THE CELL MAKES PROTEINS BY "READING" INFORMATION STORED IN DNA.
RUBISCO
ATP SYNTHASE
SODIUM CHANNEL
LET'S THINK ABOUT THE IMPLICATIONS FOR A MINUTE!

GENES ARE BLUEPRINTS FOR PROTEINS, AND PROTEINS MAKE ORGANISMS. SO IT'S FAIR TO SAY THAT **GENES MAKE AN ORGANISM WHAT IT IS.** DIFFERENT ORGANISMS HAVE DIFFERENT GENOMES.

A JELLYFISH, AN ELEPHANT, AND A CACTUS SHARE MANY GENES IN COMMON, BUT EACH CREATURE ALSO HAS UNIQUE GENES ENCODING ITS OWN SPECIAL CHARACTERISTICS.

GENETIC VARIATIONS BETWEEN PEOPLE ACCOUNT FOR DIFFERENCES IN SKIN COLOR, EYE COLOR, AND HAIR COLOR. GENES CONTROL AN ORGANISM'S FEATURES.

GENES, AND PEROXIDE!

AS WE'LL SOON SEE, GENES ARE ALSO **INHERITED** BY THE NEXT GENERATION: GENES MAKE OFFSPRING LOOK LIKE THEIR PARENT(S). NO WONDER THE SCIENCE OF GENOMICS HAS BECOME SUCH A BIG DEAL!!

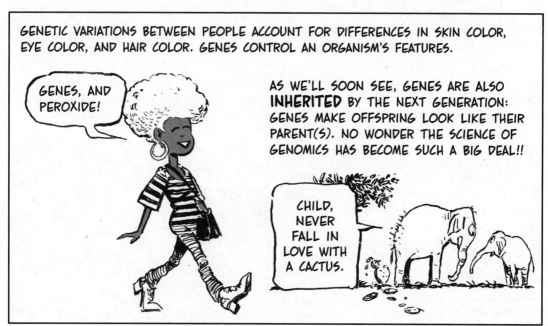

CHILD, NEVER FALL IN LOVE WITH A CACTUS.

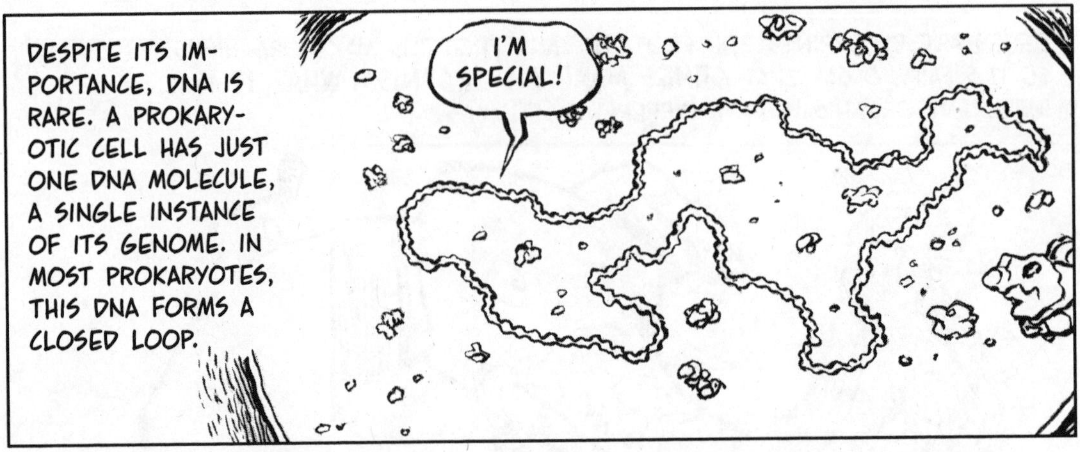
DESPITE ITS IMPORTANCE, DNA IS RARE. A PROKARYOTIC CELL HAS JUST ONE DNA MOLECULE, A SINGLE INSTANCE OF ITS GENOME. IN MOST PROKARYOTES, THIS DNA FORMS A CLOSED LOOP.
I'M SPECIAL!

EUKARYOTES TYPICALLY CARRY A SECOND COPY OF THEIR GENOME,* WITH EACH COPY OCCUPYING SEVERAL SEPARATE DNA MOLECULES, KNOWN AS **CHROMOSOMES.** HUMANS HAVE 46 CHROMOSOMES, TWO SETS OF 23.

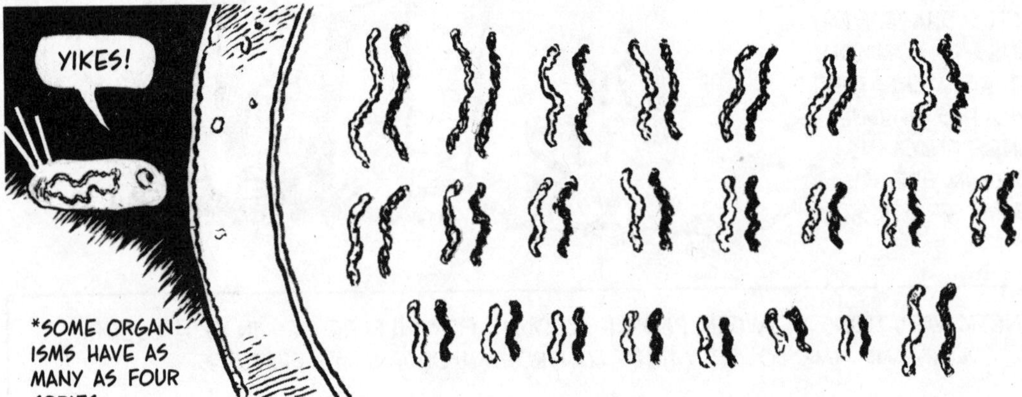
YIKES!

*SOME ORGANISMS HAVE AS MANY AS FOUR COPIES.

BECAUSE DNA IS PRECIOUS STUFF, CELLS SURROUND DNA WITH PROTECTIVE PROTEINS...
HOW MAY WE SERVE, O MASTER LIST?
UNTANGLE?
DE-KINK?
MESSAGE MASSAGE?

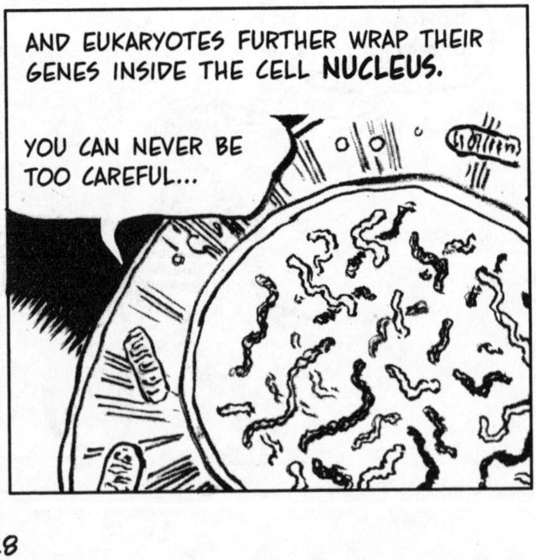
AND EUKARYOTES FURTHER WRAP THEIR GENES INSIDE THE CELL **NUCLEUS.**
YOU CAN NEVER BE TOO CAREFUL...

PROTEIN BUILDING, ON THE OTHER HAND, USES HEAVY MACHINERY THAT RUNS PRETTY MUCH NONSTOP. HOW TO KEEP THIS EQUIPMENT FROM MESSING UP DNA?

STEP BACK!

FIRST, THE CELL MAKES A TEMPORARY **WORKING COPY** OF THE GENE, A **TRANSCRIPT** IN THE FORM OF AN RNA MOLECULE.

GENE

| G C A C G C C A C A T G A G T T C A A G A G G C G A A |
| C G T G C G C T G T A C T C A A G T T C T C C G C T T |

RNA

| G C A C G C C A C A U G A G U U C A A G A G G C G A A |

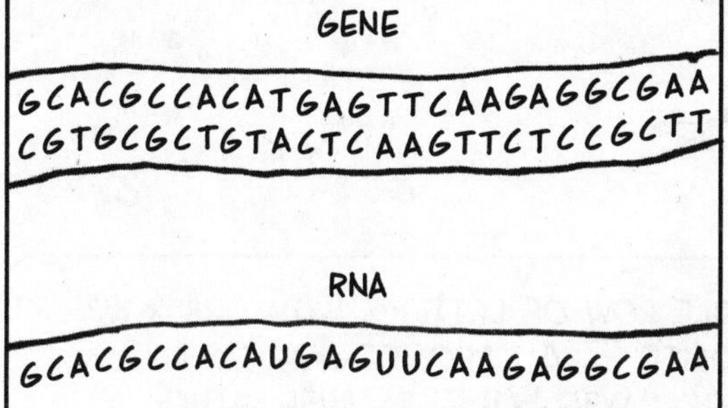

THEN THIS **MESSENGER RNA, mRNA,** RUNS THROUGH A SORT OF TAPE HEAD THAT **TRANSLATES** THE MESSAGE INTO A PROTEIN.

PROTEIN UNDER CONSTRUCTION

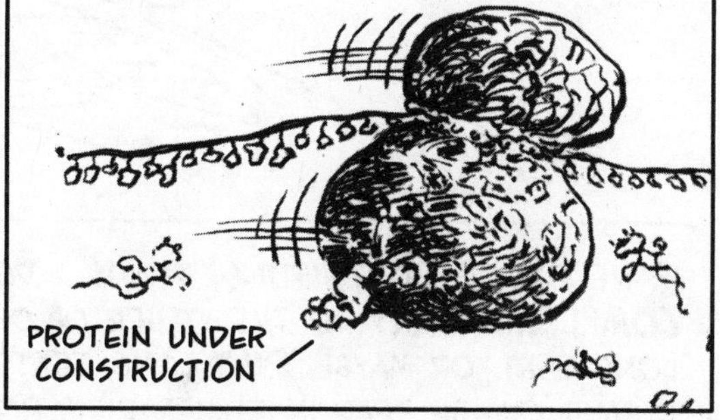

GENE EXPRESSION IS A TWO-STEP PROGRAM: FIRST TRANSCRIPTION, THEN TRANSLATION. THIS KEEPS MOST OF THE ACTION AWAY FROM THE ALL-IMPORTANT GENETIC MATERIAL.

TRANSCRIPTION OF GENE TO mRNA

DNA

mRNA

TRANSLATION TO PROTEIN

PROTEIN

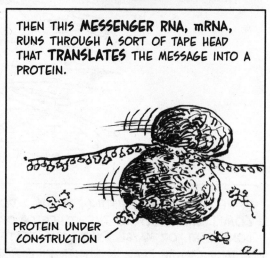

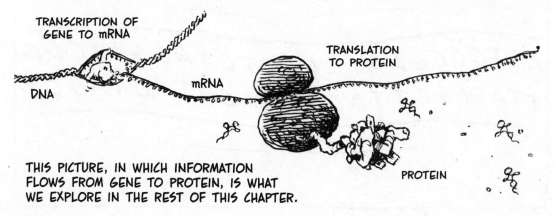

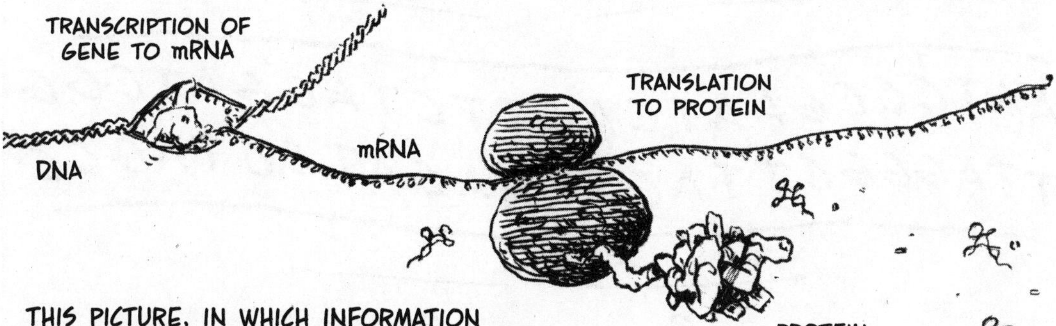

THIS PICTURE, IN WHICH INFORMATION FLOWS FROM GENE TO PROTEIN, IS WHAT WE EXPLORE IN THE REST OF THIS CHAPTER.

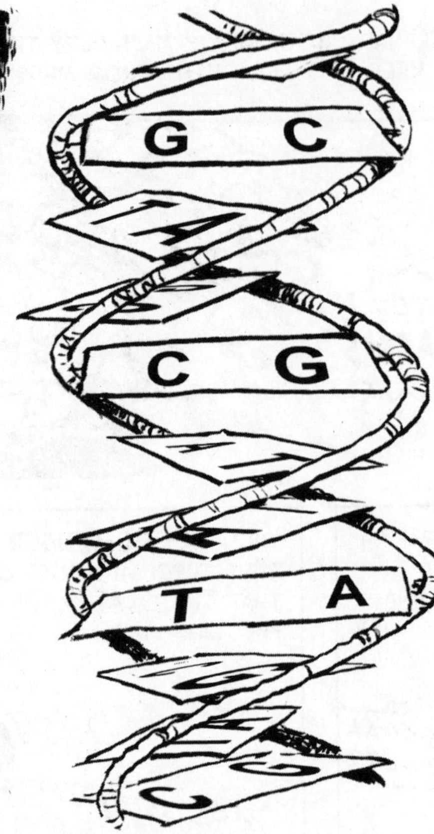

DNA, AS WE SAW ON P. 35, CONSISTS OF TWO STRANDS TWISTED INTO A DOUBLE HELIX LIKE A SPIRAL STAIRCASE. EACH "STEP" IS A **BASE PAIR.**

THE BASES ARE KNOWN AS **A, C, G,** AND **T.** **A** ALWAYS PAIRS WITH **T,** AND **C** ALWAYS PAIRS WITH **G.**

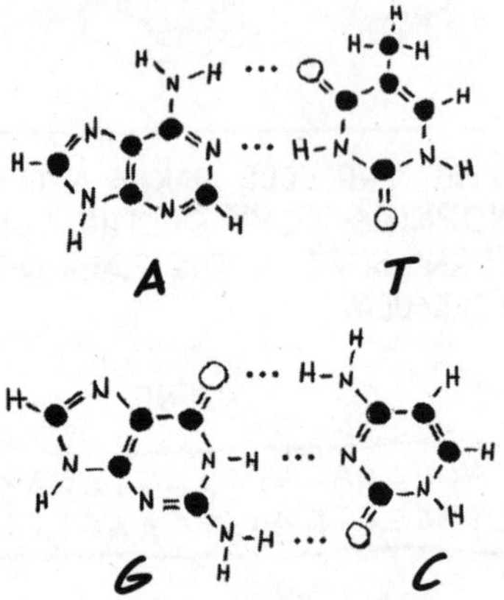

IF WE IGNORE THE TWISTING, DNA IS A DOUBLE ROW OF LETTERS, WITH ONE ROW **COMPLEMENTARY** TO THE OTHER (**A** OPPOSITE **T, C** OPPOSITE **G**). IT'S LIKE A LONG TEXT, OR MAYBE **TWO** LONG TEXTS—OR MAYBE **FOUR,** BECAUSE EITHER STRAND CAN BE READ IN EITHER DIRECTION! WHICH SEQUENCE IS THE GENE?

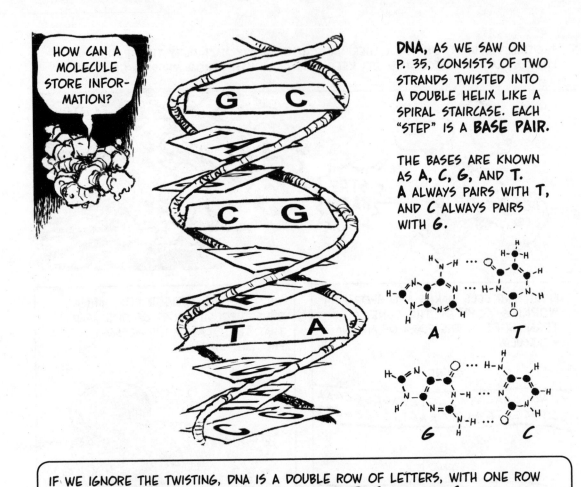

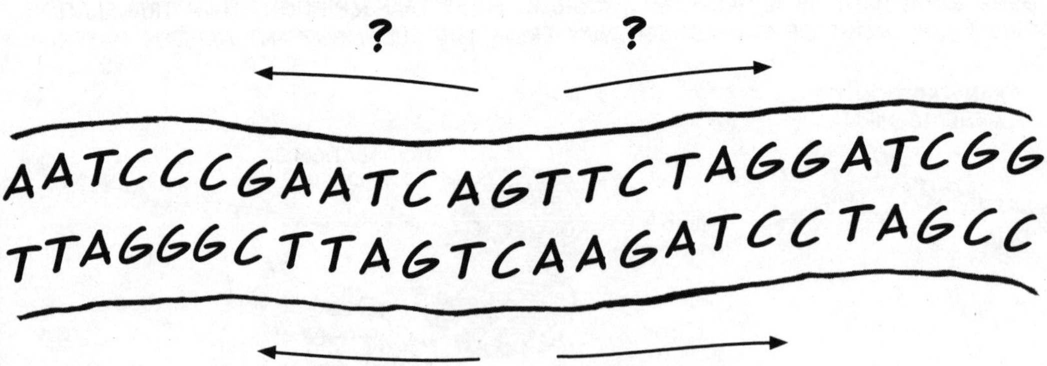

IN FACT, EVERY NUCLEIC ACID STRAND HAS A **PREFERRED DIRECTION,** BECAUSE ITS SUGARS ARE **LOPSIDED.** ONE CARBON, THE 5′, JUTS OFF THE RING.

DEOXY-RIBOSE

5′ 4′ 3′ 2′ 1′

BASE

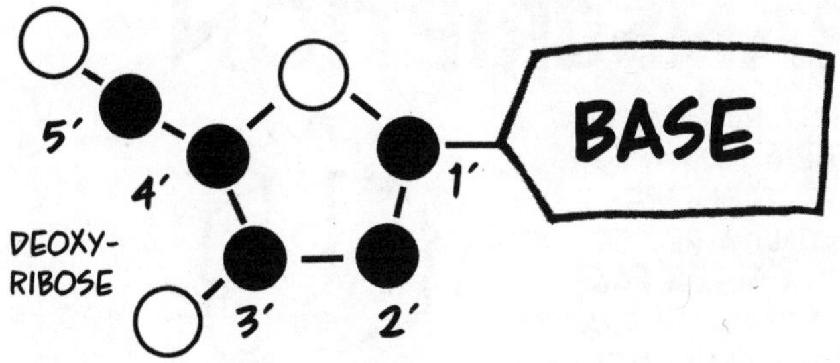

A NUCLEIC ACID BACKBONE RUNS THROUGH THE 3′, 4′, AND 5′ CARBONS, WITH THE 5′ CARBONS ALL POINTING ONE WAY. THE STRAND, THEN, HAS TWO DEFINITE ORIENTATIONS, ONE GOING FROM 5′ TO 3′, THE OTHER FROM 3′ TO 5′. DNA'S TWO STRANDS RUN IN **OPPOSITE DIRECTIONS.**

3′ → 5′

5′ → 3′

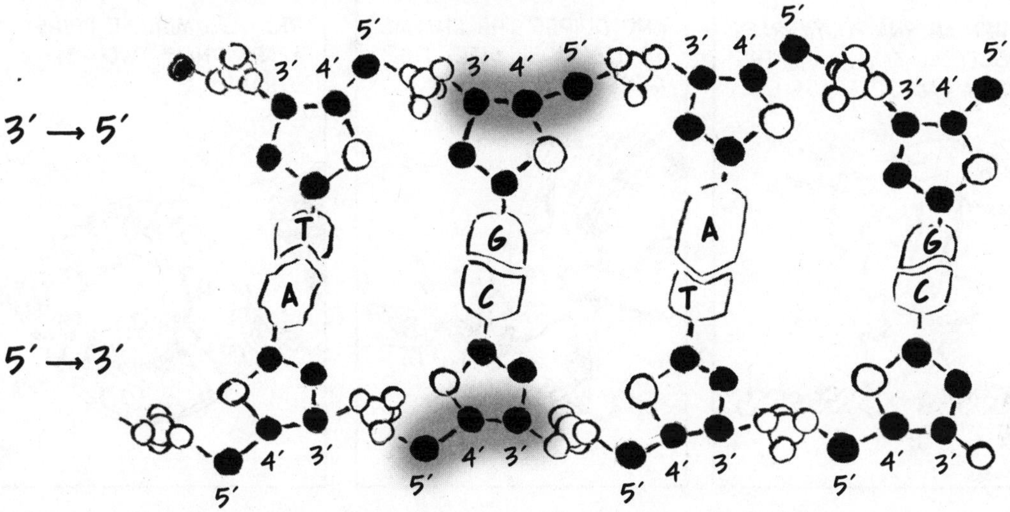

THE CELL **ALWAYS READS GENES FROM 5′ TO 3′.** GIVEN A STARTING POINT ON DNA, THE GENE CAN BE ONLY ONE OF TWO POSSIBLE SEQUENCES.

5′ C   G 3′
    A   T
    C   G
    C   G
    T   A
    T   A
    G   C
    T   A
    C   G
3′ C   G 5′

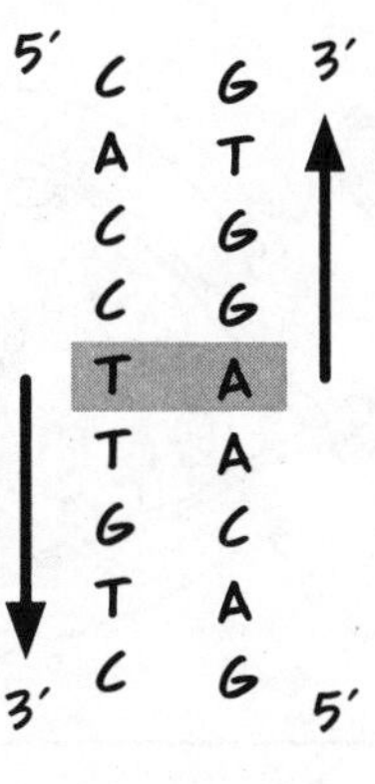

THE CHOICE IS MADE BY PROTEINS SITTING ON DNA JUST "UPSTREAM" FROM THE GENE, AS WE ARE ABOUT TO SEE.

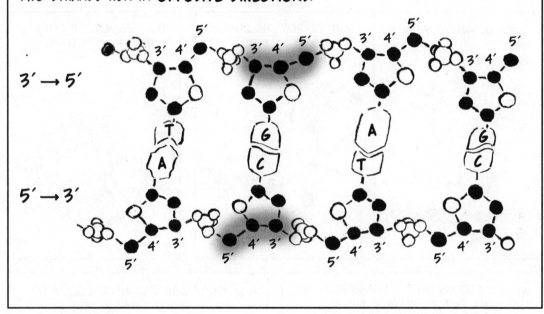

TRANSCRIPTION
WE FIRST DESCRIBE HOW BACTERIA COPY A GENE. TWO PROTEINS ARE ESSENTIAL: A HELPER CALLED A SIGMA FACTOR, SITTING ON DNA NEAR THE GENE, AND RNA POLYMERASE, WHICH BUILDS THE RNA TRANSCRIPT.
SEQUENCE WITH BOUND σ FACTOR
GENE
WHEN AN RNA POLYMERASE MOLECULE COMES BY, THE SIGMA FACTOR GRABS IT...
AND GUIDES THE ENZYME TO ITS WORK SITE, THE GENE.
RNA POLYMERASE PRIES APART DNA'S TWO STRANDS.
WHICH DIRECTION WILL POLYMERASE TAKE? THE SIGMA FACTOR SHOWS IT THE WAY.
RNA POLYMERASE WILL MAKE A COPY OF THE 5′→3′ STRAND BY BUILDING A COMPLEMENT TO THE 3′→5′ STRAND.
3′
5′
3′
5′

WHEN THE ENZYME SEES **T** ON DNA, IT ADDS **A** TO THE RNA STRAND. SEEING **G**, IT ADDS **C**; READING **C**, IT ADDS **G**. AND READING **A**, IT ADDS... **U**?

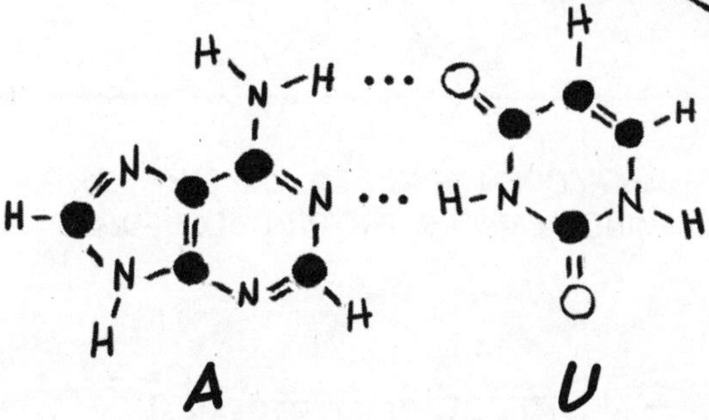

AS WE SAW ON P. 32, RNA ALWAYS USES THE BASE **URACIL** IN PLACE OF DNA'S **THYMINE.** THIS WORKS BECAUSE **U**, LIKE **T**, IS ALSO COMPLEMENTARY TO **A**, ADENINE.

DNA'S 5′→3′ STRAND IS CALLED THE **SENSE** OR **CODING** STRAND. THE OTHER STRAND IS **ANTISENSE.** MOVING BASE BY BASE ALONG THE ANTISENSE STRAND, RNA POLYMERASE MAKES AN **EXACT COPY** OF THE **SENSE** STRAND—EXCEPT THAT **U** REPLACES **T**.

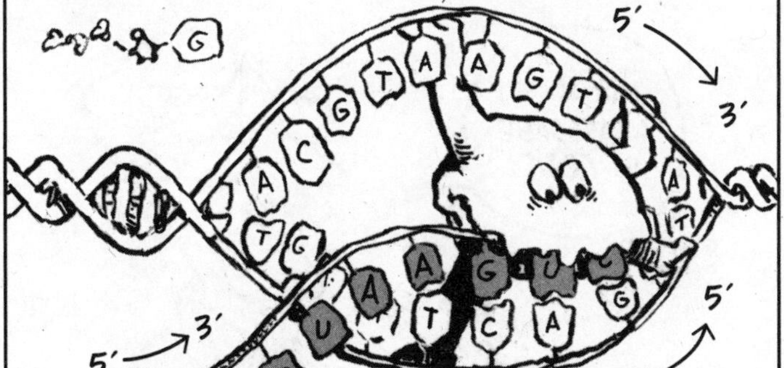

WHEN RNA POLYMERASE REACHES THE GENE'S END, THE ENZYME GETS A SIGNAL TO QUIT. THE RNA COPY IS CALLED **MESSENGER RNA,** mRNA.

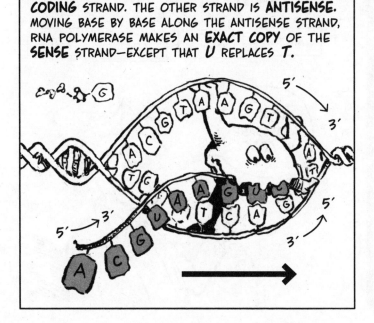

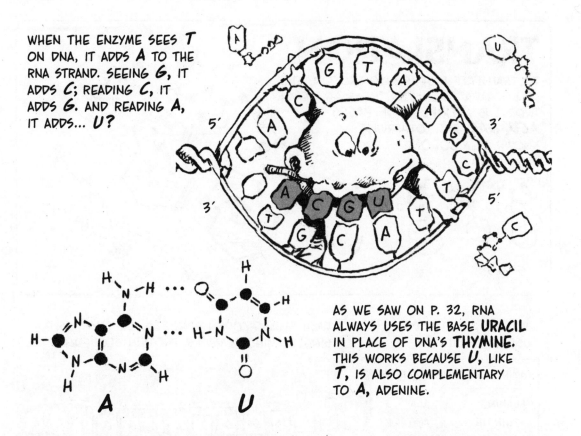

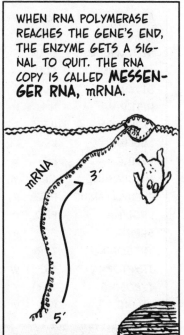

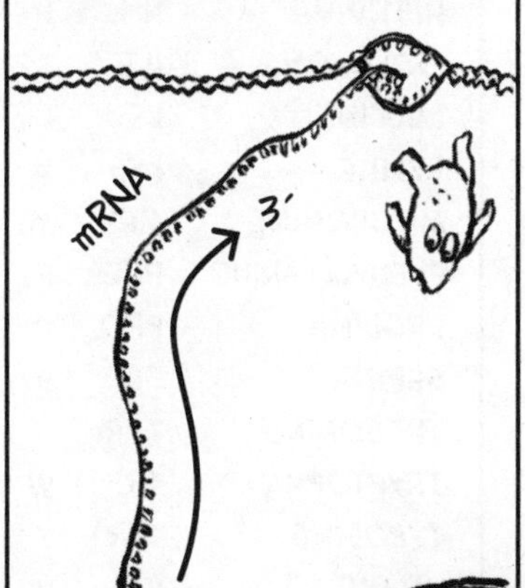

TRANSLATION

TO TRANSLATE A GENE INTO PROTEIN, THE CELL "READS" THE mRNA "MESSAGE." EACH "WORD" IS A **TRIPLET OF BASES— ACU, GAC,** ETC.—OTHERWISE KNOWN AS A **CODON.**

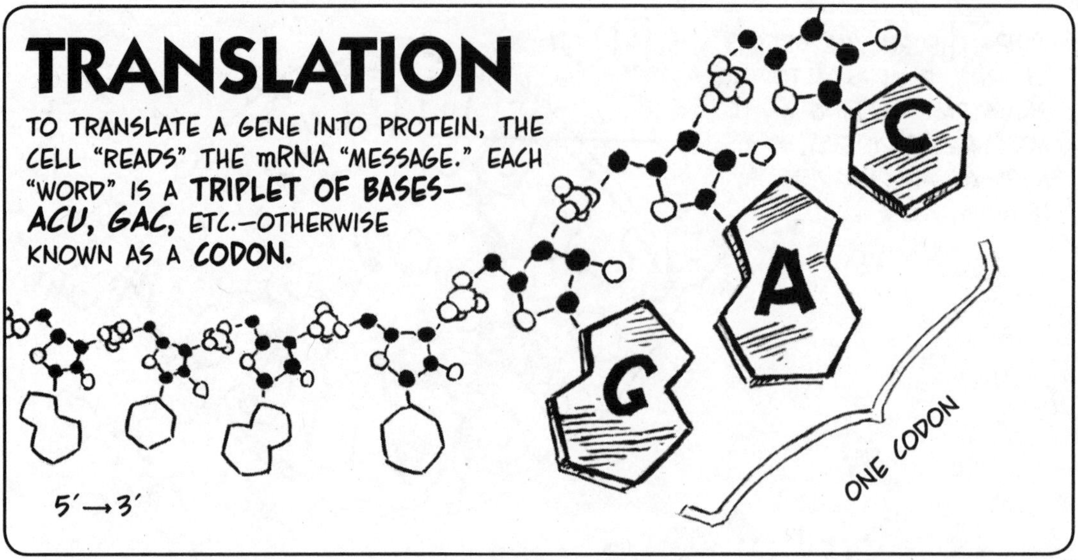

THE 20 AMINO ACIDS USED IN PROTEINS:

AMINO ACID	ABBREVIATIONS	
ALANINE	ALA	A
ARGININE	ARG	R
ASPARGINE	ASN	N
ASPARTIC ACID	ASP	D
CYSTEINE	CYS	C
GLUTAMIC ACID	GLU	E
GLUTAMINE	GLN	Q
GLYCINE	GLY	G
HISTIDINE	HIS	H
ISOLEUCINE	ILE	I
LEUCINE	LEU	L
LYSINE	LYS	K
METHIONINE	MET	M
PHENYLALANINE	PHE	F
PROLINE	PRO	P
SERINE	SER	S
THREONINE	THR	T
TRYPTOPHAN	TRP	W
TYROSINE	TYR	Y
VALINE	VAL	V

EACH 3-BASE CODON SPECIFIES, OR ENCODES, A SINGLE AMINO ACID IN A PROTEIN SEQUENCE.

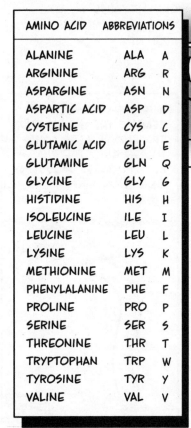

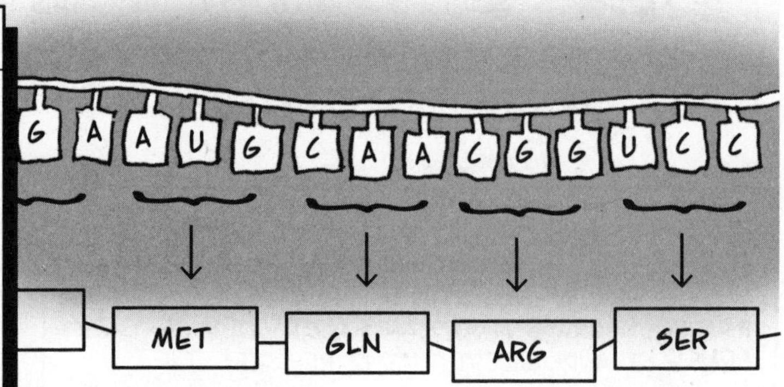

THE FULL SET OF
TRANSLATIONS IS
CALLED THE

GENETIC CODE.

UUU UUC	PHE	UCU UCC	SER	UAU UAC	TYR	UGU UGC	CYS
UUA UUG	LEU	UCA UCG		UAA UAG	STOP	UGA	STOP
						UGG	TRP
CUU CUC	LEU	CCU CCC	PRO	CAU CAC	HIS	CGU CGC	ARG
CUA CUG		CCA CCG		CAA CAG	GLN	CGA CGG	
AUU AUC	ILE	ACU ACC	THR	AAU AAC	ASN	AGU AGC	SER
AUA		ACA		AAA	LYS	AGA AGG	ARG
AUG	MET	ACG		AAG			
GUU GUC	VAL	GCU GCC	ALA	GAU GAC	ASP	GGU GGC	GLY
GUA GUG		GCA GCG		GAA GAG	GLU	GGA GGG	

AUGUCCACAUGGUC

OOK!

THE GENETIC CODE IS **REDUNDANT.** WITH 64
(4 × 4 × 4) CODONS AND ONLY 20 AMINO ACIDS,
SEVERAL DIFFERENT CODONS WILL BE "SYNONYMS,"
ENCODING THE SAME AMINO ACID.

THREE CODONS ARE **STOP** SIGNALS, ENCODING NO
AMINO ACID.

THE CODE IS **NON-OVERLAPPING.** "WORDS" FOL-
LOW EACH OTHER WITHOUT GAPS OR OVERLAPS.

THE CODE IS **UNIVERSAL.** EVERY ORGANISM USES IT.

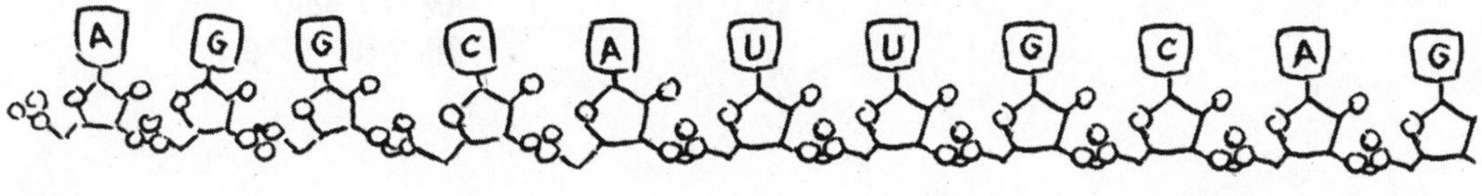

HOW DOES THE CELL ACTUALLY "GET FROM" CODON TO AMINO ACID? BY USING SPECIAL TRANSLATOR MOLECULES THAT EMBODY THE CODE. **TRANSFER RNA, tRNA,** MAKES THE **PHYSICAL LINK** BETWEEN MESSAGE AND PROTEIN.

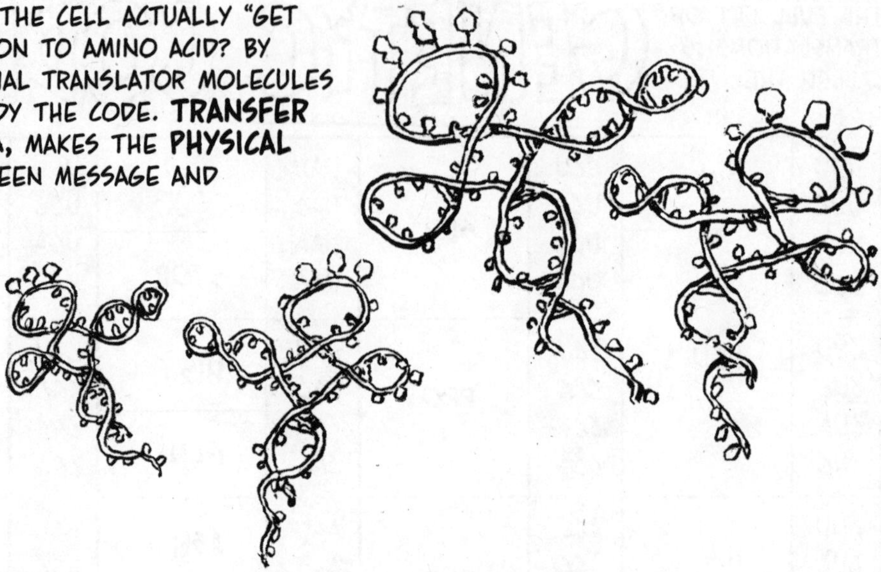

THESE KEY tRNA MOLECULES ARE SHAPED, APPROPRIATELY ENOUGH, LIKE KEYS. THANKS TO HYDROGEN BONDS BETWEEN THEIR OWN BASES, tRNA TWISTS INTO THIS SHAPE, WITH A 3-BASE **ANTICODON** AT THE HEAD AND AN AMINO ACID BINDING SITE AT THE TAIL.

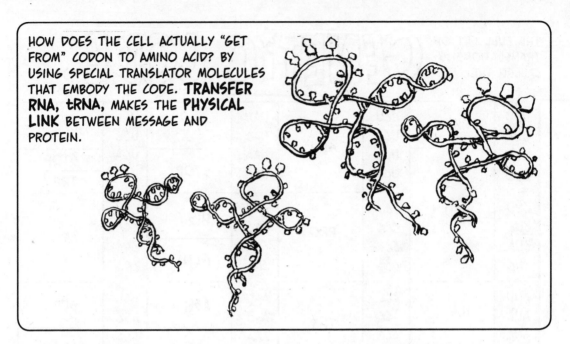

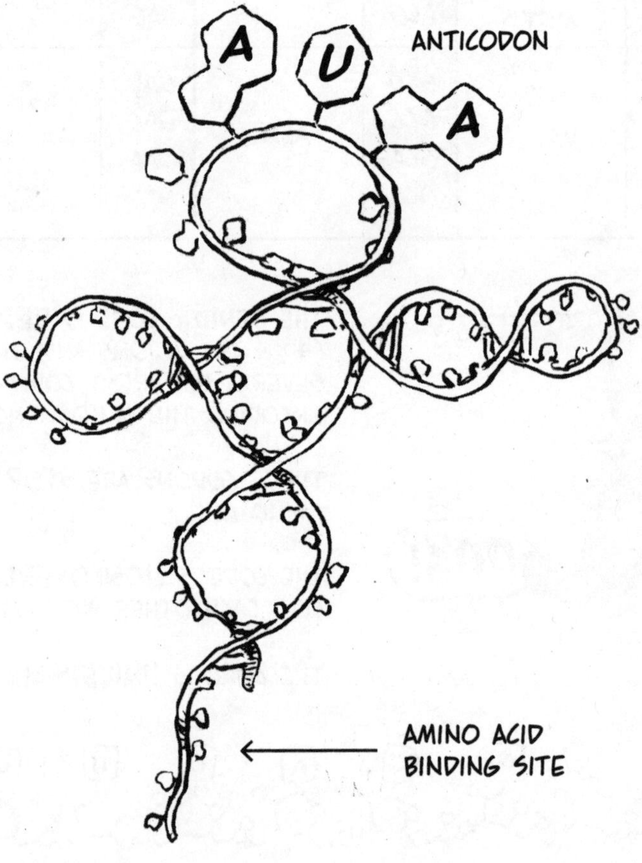

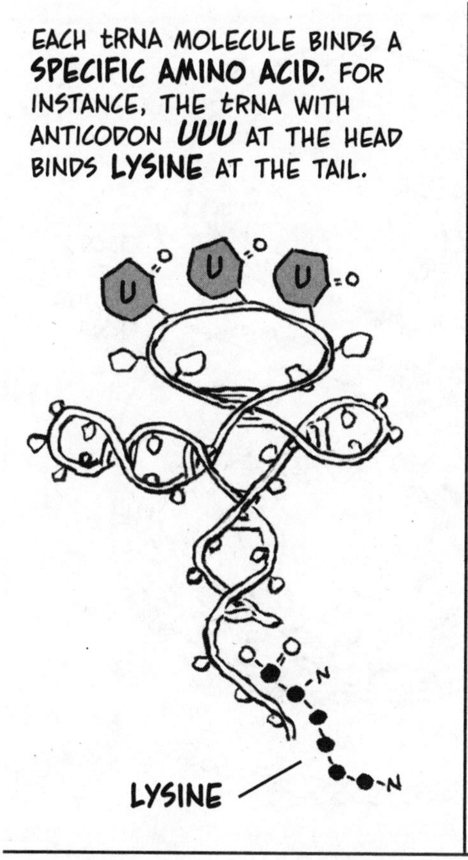
EACH tRNA MOLECULE BINDS A SPECIFIC AMINO ACID. FOR INSTANCE, THE tRNA WITH ANTICODON UUU AT THE HEAD BINDS LYSINE AT THE TAIL.
U U U
N
N
LYSINE

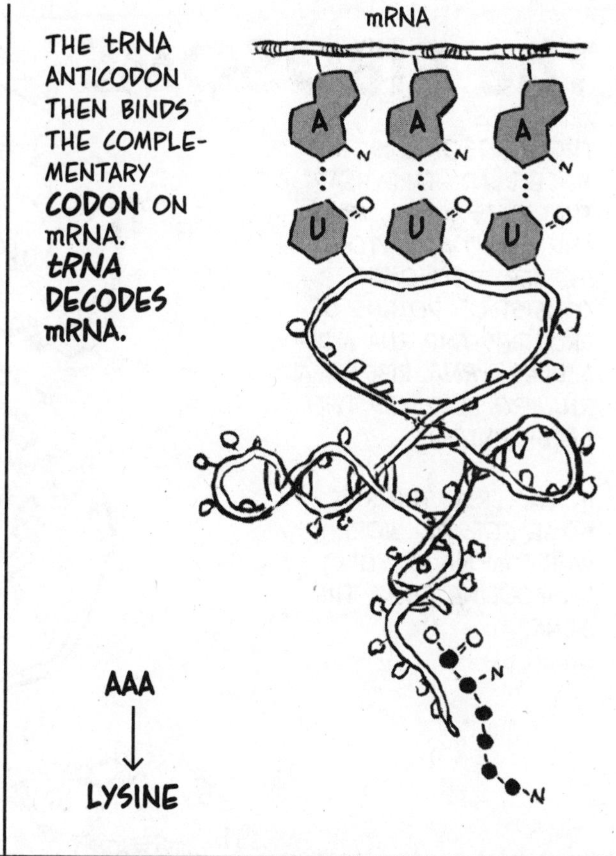
THE tRNA ANTICODON THEN BINDS THE COMPLEMENTARY CODON ON mRNA. tRNA DECODES mRNA.
mRNA
A A A
U U U
N
N
AAA
LYSINE

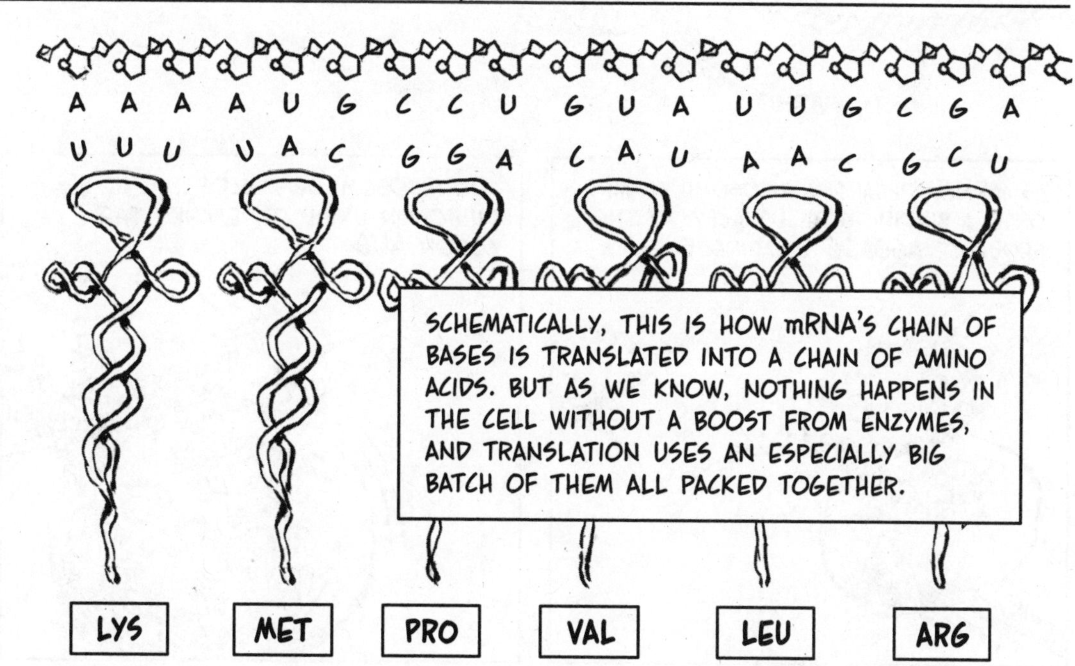
A A A A U G C C U G U A U U G C G A
U U U U A C G G A C A U A A C G C U
SCHEMATICALLY, THIS IS HOW mRNA'S CHAIN OF BASES IS TRANSLATED INTO A CHAIN OF AMINO ACIDS. BUT AS WE KNOW, NOTHING HAPPENS IN THE CELL WITHOUT A BOOST FROM ENZYMES, AND TRANSLATION USES AN ESPECIALLY BIG BATCH OF THEM ALL PACKED TOGETHER.
LYS
MET
PRO
VAL
LEU
ARG

The RIBOSOME

THE **RIBOSOME** IS THE MOLECULAR "TAPE HEAD" THAT PUTS *m*RNA, *t*RNA, AND AMINO ACIDS TOGETHER. RIBOSOMES CONSIST OF DOZENS OF PROTEINS AND RNA MOLECULES (*r*RNA, RIBOSOMAL RNA) ROLLED INTO TWO LARGE UNITS.

IN BACTERIA, THE RIBOSOME GETS TO WORK WHILE *m*RNA IS STILL UNSPOOLING FROM THE GENE.

"SLOTS" FOR BINDING tRNA

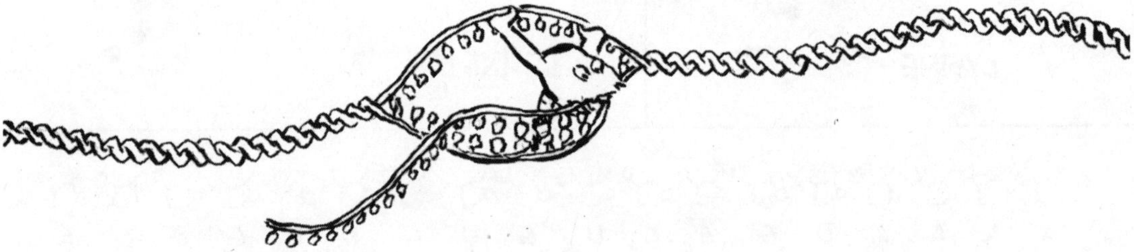

AS mRNA APPEARS, A RIBOSOME CLAMPS ONTO A BINDING SITE, USUALLY TO THE SEQUENCE AGGAGG (READING 5′→3′).

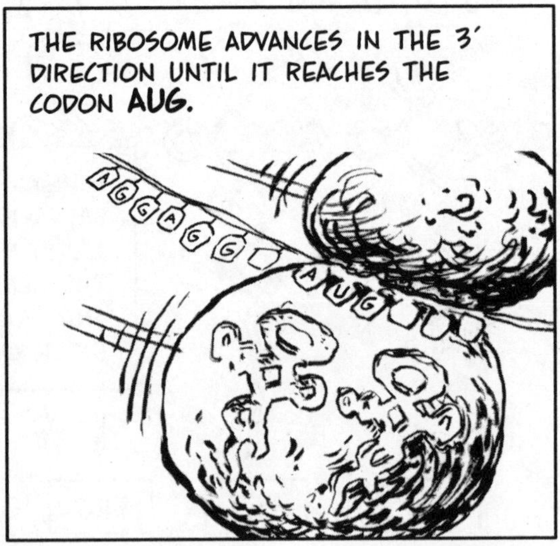
THE RIBOSOME ADVANCES IN THE 3′ DIRECTION UNTIL IT REACHES THE CODON AUG.
AGGAGG
AUG

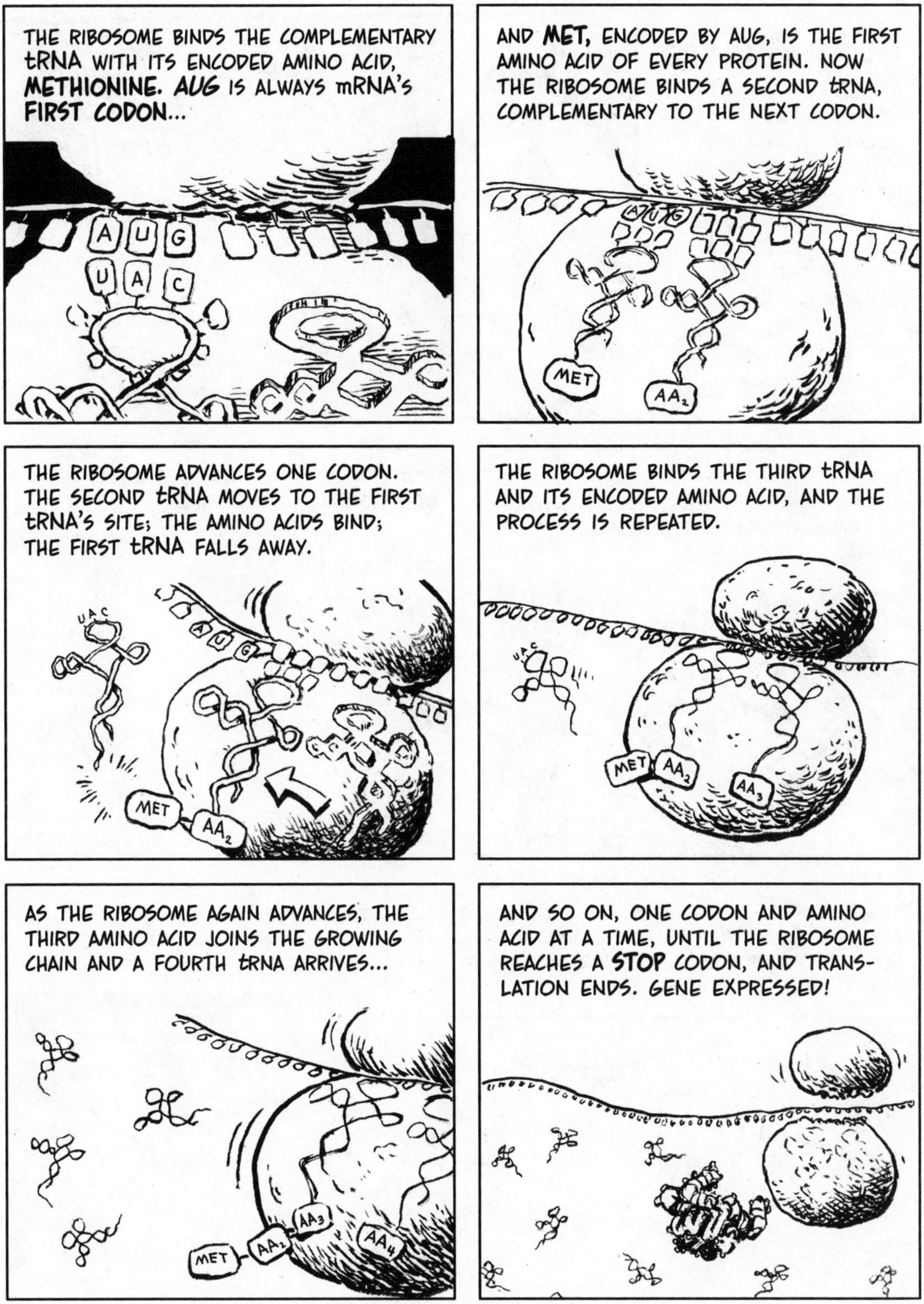

THE RIBOSOME BINDS THE COMPLEMENTARY tRNA WITH ITS ENCODED AMINO ACID, **METHIONINE**. AUG IS ALWAYS mRNA'S **FIRST CODON**...

AND **MET**, ENCODED BY AUG, IS THE FIRST AMINO ACID OF EVERY PROTEIN. NOW THE RIBOSOME BINDS A SECOND tRNA, COMPLEMENTARY TO THE NEXT CODON.

THE RIBOSOME ADVANCES ONE CODON. THE SECOND tRNA MOVES TO THE FIRST tRNA'S SITE; THE AMINO ACIDS BIND; THE FIRST tRNA FALLS AWAY.

THE RIBOSOME BINDS THE THIRD tRNA AND ITS ENCODED AMINO ACID, AND THE PROCESS IS REPEATED.

AS THE RIBOSOME AGAIN ADVANCES, THE THIRD AMINO ACID JOINS THE GROWING CHAIN AND A FOURTH tRNA ARRIVES...

AND SO ON, ONE CODON AND AMINO ACID AT A TIME, UNTIL THE RIBOSOME REACHES A **STOP** CODON, AND TRANSLATION ENDS. GENE EXPRESSED!

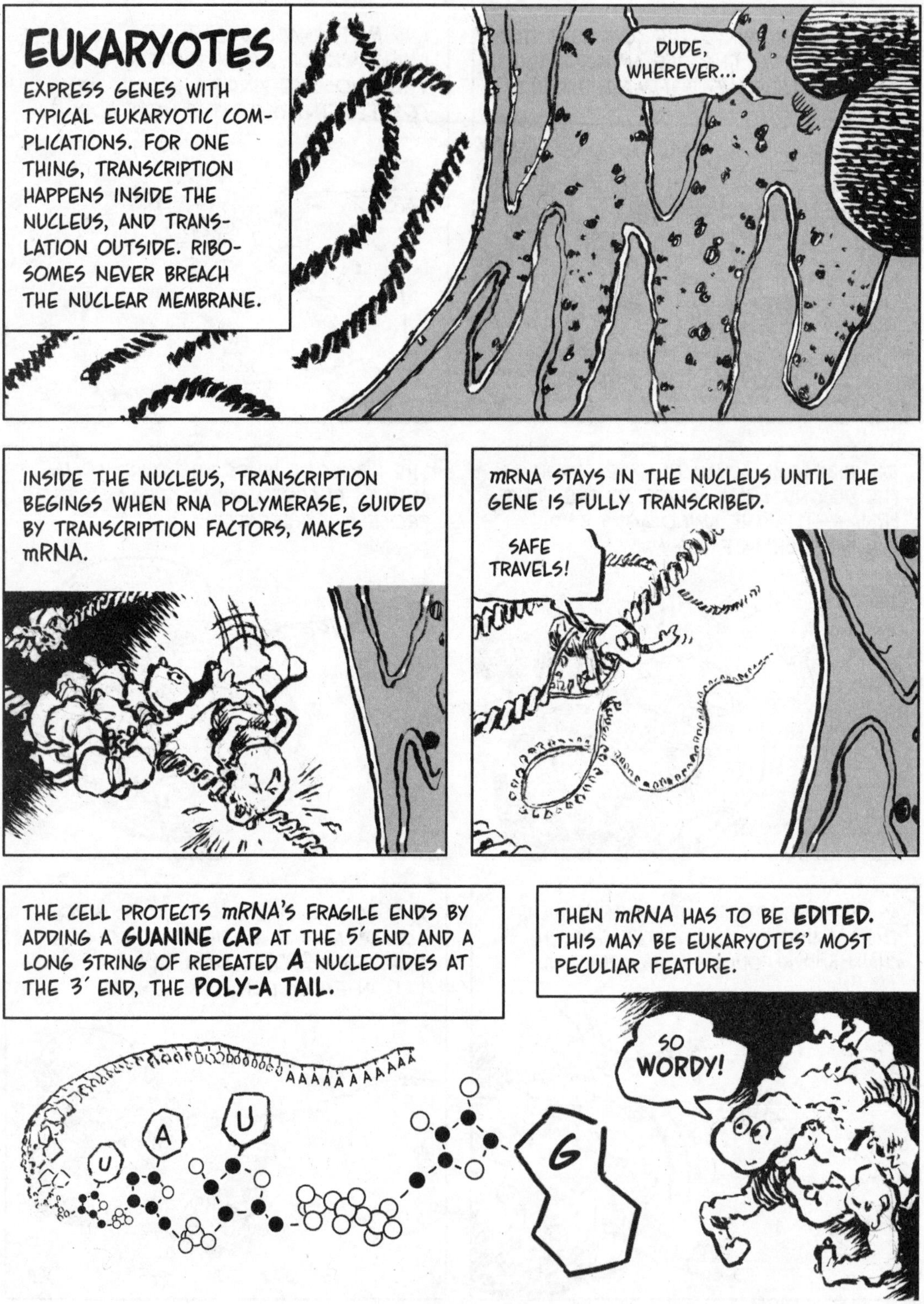
EUKARYOTES
EXPRESS GENES WITH TYPICAL EUKARYOTIC COMPLICATIONS. FOR ONE THING, TRANSCRIPTION HAPPENS INSIDE THE NUCLEUS, AND TRANSLATION OUTSIDE. RIBOSOMES NEVER BREACH THE NUCLEAR MEMBRANE.
DUDE, WHEREVER...
INSIDE THE NUCLEUS, TRANSCRIPTION BEGINGS WHEN RNA POLYMERASE, GUIDED BY TRANSCRIPTION FACTORS, MAKES mRNA.
mRNA STAYS IN THE NUCLEUS UNTIL THE GENE IS FULLY TRANSCRIBED.
SAFE TRAVELS!
THE CELL PROTECTS mRNA'S FRAGILE ENDS BY ADDING A GUANINE CAP AT THE 5′ END AND A LONG STRING OF REPEATED A NUCLEOTIDES AT THE 3′ END, THE POLY-A TAIL.
AAAAAAAAAA
U
A
U
G
THEN mRNA HAS TO BE EDITED. THIS MAY BE EUKARYOTES' MOST PECULIAR FEATURE.
SO WORDY!

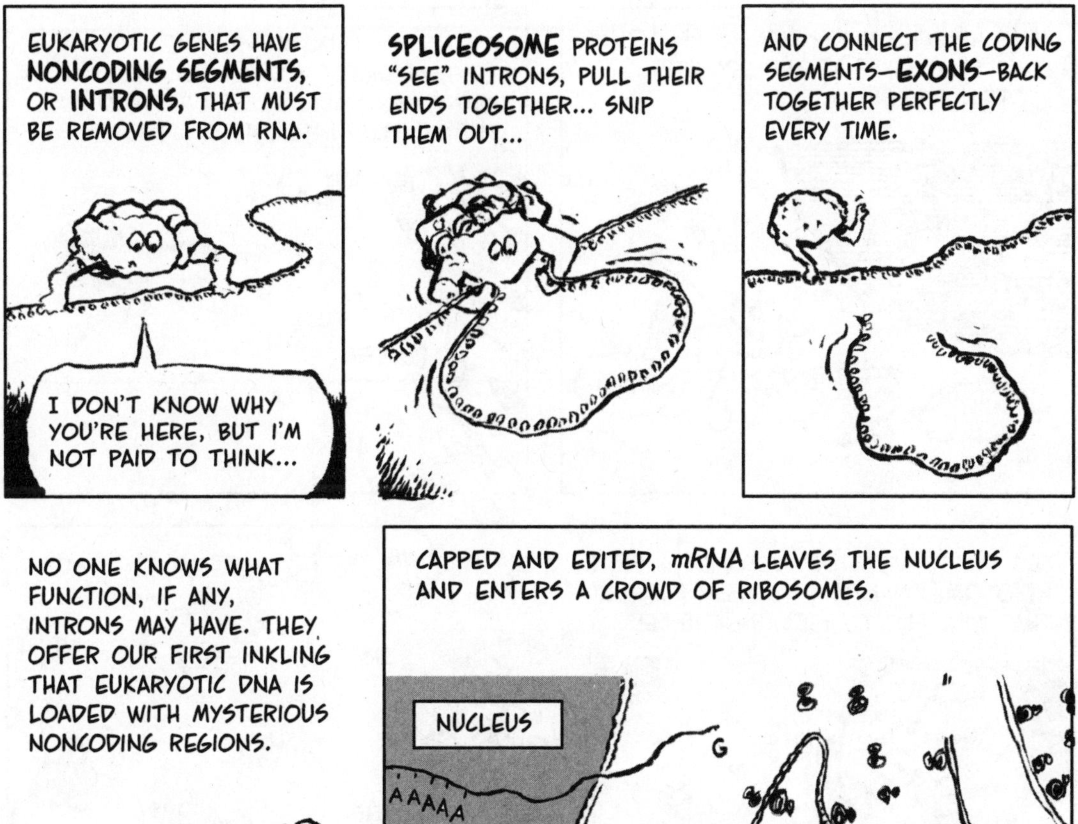
EUKARYOTIC GENES HAVE **NONCODING SEGMENTS,** OR **INTRONS,** THAT MUST BE REMOVED FROM RNA.
I DON'T KNOW WHY YOU'RE HERE, BUT I'M NOT PAID TO THINK...
SPLICEOSOME PROTEINS "SEE" INTRONS, PULL THEIR ENDS TOGETHER... SNIP THEM OUT...
AND CONNECT THE CODING SEGMENTS—**EXONS**—BACK TOGETHER PERFECTLY EVERY TIME.
NO ONE KNOWS WHAT FUNCTION, IF ANY, INTRONS MAY HAVE. THEY OFFER OUR FIRST INKLING THAT EUKARYOTIC DNA IS LOADED WITH MYSTERIOUS NONCODING REGIONS.
?
CAPPED AND EDITED, *m*RNA LEAVES THE NUCLEUS AND ENTERS A CROWD OF RIBOSOMES.
NUCLEUS
G
AAAAA
ENDOPLASMIC RETICULUM (E.R.)
A RIBOSOME GRABS *m*RNA JUST DOWN-STREAM FROM THE GUANINE CAP (RATHER THAN AT A SPECIAL BINDING SEQUENCE, AS IN PROKARYOTES).
G
AFTER THAT, TRANSLATION RUNS DOWN-STREAM AS BEFORE, STARTING WITH THE FIRST **AUG** CODON ENCOUNTERED.
5′
3′
AUG

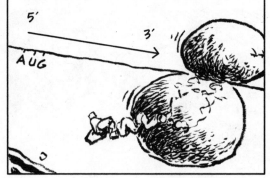

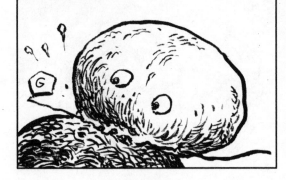

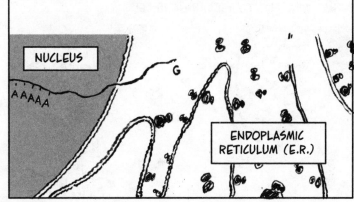

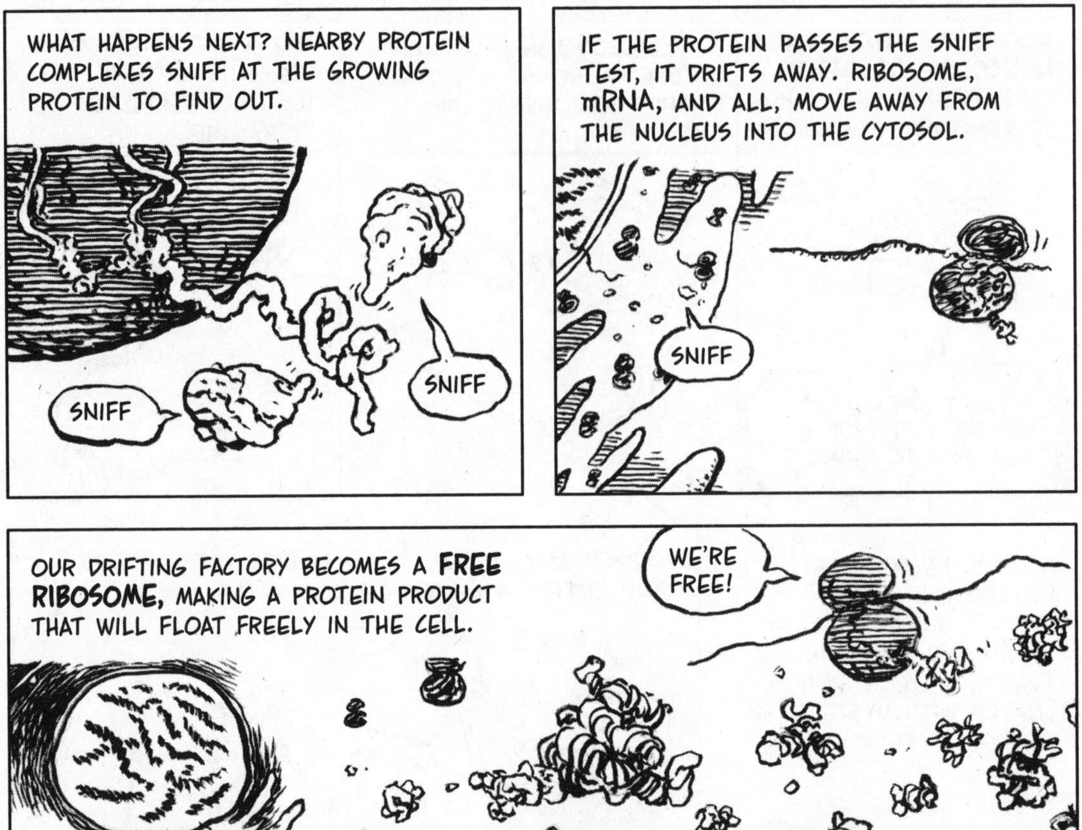
WHAT HAPPENS NEXT? NEARBY PROTEIN COMPLEXES SNIFF AT THE GROWING PROTEIN TO FIND OUT.
SNIFF
SNIFF
IF THE PROTEIN PASSES THE SNIFF TEST, IT DRIFTS AWAY. RIBOSOME, mRNA, AND ALL, MOVE AWAY FROM THE NUCLEUS INTO THE CYTOSOL.
SNIFF
OUR DRIFTING FACTORY BECOMES A **FREE RIBOSOME,** MAKING A PROTEIN PRODUCT THAT WILL FLOAT FREELY IN THE CELL.
WE'RE FREE!
ON THE OTHER HAND, SOME PROTEINS DON'T MAKE IT SO FAR. THEY HAVE A **SIGNAL SEQUENCE** THAT BINDS A **SIGNAL RECOGNITION PARTICLE.**
SNIFF!
THE SIGNAL RECOGNITION PARTICLE ATTACHES THE RIBOSOME TO A WALL OF THE **ENDOPLASMIC RETICULUM** (E.R.).

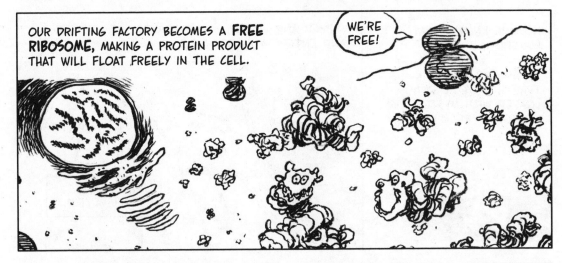

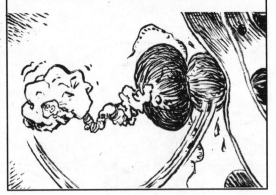

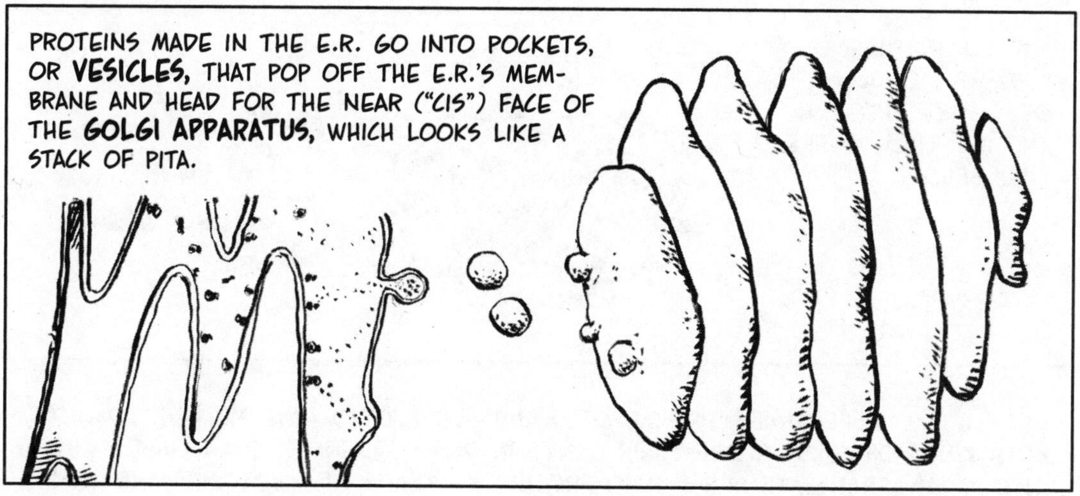
PROTEINS MADE IN THE E.R. GO INTO POCKETS, OR **VESICLES,** THAT POP OFF THE E.R.'S MEMBRANE AND HEAD FOR THE NEAR ("CIS") FACE OF THE **GOLGI APPARATUS,** WHICH LOOKS LIKE A STACK OF PITA.

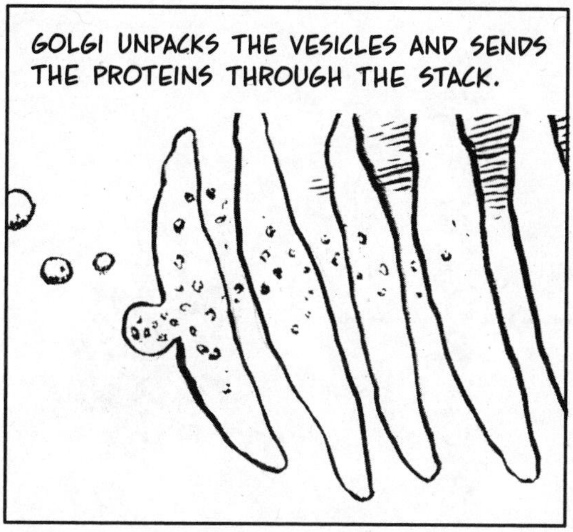
GOLGI UNPACKS THE VESICLES AND SENDS THE PROTEINS THROUGH THE STACK.

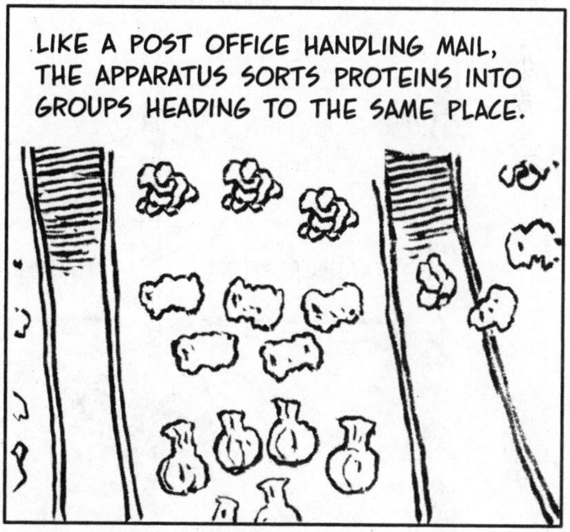
LIKE A POST OFFICE HANDLING MAIL, THE APPARATUS SORTS PROTEINS INTO GROUPS HEADING TO THE SAME PLACE.

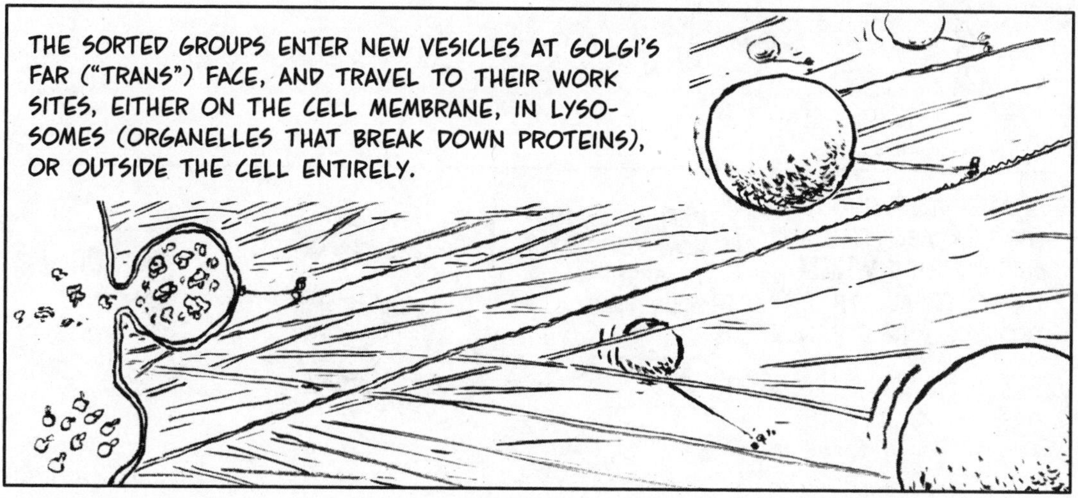
THE SORTED GROUPS ENTER NEW VESICLES AT GOLGI'S FAR ("TRANS") FACE, AND TRAVEL TO THEIR WORK SITES, EITHER ON THE CELL MEMBRANE, IN LYSOSOMES (ORGANELLES THAT BREAK DOWN PROTEINS), OR OUTSIDE THE CELL ENTIRELY.

AND SO, IN THEIR BUSY WAY, EUKARYOTES MAKE PROTEINS AND PUT THEM WHERE THEY BELONG.
CAP
EDIT
SPLICE

DESPITE THE DIFFERENCES, PROKARYOTES AND EUKARYOTES HAVE MUCH IN COMMON: BOTH KINDS OF CELLS STORE THEIR GENES IN DNA SEQUENCES, TRANSCRIBE GENES INTO mRNA, TRANSLATE mRNA INTO PROTEIN BY MEANS OF RIBOSOMES AND tRNA, AND USE THE SAME GENETIC CODE.

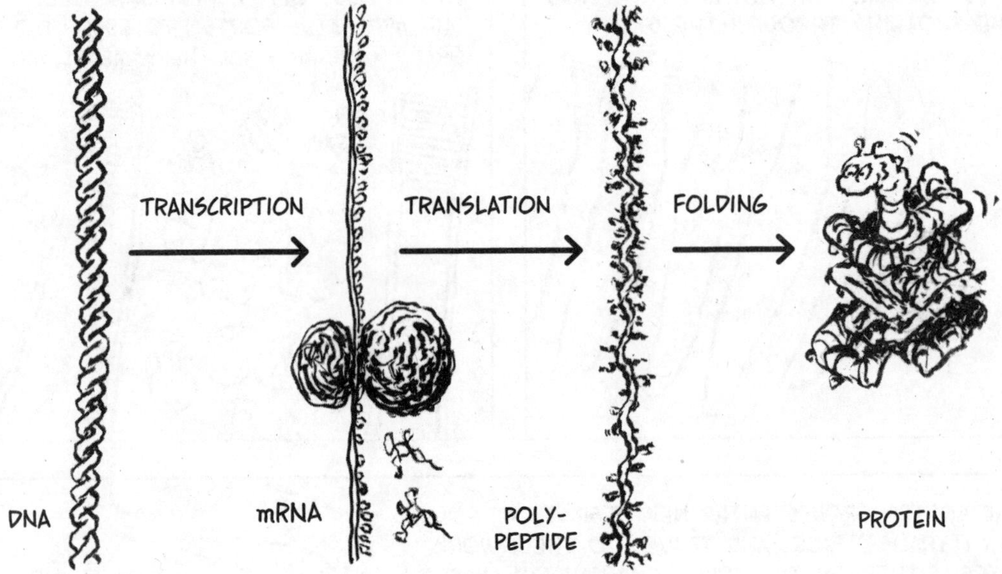
TRANSCRIPTION
TRANSLATION
FOLDING
DNA
mRNA
POLY-PEPTIDE
PROTEIN

IN THE NEXT CHAPTER, WE TALK ABOUT HOW CELLS DECIDE **WHEN** TO GO TO ALL THAT TROUBLE AND EXPENSE.
DEFINITELY **NOT** HOW YOU WANT TO SPEND ALL YOUR TIME...
NUH-**UH**!

Chapter 10
GENE REGULATION

I'M SO EXPRES-SIVE I MAKE MYSELF SICK!

LIVING THINGS, LIKE ARTISTS, DEVOTE THEMSELVES TO **EXPRESSION,** AND, LIKE ANY ARTIST OF TALENT, THEY HAVE TO KNOW WHEN TO **STOP** EXPRESSING THEMSELVES.

NO ORGANISM EXPRESSES ALL ITS GENES ALL THE TIME. NONSTOP PROTEIN PRODUCTION WOULD WASTE ENERGY AND OVERSTUFF AN ALREADY CROWDED CELL.

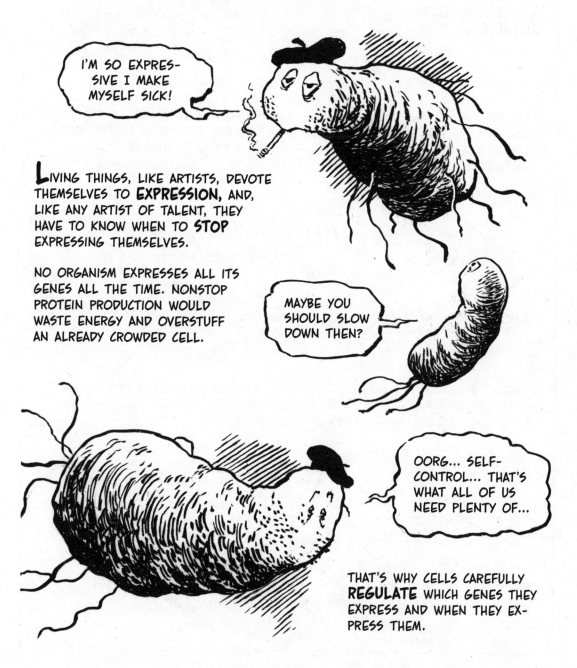

MAYBE YOU SHOULD SLOW DOWN THEN?

OORG... SELF-CONTROL... THAT'S WHAT ALL OF US NEED PLENTY OF...

THAT'S WHY CELLS CAREFULLY **REGULATE** WHICH GENES THEY EXPRESS AND WHEN THEY EX-PRESS THEM.

CELLS MANAGE GENE EXPRESSION BY TURNING **TRANSCRIPTION** ON AND OFF. JUST AHEAD OF A GENE ARE DNA SEQUENCES WHERE **REGULATORY PROTEINS** CAN BIND.

IN BACTERIA, AS WE SAW ON P. 132, AN UPSTREAM **SIGMA FACTOR** GUIDES RNA POLYMERASE TO THE GENE. THE DNA SEQUENCE WHERE THE SIGMA FACTOR BINDS IS CALLED THE **PROMOTER** SEQUENCE. A COMMON SIGMA FACTOR IS σ-70, SO NAMED BECAUSE ITS MASS IS AROUND 70,000 Da.*

HERE!

BETWEEN PROMOTER AND GENE MAY BE ANOTHER REGULATORY SEQUENCE, THE **OPERATOR,** WHERE A **REPRESSOR** PROTEIN CAN BIND, TOTALLY BLOCKING RNA POLYMERASE'S ACCESS, REGARDLESS OF ACTIVATION.

OOP!
NOT SO FAST, PAL!

*ONE DALTON IS APPROXIMATELY THE MASS OF A HYDROGEN ATOM.

THE

LAC OPERON

FOUND IN *E. COLI* IS A PRIME EXAMPLE.

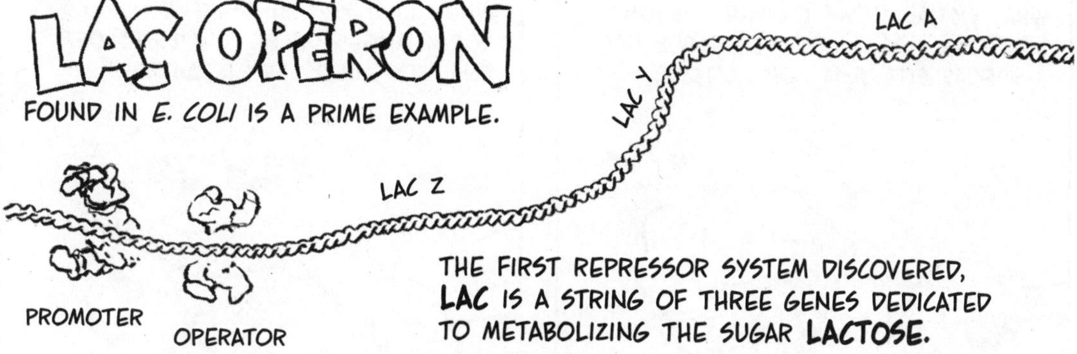

THE FIRST REPRESSOR SYSTEM DISCOVERED, **LAC** IS A STRING OF THREE GENES DEDICATED TO METABOLIZING THE SUGAR **LACTOSE.**

THE GENES ARE: ***LAC Z,*** ENCODING AN ENZYME, BETA-GALACTOSIDASE, THAT BREAKS DOWN LACTOSE; ***LAC Y,*** ENCODING PERMEASE, WHICH OPENS THE CELL MEMBRANE TO LACTOSE; AND ***LAC A,*** ENCODING A PROTEIN OF UNKNOWN IMPORTANCE.

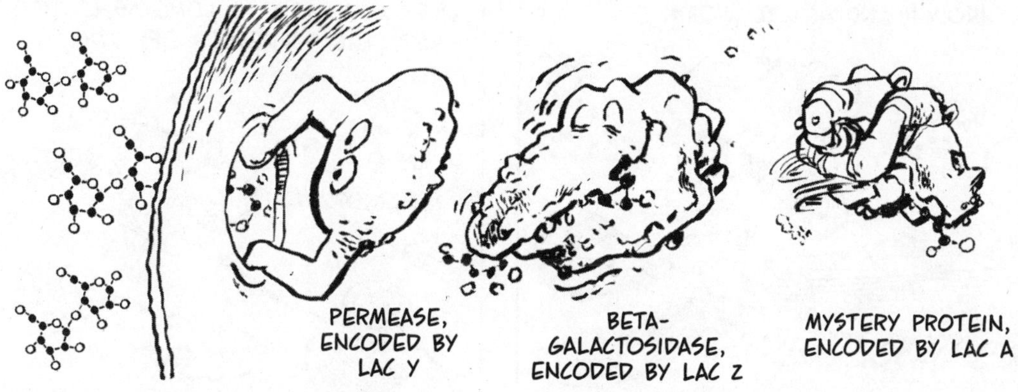

A GROUP OF GENES, ALL REGULATED TOGETHER, IS CALLED AN **OPERON.** OPERONS ARE COMMON IN PROKARYOTES. IN **LAC,** A SINGLE PROMOTER GOVERNS ALL THREE GENES, AND SO DOES AN **OPERATOR** SEQUENCE WHERE A REPRESSOR PROTEIN IS BOUND.* THE QUESTION IS—

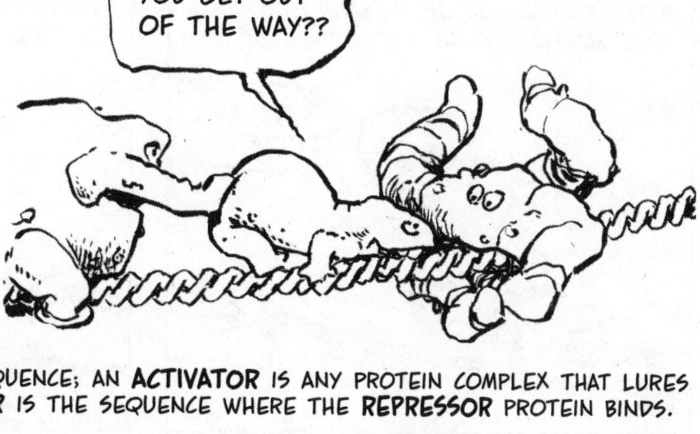

WHEN **WILL** YOU GET OUT OF THE WAY??

*THE **PROMOTER** IS THE DNA SEQUENCE; AN **ACTIVATOR** IS ANY PROTEIN COMPLEX THAT LURES RNA POLYMERASE; THE **OPERATOR** IS THE SEQUENCE WHERE THE **REPRESSOR** PROTEIN BINDS.

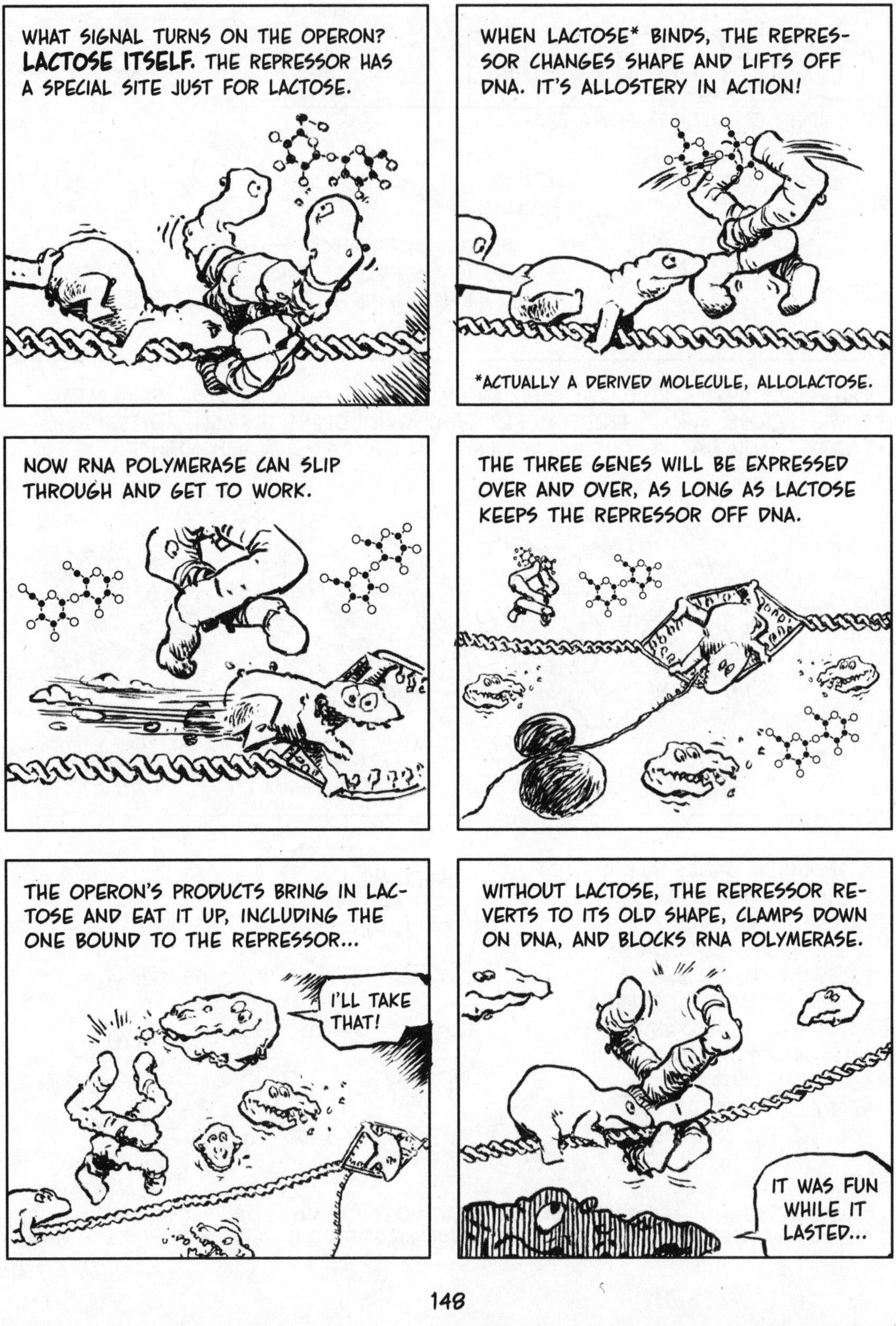
WHAT SIGNAL TURNS ON THE OPERON? **LACTOSE ITSELF.** THE REPRESSOR HAS A SPECIAL SITE JUST FOR LACTOSE.
WHEN LACTOSE* BINDS, THE REPRESSOR CHANGES SHAPE AND LIFTS OFF DNA. IT'S ALLOSTERY IN ACTION!
*ACTUALLY A DERIVED MOLECULE, ALLOLACTOSE.
NOW RNA POLYMERASE CAN SLIP THROUGH AND GET TO WORK.
THE THREE GENES WILL BE EXPRESSED OVER AND OVER, AS LONG AS LACTOSE KEEPS THE REPRESSOR OFF DNA.
THE OPERON'S PRODUCTS BRING IN LACTOSE AND EAT IT UP, INCLUDING THE ONE BOUND TO THE REPRESSOR...
I'LL TAKE THAT!
WITHOUT LACTOSE, THE REPRESSOR REVERTS TO ITS OLD SHAPE, CLAMPS DOWN ON DNA, AND BLOCKS RNA POLYMERASE.
IT WAS FUN WHILE IT LASTED...

CHICKEN-AND-EGG QUESTION: IF THE LAC OPERON BRINGS LACTOSE INTO THE CELL, WHERE DID THE REPRESSOR-BINDING LACTOSE MOLECULE COME FROM? ANSWER: THE LAC REPRESSOR IS SOMEWHAT "LEAKY." THE OPERON IS OCCASIONALLY EXPRESSED, SO IF LACTOSE IS OUTSIDE THE CELL, A TRACE WILL COME IN.

THE LAC OPERON SWITCHES **OFF** WHEN LACTOSE IS ABSENT. BY CONTRAST, THE **TRP** OPERON, FIVE GENES CODING FOR PROTEINS THAT SYTHESIZE THE AMINO ACID TRYPTOPHAN, MUST BE SWITCHED **ON** WHEN TRYPTOPHAN IS ABSENT, TO MAKE THE STUFF. THE **TRP REPRESSOR** STAYS OFF DNA IN THE ABSENCE OF TRYPTOPHAN.

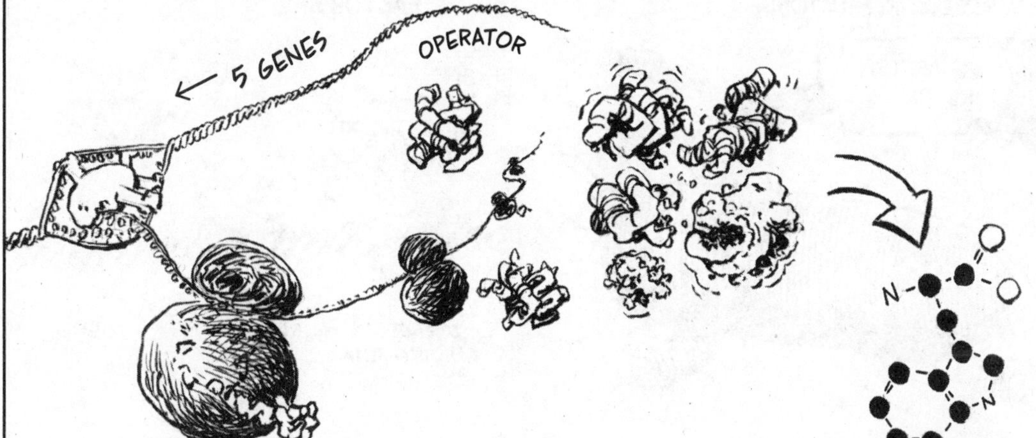

AS TRYPTOPHAN LEVELS RISE, THE TRP REPRESSOR BINDS THE AMINO ACID **AND** DNA. TRANSCRIPTION STOPS WHEN THE CELL HAS ENOUGH TRYPTOPHAN.

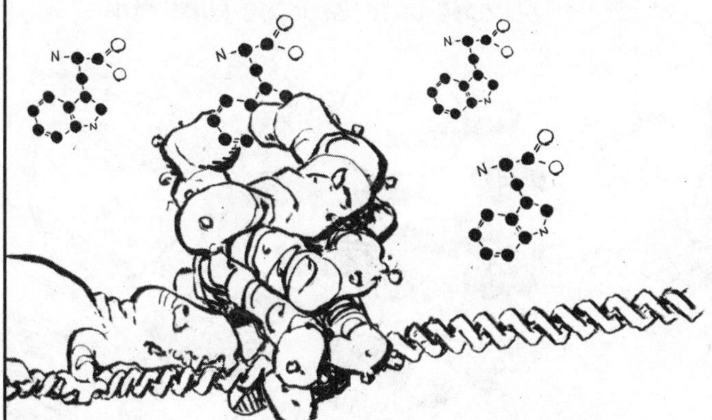

IN BIO-JARGON, LAC IS **CATABOLIC** AND LACTOSE IS ITS **INDUCER**; TRP IS **ANABOLIC** AND TRYPTOPHAN IS A **CO-REPRESSOR.**

AN OPERON, WITH ONLY A FEW GENES LAID END TO END, CAN BE SWITCHED BY A SINGLE REPRESSOR. PROKARYOTES ALSO HAVE WAYS TO CONTROL **MANY OPERONS AT ONCE.**

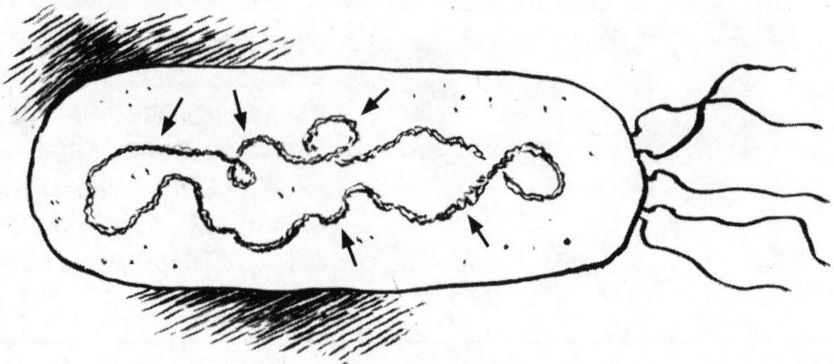

FOR INSTANCE, IF GLUCOSE IS AVAILABLE, *E. COLI* "WANTS" TO EAT THAT SUGAR BEFORE ANY OTHER, SUCH AS LACTOSE, SUCROSE, OR FRUCTOSE.

WE ARTISTS ARE PICKY...

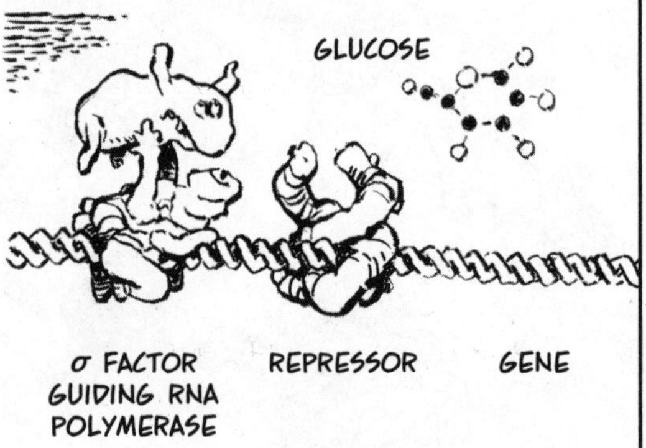

THE PRESENCE OF GLUCOSE DISABLES THE SIGMA FACTORS OF THESE OTHER SUGAR OPERONS AND SO SILENCES THEM.

IN THAT CASE, *E. COLI* WILL SHUT DOWN **ALL** OPERONS THAT GOVERN ALTERNATIVE SUGAR DIGESTION BY MESSING WITH THEIR **SIGMA FACTORS.**

GLUCOSE

σ FACTOR GUIDING RNA POLYMERASE   REPRESSOR   GENE

RESULT: EVEN IF ANOTHER SUGAR (LIKE LACTOSE) IS PRESENT, *E. COLI* WILL EAT ONLY GLUCOSE UNTIL GLUCOSE RUNS OUT.

SO TRUE...

E. COLI LOVES GLUCOSE

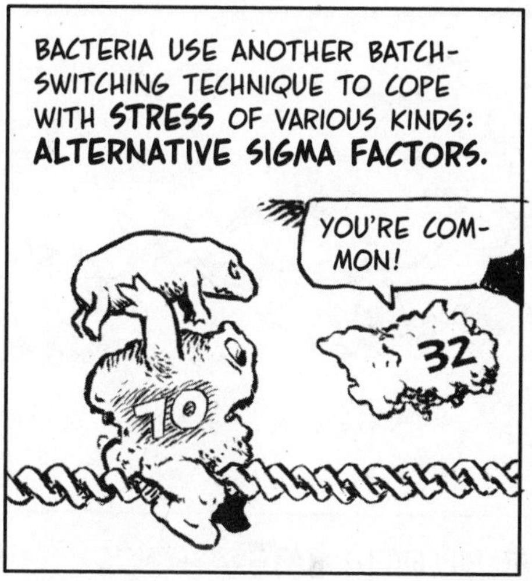
BACTERIA USE ANOTHER BATCH-SWITCHING TECHNIQUE TO COPE WITH **STRESS** OF VARIOUS KINDS: **ALTERNATIVE SIGMA FACTORS.**
YOU'RE COM-MON!
70
32

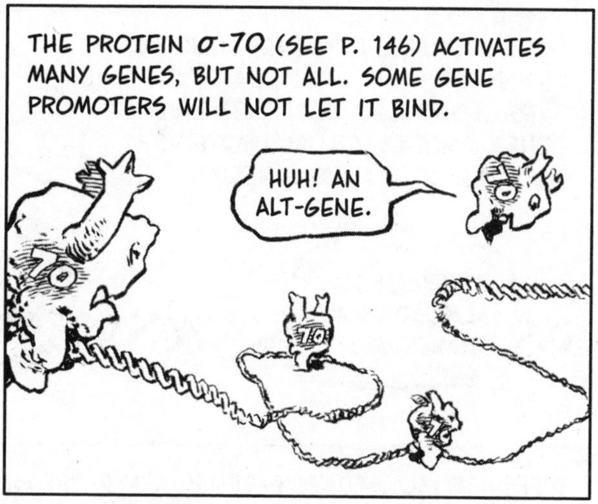
THE PROTEIN σ-70 (SEE P. 146) ACTIVATES MANY GENES, BUT NOT ALL. SOME GENE PROMOTERS WILL NOT LET IT BIND.
HUH! AN ALT-GENE.
70

STRESSFUL CONDITIONS—FOOD SHORTAGE, DNA BREAKAGE, ETC.—ACTIVATE GENES THAT ENCODE OTHER σ FACTORS. WHEN **TEMPERATURE** RISES, FOR INSTANCE, THE CELL EXPRESSES A GENE THAT ENCODES **σ-32.**

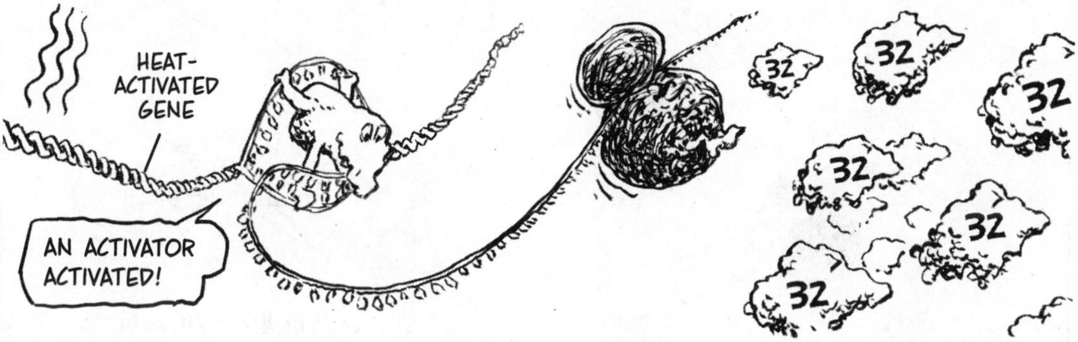

AN ACTIVATOR ACTIVATED!

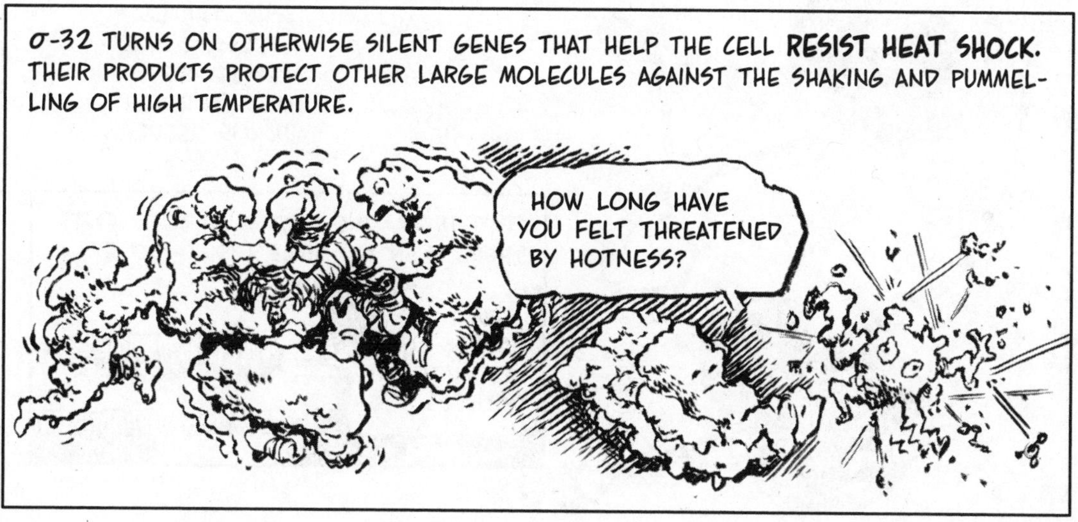
σ-32 TURNS ON OTHERWISE SILENT GENES THAT HELP THE CELL **RESIST HEAT SHOCK.** THEIR PRODUCTS PROTECT OTHER LARGE MOLECULES AGAINST THE SHAKING AND PUMMEL-LING OF HIGH TEMPERATURE.
HOW LONG HAVE YOU FELT THREATENED BY HOTNESS?

SOME PROKARYOTIC GENES HAVE NO REPRESSORS AND ARE ALWAYS "ON." THESE SO-CALLED **CONSTITUTIVE** GENES MAKE ESSENTIAL PROTEINS THAT THE CELL ALWAYS NEEDS.

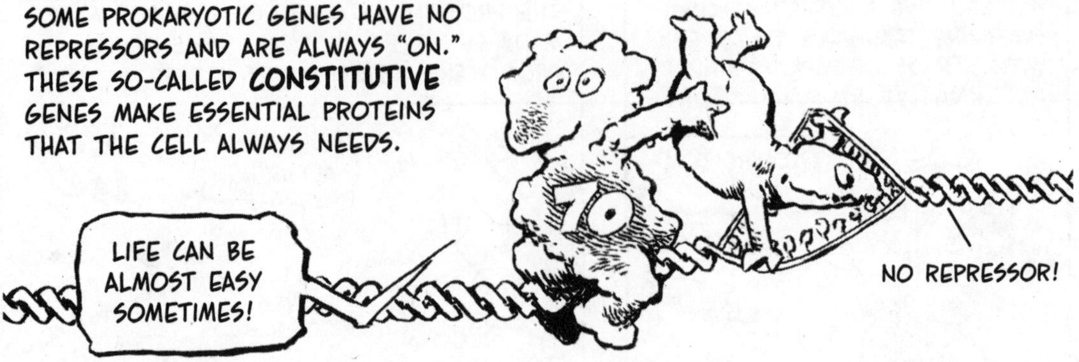
70
LIFE CAN BE ALMOST EASY SOMETIMES!
NO REPRESSOR!

DIFFERENT CONSTITUTIVE GENES ARE EXPRESSED AT DIFFERENT **RATES,** BECAUSE THEIR **PROMOTER** REGIONS ARE NOT THE SAME. THE GENE ENCODING **RNA POLYMERASE** ITSELF, FOR INSTANCE, HAS AN EFFICIENT PROMOTER THAT ACTIVATES PLENTY OF THIS MUCH-NEEDED ENZYME. THE **LAC REPRESSOR** GENE'S PROMOTER BINDS WEAKLY TO A SIGMA FACTOR, SO EXPRESSION HAPPENS RARELY.

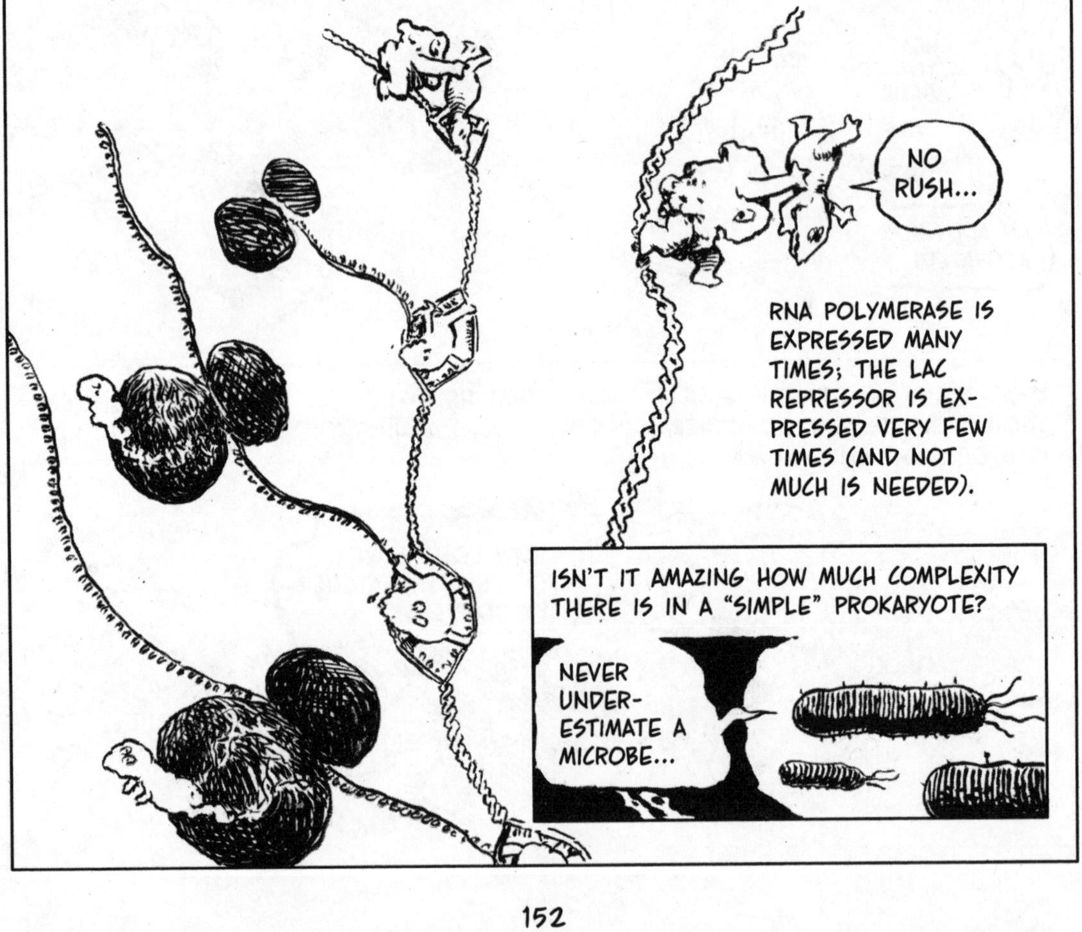
NO RUSH...
RNA POLYMERASE IS EXPRESSED MANY TIMES; THE LAC REPRESSOR IS EXPRESSED VERY FEW TIMES (AND NOT MUCH IS NEEDED).
ISN'T IT AMAZING HOW MUCH COMPLEXITY THERE IS IN A "SIMPLE" PROKARYOTE?
NEVER UNDER-ESTIMATE A MICROBE...

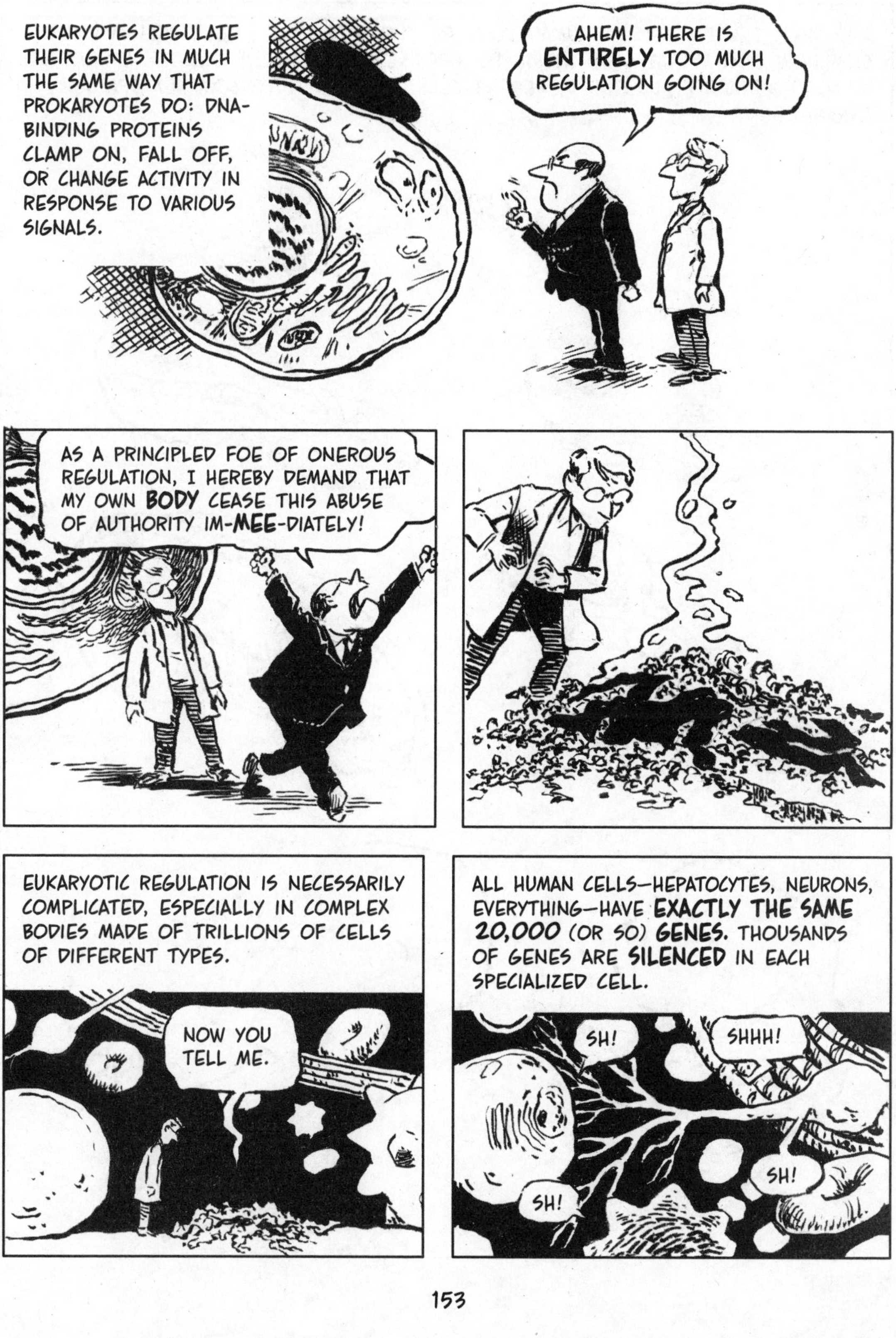
EUKARYOTES REGULATE THEIR GENES IN MUCH THE SAME WAY THAT PROKARYOTES DO: DNA-BINDING PROTEINS CLAMP ON, FALL OFF, OR CHANGE ACTIVITY IN RESPONSE TO VARIOUS SIGNALS.
AHEM! THERE IS ENTIRELY TOO MUCH REGULATION GOING ON!
AS A PRINCIPLED FOE OF ONEROUS REGULATION, I HEREBY DEMAND THAT MY OWN BODY CEASE THIS ABUSE OF AUTHORITY IM-MEE-DIATELY!
EUKARYOTIC REGULATION IS NECESSARILY COMPLICATED, ESPECIALLY IN COMPLEX BODIES MADE OF TRILLIONS OF CELLS OF DIFFERENT TYPES.
NOW YOU TELL ME.
ALL HUMAN CELLS—HEPATOCYTES, NEURONS, EVERYTHING—HAVE EXACTLY THE SAME 20,000 (OR SO) GENES. THOUSANDS OF GENES ARE SILENCED IN EACH SPECIALIZED CELL.
SH!
SHHH!
SH!
SH!

AND THEN THERE'S THIS: A HUMAN BODY DEVELOPS FROM A SINGLE CELL, WHICH SOMEHOW ORCHESTRATES AN INTRICATE, PRECISE SYMPHONY OF GENETIC EXPRESSION TO GENERATE ALL OUR DIFFERENT CELL TYPES IN THEIR COMPLEX SPACIAL ARRANGEMENT.

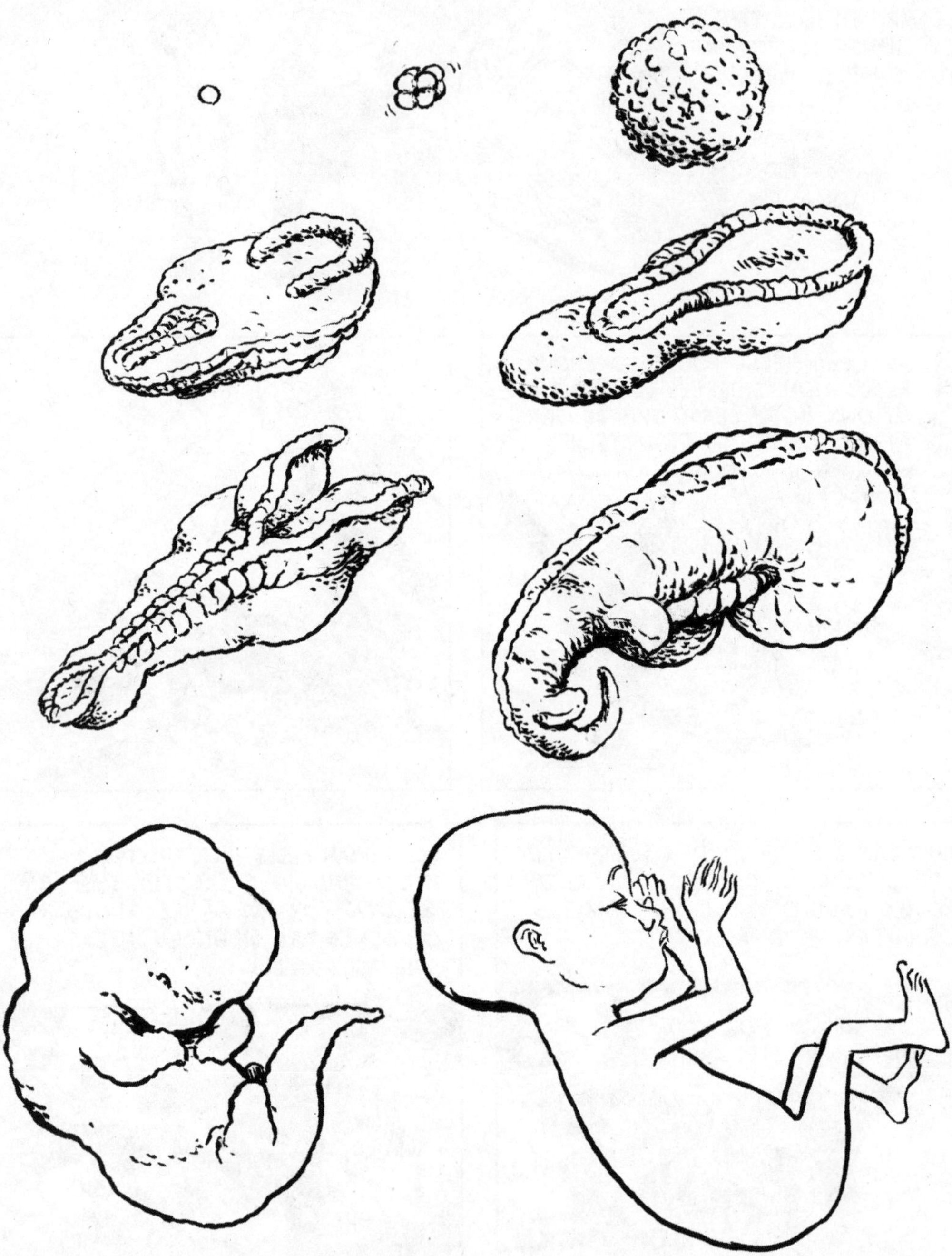

THERE'S FAR TOO MUCH GOING ON HERE FOR A GENERAL BIOLOGY BOOK, SO WE'LL JUST OUTLINE ONE IMPORTANT PROCESS. GENES CAN BE TURNED ON ACCORDING TO **WHERE** THEY ARE IN THE BODY.

ANIMALS DO THIS BY MEANS OF A SET OF GENES CALLED ***HOX,*** SHORT FOR **HOMEOBOX.**

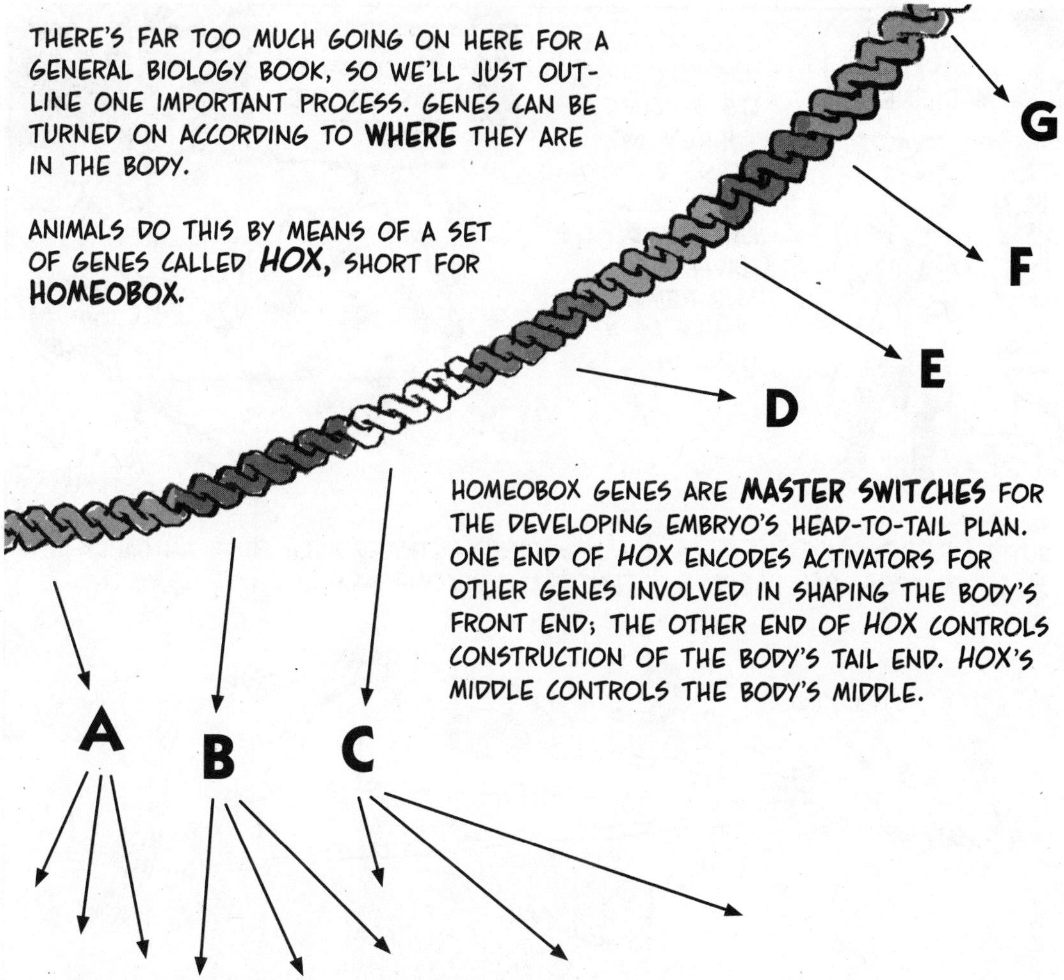

HOMEOBOX GENES ARE **MASTER SWITCHES** FOR THE DEVELOPING EMBRYO'S HEAD-TO-TAIL PLAN. ONE END OF *HOX* ENCODES ACTIVATORS FOR OTHER GENES INVOLVED IN SHAPING THE BODY'S FRONT END; THE OTHER END OF *HOX* CONTROLS CONSTRUCTION OF THE BODY'S TAIL END. *HOX*'S MIDDLE CONTROLS THE BODY'S MIDDLE.

THE ACTIVATED GENES CAN THEMSELVES ACTIVATE OTHER GENES, SO PRODUCTS APPEAR IN PRECISE ORDER. FINGERS, FOR INSTANCE, FORM WHEN SOME CELLS TURN ON GENES FOR **COMMITTING SUICIDE,** LEAVING GAPS IN THE DEVELOPING HAND.

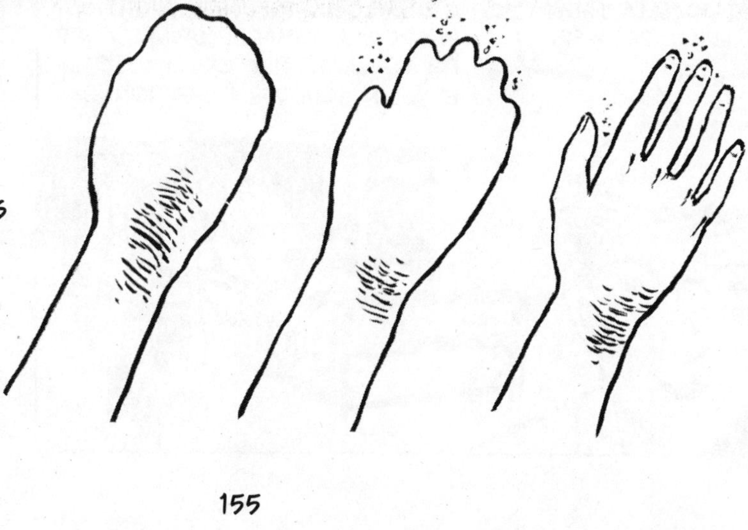

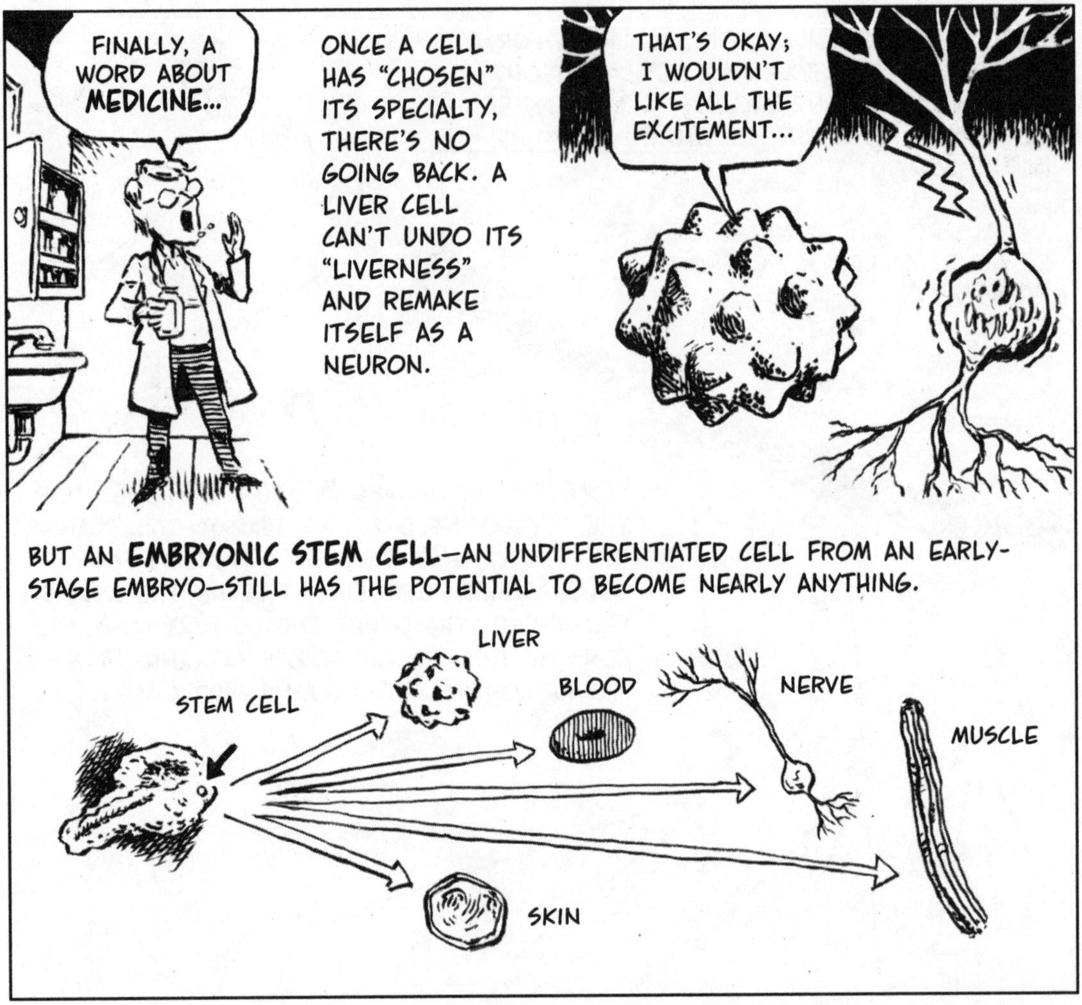
FINALLY, A WORD ABOUT **MEDICINE**...
ONCE A CELL HAS "CHOSEN" ITS SPECIALTY, THERE'S NO GOING BACK. A LIVER CELL CAN'T UNDO ITS "LIVERNESS" AND REMAKE ITSELF AS A NEURON.
THAT'S OKAY; I WOULDN'T LIKE ALL THE EXCITEMENT...
BUT AN **EMBRYONIC STEM CELL**—AN UNDIFFERENTIATED CELL FROM AN EARLY-STAGE EMBRYO—STILL HAS THE POTENTIAL TO BECOME NEARLY ANYTHING.
STEM CELL
LIVER
BLOOD
NERVE
MUSCLE
SKIN

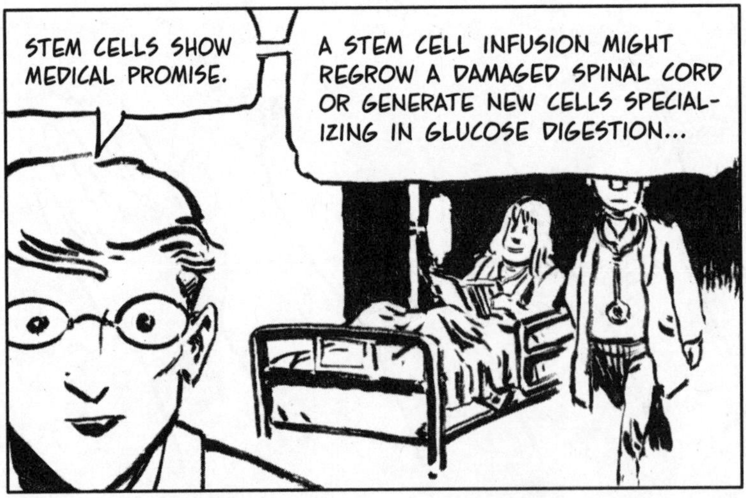
STEM CELLS SHOW MEDICAL PROMISE.
A STEM CELL INFUSION MIGHT REGROW A DAMAGED SPINAL CORD OR GENERATE NEW CELLS SPECIALIZING IN GLUCOSE DIGESTION...

AND IT'S ALL ABOUT REGULATING EXPRESSION!

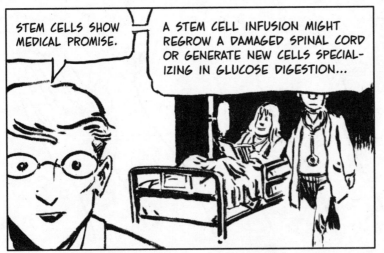

Chapter 11
MULTICELLULARITY

HOW CELLS COOPERATE

WHY DO LARGE EUKARYOTIC ORGANISMS NEED SO MANY CELLULAR SPECIALISTS? HOW DO DIFFERENT CELL TYPES WORK WITH EACH OTHER TO PROMOTE AN ORGANISM'S HEALTH AND WELL-BEING? IN THIS CHAPTER, WE DESCRIBE HOW CELLS COOPERATE TO MAKE MULTICELLULAR LIFE POSSIBLE. WE'LL MOSTLY USE HUMAN EXAMPLES.

FIRST OF ALL, SIMILAR CELLS GROW TOGETHER (OFTEN JOINED BY GROUPS OF DISSIMILAR CELLS) TO FORM **TISSUE.** TISSUE COMES IN FOUR BASIC TYPES:

EPITHELIUM IS A SHEET (OR STACKED SHEETS) OF CELLS SERVING AS A PROTECTIVE COATING.

THIN EPITHELIUM

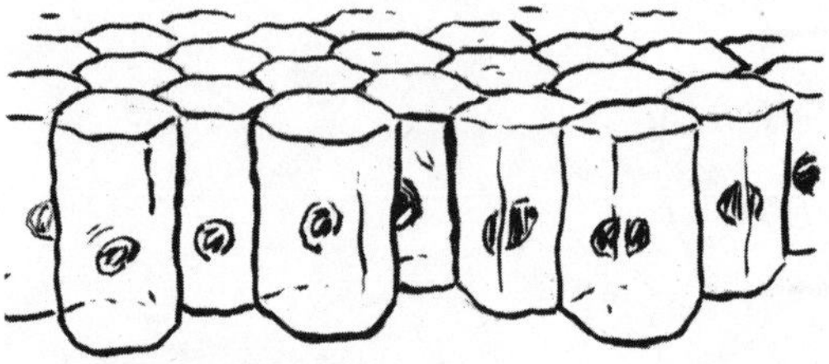

COLUMNAR EPITHELIUM

CONNECTIVE TISSUE MAKES MANY KINDS OF MOLECULAR MESH THAT BIND OTHER TISSUES TOGETHER.

NERVES, THE BODY'S WIRING, ARE BUNDLES OF NEURONS AND SUPPORTING **GLIAL** ("GLUING") CELLS THAT HELP CHANNEL ELECTRICITY.

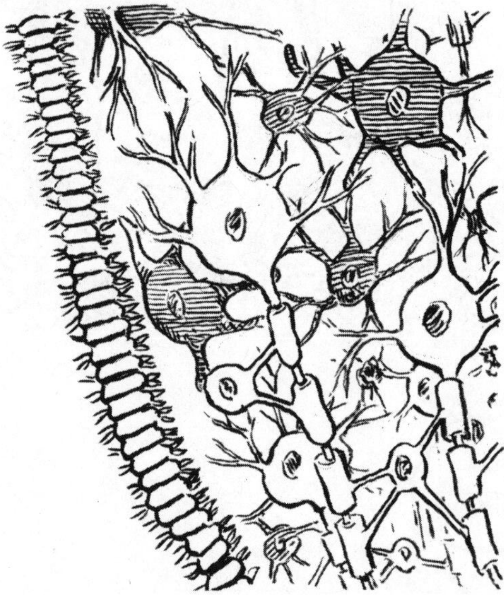

MUSCLE IS MEATY, DENSE TISSUE THAT MAKES MOVEMENT POSSIBLE.

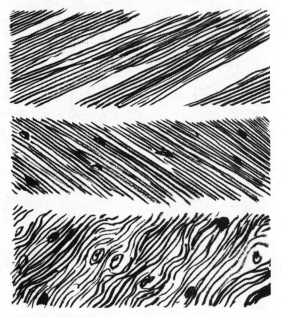

TISSUES ARRANGE THEMSELVES INTO

ORGANS.

MOST ORGANS DEVELOP WITHIN THE EMBRYO AND MAINTAIN THEIR STRUCTURE THROUGHOUT LIFE.

YOU **SAID** IT'S A SYMPHONY!

HERE'S ONE NOW: A HOLLOW TUBE WRAPPED IN A LAYER OF EPITHELIUM INSIDE AN ELASTIC SHEATH OF CONNECTIVE TISSUE, INSIDE A LAYER OF MUSCLE, INSIDE AN ELASTIC COMPOSITE OF CONNECTIVE AND EPITHELIAL TISSUE—IT'S A **BLOOD VESSEL.**

THE TUBE CHANNELS THE FLOW OF BLOOD, ITSELF A MIX OF ASSORTED CELLS, MOLECULES, AND IONS CARRIED BY A FLUID CALLED **PLASMA.**

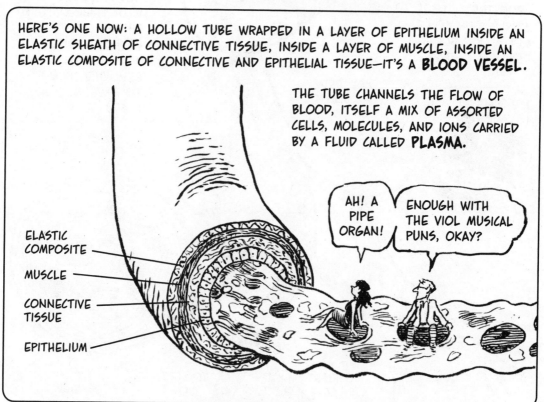

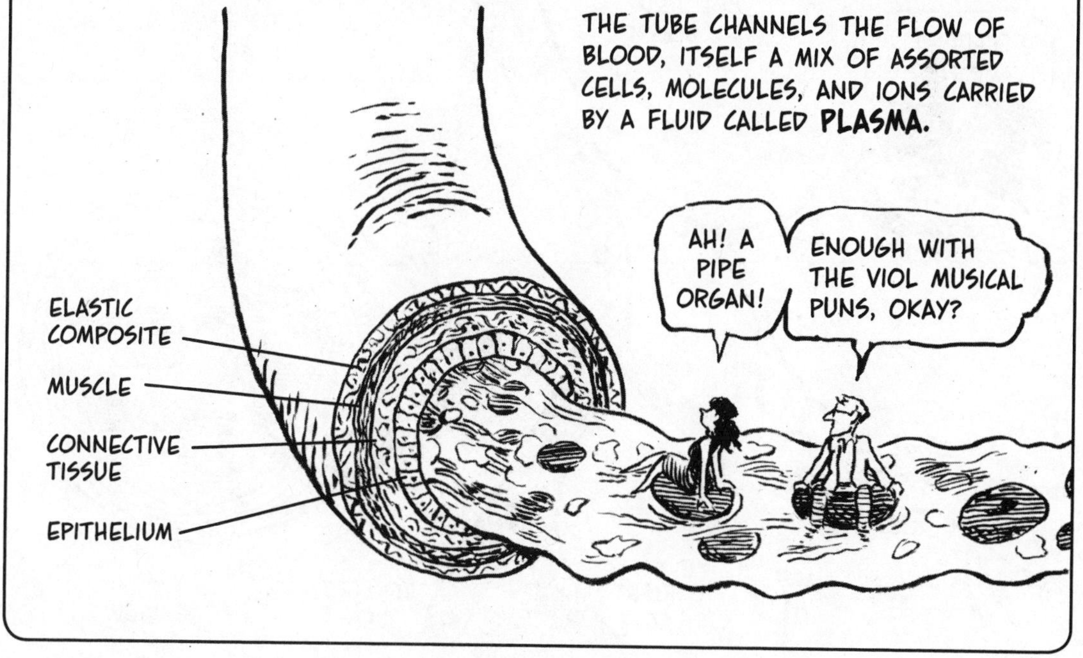
AH! A PIPE ORGAN!
ENOUGH WITH THE VIOL MUSICAL PUNS, OKAY?
ELASTIC COMPOSITE
MUSCLE
CONNECTIVE TISSUE
EPITHELIUM

THE **HEART**—A CHAMBERED ORGAN, MOSTLY MUSCLE TISSUE DRIVEN BY NEURAL SIGNALS—PUMPS OUT OXYGENATED BLOOD IN A SERIES OF PRESSURIZED PULSES. OUTBOUND VESSELS, THE **ARTERIES,** BRANCH INTO EVER-FINER PASSAGES THAT THREAD THEIR WAY THROUGHOUT EVERY TISSUE.

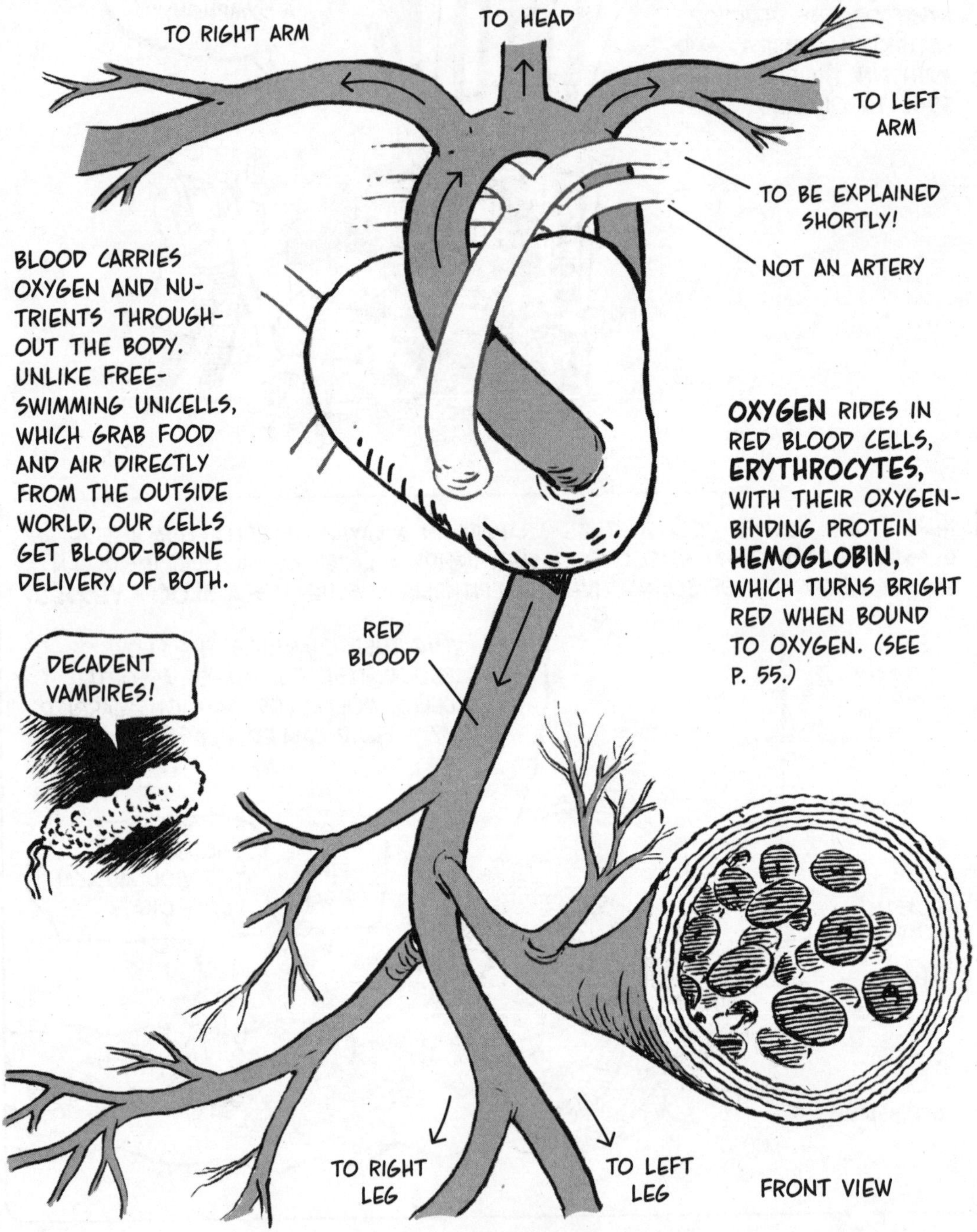

BLOOD CARRIES OXYGEN AND NUTRIENTS THROUGHOUT THE BODY. UNLIKE FREE-SWIMMING UNICELLS, WHICH GRAB FOOD AND AIR DIRECTLY FROM THE OUTSIDE WORLD, OUR CELLS GET BLOOD-BORNE DELIVERY OF BOTH.

OXYGEN RIDES IN RED BLOOD CELLS, **ERYTHROCYTES,** WITH THEIR OXYGEN-BINDING PROTEIN **HEMOGLOBIN,** WHICH TURNS BRIGHT RED WHEN BOUND TO OXYGEN. (SEE P. 55.)

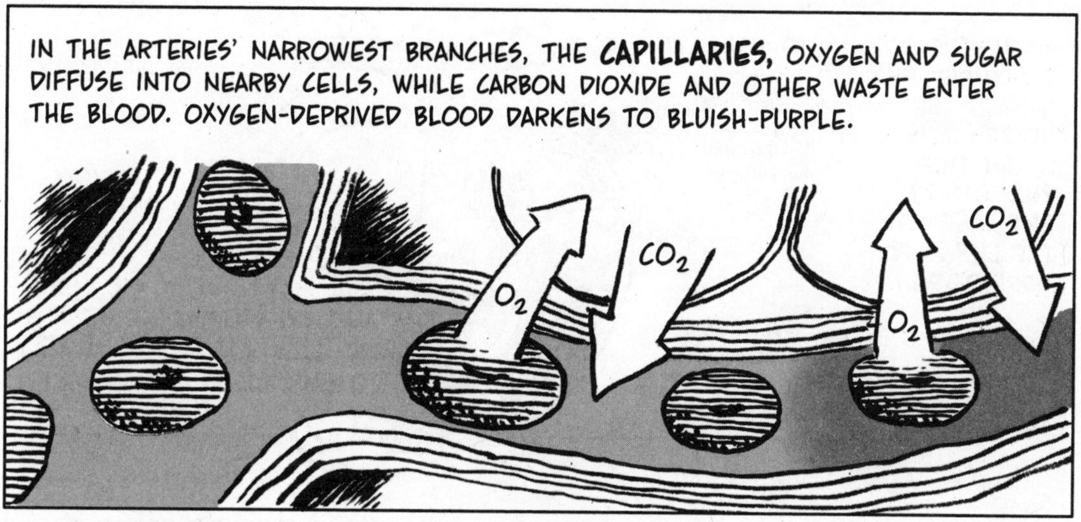
IN THE ARTERIES' NARROWEST BRANCHES, THE **CAPILLARIES,** OXYGEN AND SUGAR DIFFUSE INTO NEARBY CELLS, WHILE CARBON DIOXIDE AND OTHER WASTE ENTER THE BLOOD. OXYGEN-DEPRIVED BLOOD DARKENS TO BLUISH-PURPLE.
O_2
CO_2
O_2
CO_2

FROM THE CAPILLARIES, "BLUE" BLOOD FLOWS BACK THROUGH **VEINS** TO THE HEART. VEINS, WHICH HAVE NO PULSE, COME EQUIPPED WITH SPECIAL VALVES TO PREVENT BACKFLOW.

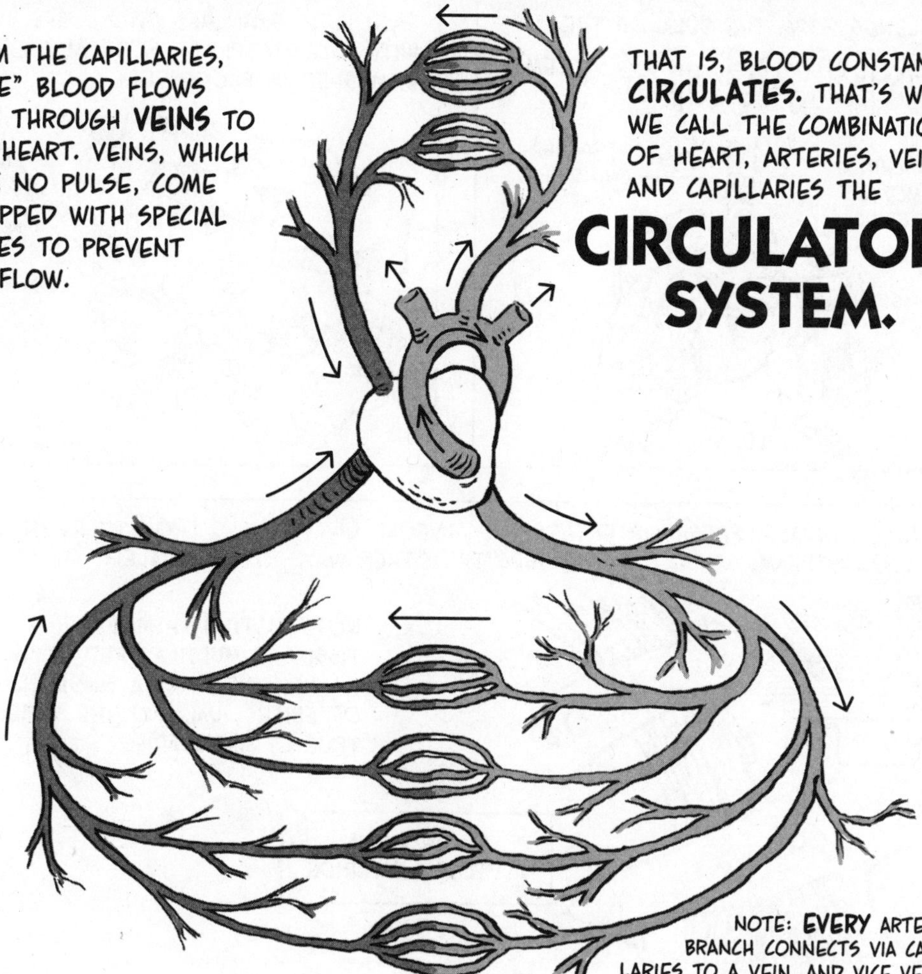

THAT IS, BLOOD CONSTANTLY **CIRCULATES.** THAT'S WHY WE CALL THE COMBINATION OF HEART, ARTERIES, VEINS, AND CAPILLARIES THE

CIRCULATORY SYSTEM.

NOTE: **EVERY** ARTERIAL BRANCH CONNECTS VIA CAPILLARIES TO A VEIN, AND VICE VERSA.

WHEN BLUE BLOOD RETURNS "HOME," THE HEART PUMPS IT OUT THE **PULMONARY ARTERIES** TO THE **LUNGS** FOR MORE OXYGEN.

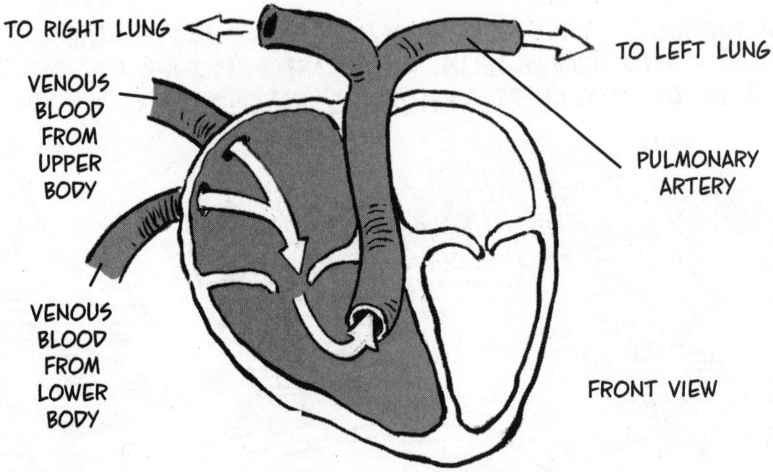
TO RIGHT LUNG
TO LEFT LUNG
VENOUS BLOOD FROM UPPER BODY
PULMONARY ARTERY
VENOUS BLOOD FROM LOWER BODY
FRONT VIEW

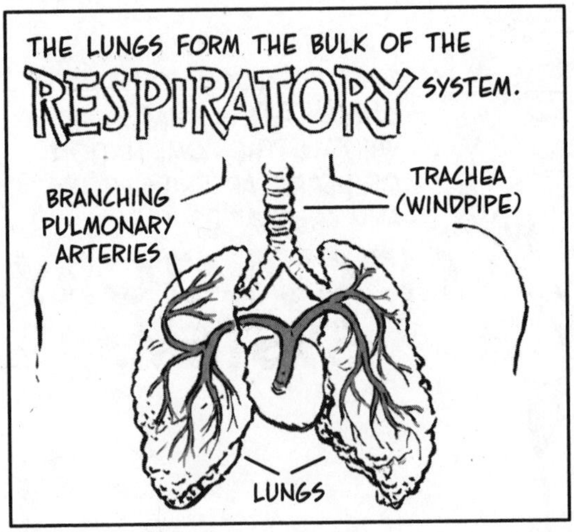
THE LUNGS FORM THE BULK OF THE RESPIRATORY SYSTEM.
BRANCHING PULMONARY ARTERIES
TRACHEA (WINDPIPE)
LUNGS

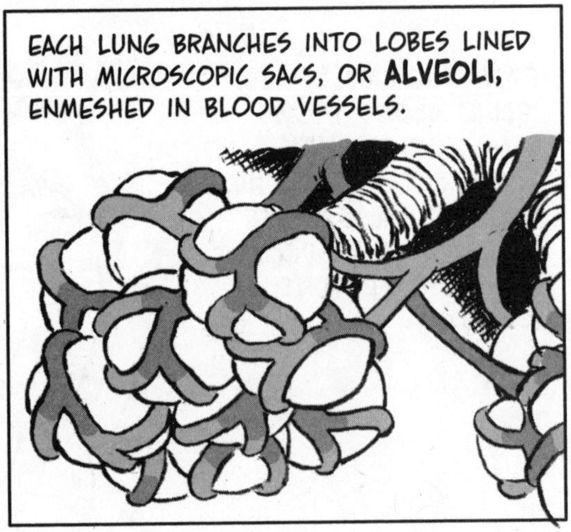
EACH LUNG BRANCHES INTO LOBES LINED WITH MICROSCOPIC SACS, OR ALVEOLI, ENMESHED IN BLOOD VESSELS.

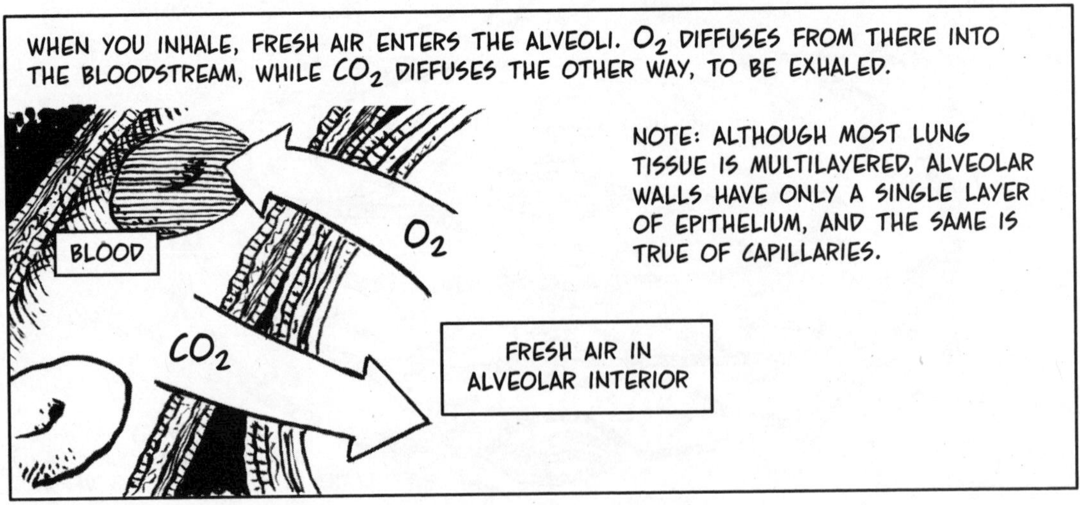
WHEN YOU INHALE, FRESH AIR ENTERS THE ALVEOLI. O_2 DIFFUSES FROM THERE INTO THE BLOODSTREAM, WHILE CO_2 DIFFUSES THE OTHER WAY, TO BE EXHALED.
NOTE: ALTHOUGH MOST LUNG TISSUE IS MULTILAYERED, ALVEOLAR WALLS HAVE ONLY A SINGLE LAYER OF EPITHELIUM, AND THE SAME IS TRUE OF CAPILLARIES.
BLOOD
O_2
CO_2
FRESH AIR IN ALVEOLAR INTERIOR

ENRICHED, RED BLOOD RETURNS FROM THE LUNGS TO THE HEART THROUGH THE **PULMONARY VEINS,** WHICH PARALLEL THE PULMONARY ARTERIES.

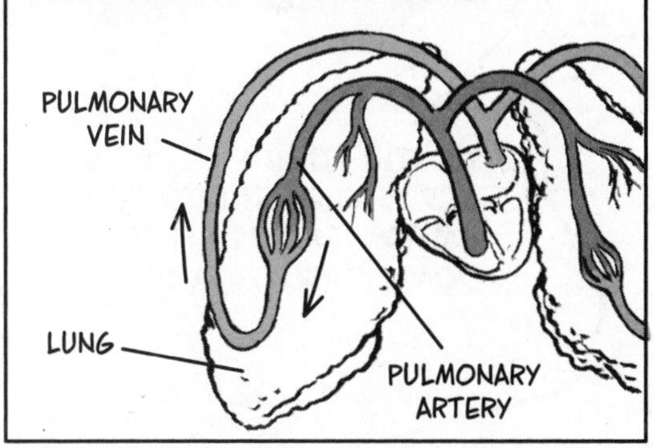
PULMONARY VEIN
LUNG
PULMONARY ARTERY

DESCENDING TO THE FOURTH CHAMBER, RED BLOOD EXITS THROUGH THE AORTA AND STARTS A NEW CIRCUIT THROUGH THE BODY.

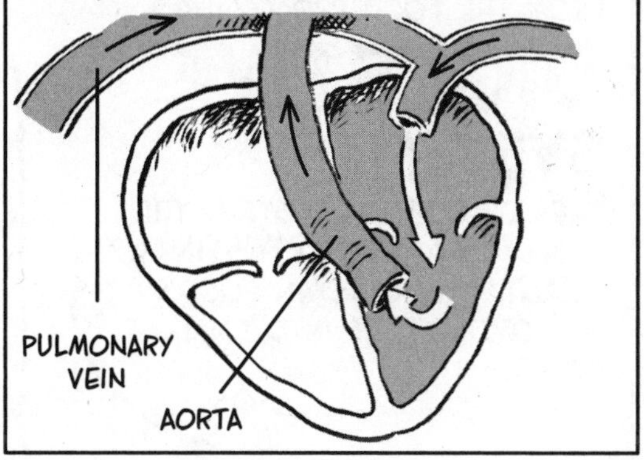
PULMONARY VEIN
AORTA

HERE'S HOW IT LOOKS WHEN YOU PUT IT ALL TOGETHER:

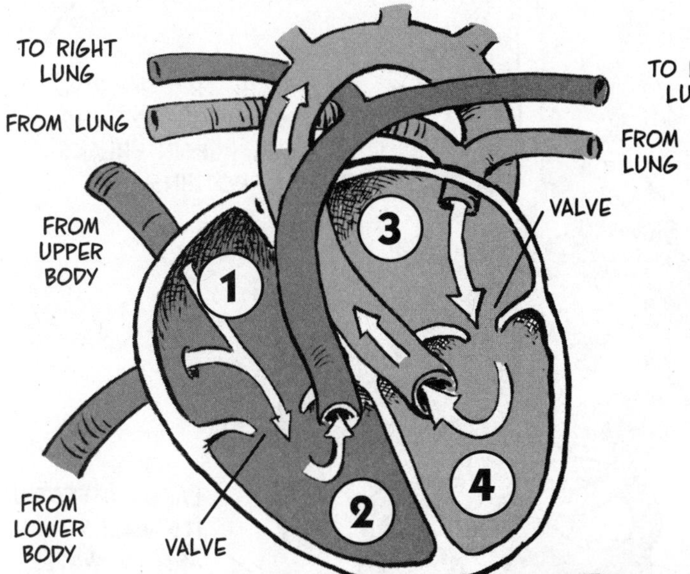
TO RIGHT LUNG
FROM LUNG
TO LEFT LUNG
FROM LUNG
FROM UPPER BODY
VALVE
1
3
4
2
FROM LOWER BODY
VALVE

VENOUS BLOOD ENTERS **(1)** THE RIGHT ATRIUM;

PASSES DOWN TO **(2)** THE RIGHT VENTRICLE, WHICH SENDS IT TO THE LUNGS;

RETURNS FROM THE LUNGS TO **(3)** THE LEFT ATRIUM;

DESCENDS TO **(4)** THE LEFT VENTRICLE, AND IS PUMPED OUT TO THE BODY.

BLOOD ACTUALLY CIRCULATES **TWICE,** ONCE THROUGH THE LUNGS TO GET O_2 AND DUMP CO_2, AND ONCE THROUGH THE BODY TO DELIVER O_2 AND COLLECT WASTE, AS IN THIS SIMPLIFIED SCHEMATIC DIAGRAM. THE UPPER CHAMBERS, THE ATRIA, RECEIVE; THE BOTTOM CHAMBERS, THE VENTRICLES, PUMP.

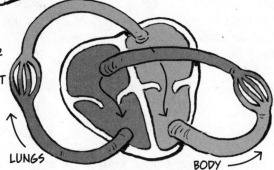

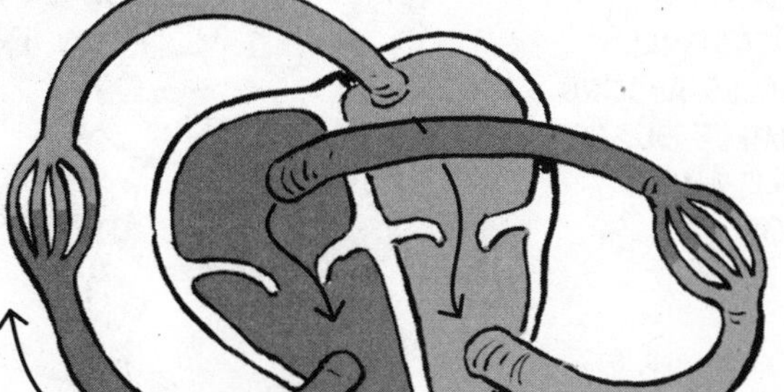
LUNGS
BODY

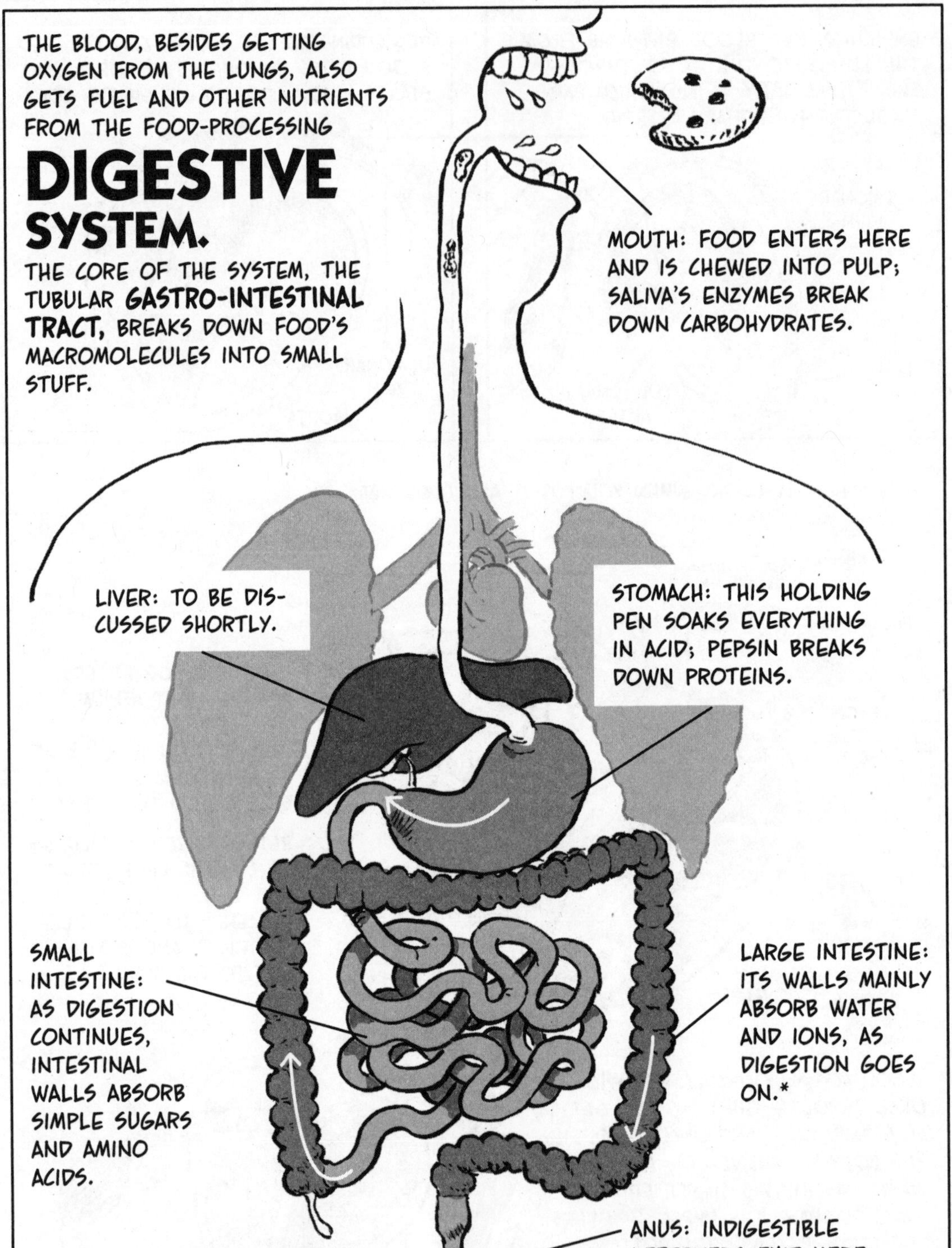
THE BLOOD, BESIDES GETTING OXYGEN FROM THE LUNGS, ALSO GETS FUEL AND OTHER NUTRIENTS FROM THE FOOD-PROCESSING
DIGESTIVE SYSTEM.
THE CORE OF THE SYSTEM, THE TUBULAR GASTRO-INTESTINAL TRACT, BREAKS DOWN FOOD'S MACROMOLECULES INTO SMALL STUFF.
MOUTH: FOOD ENTERS HERE AND IS CHEWED INTO PULP; SALIVA'S ENZYMES BREAK DOWN CARBOHYDRATES.
LIVER: TO BE DISCUSSED SHORTLY.
STOMACH: THIS HOLDING PEN SOAKS EVERYTHING IN ACID; PEPSIN BREAKS DOWN PROTEINS.
SMALL INTESTINE: AS DIGESTION CONTINUES, INTESTINAL WALLS ABSORB SIMPLE SUGARS AND AMINO ACIDS.
LARGE INTESTINE: ITS WALLS MAINLY ABSORB WATER AND IONS, AS DIGESTION GOES ON.*
ANUS: INDIGESTIBLE LEFTOVERS EXIT HERE.
*WITH THE HELP OF A HUGE NUMBER OF RESIDENT BACTERIA. SEE CHAPTER 16.

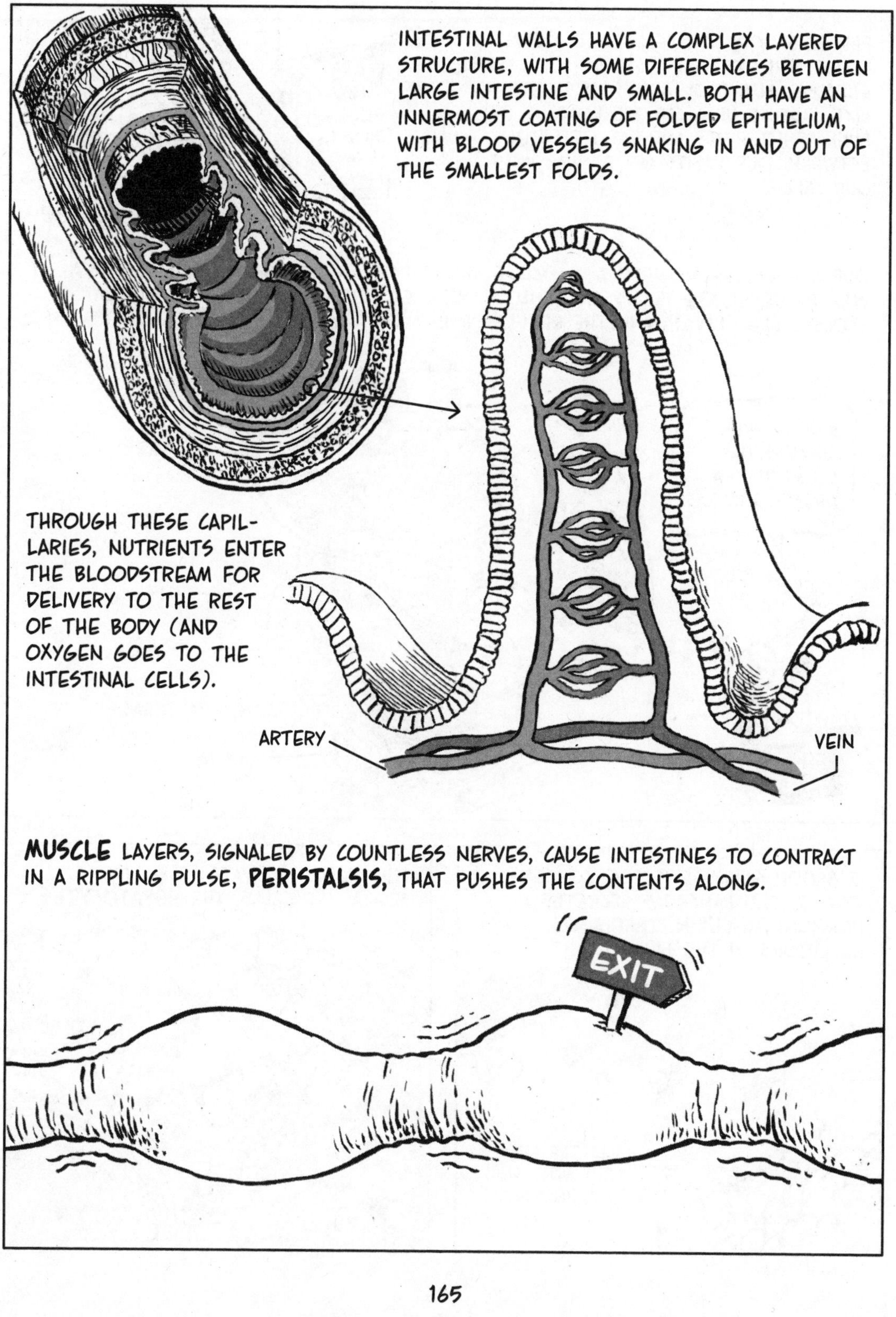
INTESTINAL WALLS HAVE A COMPLEX LAYERED STRUCTURE, WITH SOME DIFFERENCES BETWEEN LARGE INTESTINE AND SMALL. BOTH HAVE AN INNERMOST COATING OF FOLDED EPITHELIUM, WITH BLOOD VESSELS SNAKING IN AND OUT OF THE SMALLEST FOLDS.
THROUGH THESE CAPILLARIES, NUTRIENTS ENTER THE BLOODSTREAM FOR DELIVERY TO THE REST OF THE BODY (AND OXYGEN GOES TO THE INTESTINAL CELLS).
ARTERY
VEIN
MUSCLE LAYERS, SIGNALED BY COUNTLESS NERVES, CAUSE INTESTINES TO CONTRACT IN A RIPPLING PULSE, **PERISTALSIS,** THAT PUSHES THE CONTENTS ALONG.
EXIT

FED BY BLOOD, OUR CELLS FACE A CHALLENGE. THEY PREFER A STEADY SUPPLY OF GLUCOSE, BUT FOOD COMES IN SPURTS. LIKE GASSING UP A CAR, WE LOAD UP ON FUEL AT INTERVALS AND EAT LITTLE IN BETWEEN. OUR HABITS ARE AT WAR WITH OUR CELLS.

Glucose
Diesel
Hi-Test
SLOW DOWN!

OUR BODIES, THEN, NEED A "GLUCOSE TANK" FOR SHORT-TERM STORAGE AND A WAY TO **REGULATE** FUEL'S ENTRY INTO THE BLOODSTREAM. THAT TANK IS THE BLOOD-FILLED **LIVER,** AND THE REGULATOR IS THE **PANCREAS.**

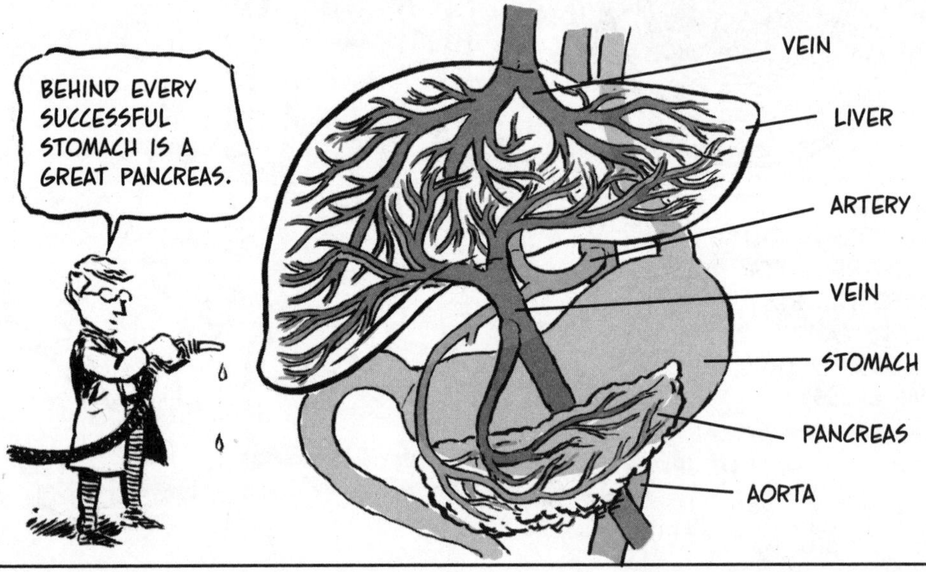
BEHIND EVERY SUCCESSFUL STOMACH IS A GREAT PANCREAS.
VEIN
LIVER
ARTERY
VEIN
STOMACH
PANCREAS
AORTA

IF BLOOD GLUCOSE IS TOO CONCENTRATED, THE PANCREAS SECRETES A HORMONE, **INSULIN,** THAT BINDS RECEPTORS IN THE LIVER.

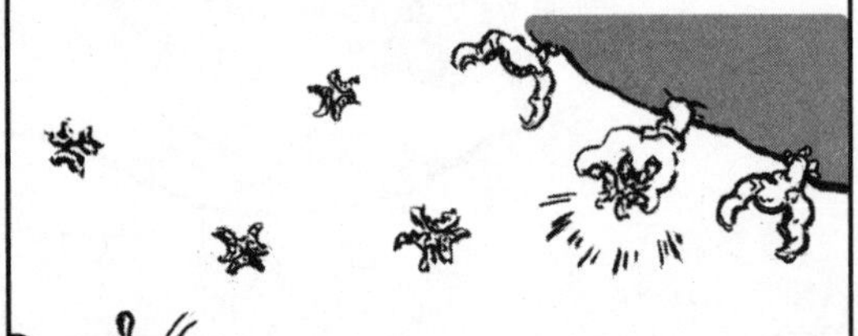
A LITTLE PROTEIN WITH ONLY 51 AMINO ACIDS

THE SIGNAL ACTIVATES THE LIVER'S SPECIALIZED CELLS, THE **HEPATOCYTES.**

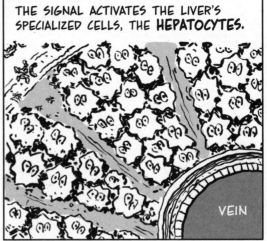

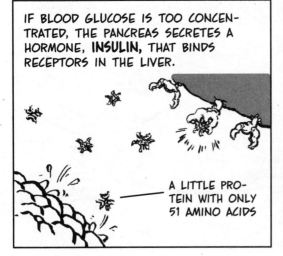

VEIN

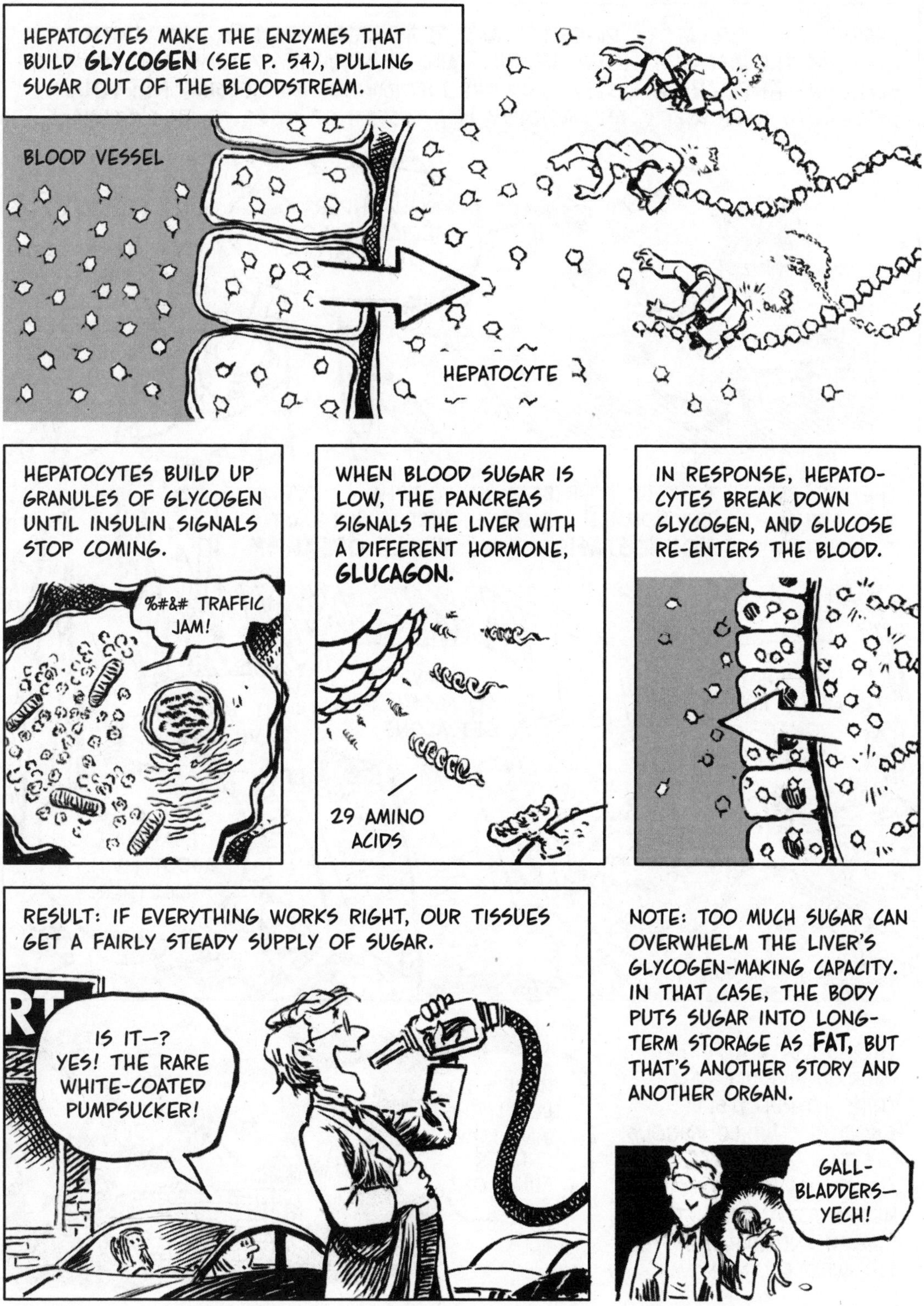
HEPATOCYTES MAKE THE ENZYMES THAT BUILD **GLYCOGEN** (SEE P. 54), PULLING SUGAR OUT OF THE BLOODSTREAM.
BLOOD VESSEL
HEPATOCYTE
HEPATOCYTES BUILD UP GRANULES OF GLYCOGEN UNTIL INSULIN SIGNALS STOP COMING.
%#&# TRAFFIC JAM!
WHEN BLOOD SUGAR IS LOW, THE PANCREAS SIGNALS THE LIVER WITH A DIFFERENT HORMONE, **GLUCAGON.**
29 AMINO ACIDS
IN RESPONSE, HEPATOCYTES BREAK DOWN GLYCOGEN, AND GLUCOSE RE-ENTERS THE BLOOD.
RESULT: IF EVERYTHING WORKS RIGHT, OUR TISSUES GET A FAIRLY STEADY SUPPLY OF SUGAR.
IS IT—? YES! THE RARE WHITE-COATED PUMPSUCKER!
NOTE: TOO MUCH SUGAR CAN OVERWHELM THE LIVER'S GLYCOGEN-MAKING CAPACITY. IN THAT CASE, THE BODY PUTS SUGAR INTO LONG-TERM STORAGE AS **FAT,** BUT THAT'S ANOTHER STORY AND ANOTHER ORGAN.
GALL-BLADDERS—YECH!

BESIDES THE CIRCULATORY, DIGESTIVE, AND RESPIRATORY SYSTEMS, THE HUMAN BODY ALSO HAS THE DISEASE-FIGHTING **IMMUNE** AND **LYMPHATIC** SYSTEMS, THE HORMONE-SECRETING **ENDOCRINE** SYSTEM, AND THE INFORMATION-PROCESSING **NERVOUS** SYSTEM, TO NAME A FEW. AN ORGANISM IS A SYSTEM OF INTERWOVEN SYSTEMS.
NO WONDER MEDICAL SCHOOL LASTS FOREVER.
THEY COOPERATE TO KEEP YOUR BODY IN WORKING CONDITION AMID A CHANGING ENVIRONMENT. ORGANS, THAT IS, INTERACT TO MAINTAIN THE **HOMEOSTASIS** OF THE **WHOLE ORGANISM.**
HEY, WHAT ABOUT **MY** HOMEO-STASIS?
YOU HAVE TO GO ALONG TO GET ALONG.
IT TAKES A VILLAGE.
SOME OTHER PLATITUDE ABOUT LIVING TOGETHER.

THIS IS QUITE A FEAT. EVERY INDIVIDUAL CELL TRIES TO RUN ITSELF SMOOTHLY, WHILE VARIOUS SYSTEMS **REGULATE** COMPETING CELLULAR NEEDS TO MAINTAIN THE HEALTH AND HAPPINESS OF THE WHOLE ORGANISM.

HERE'S TO HIGH LIVING AND LOW LIVER FUNCTION!

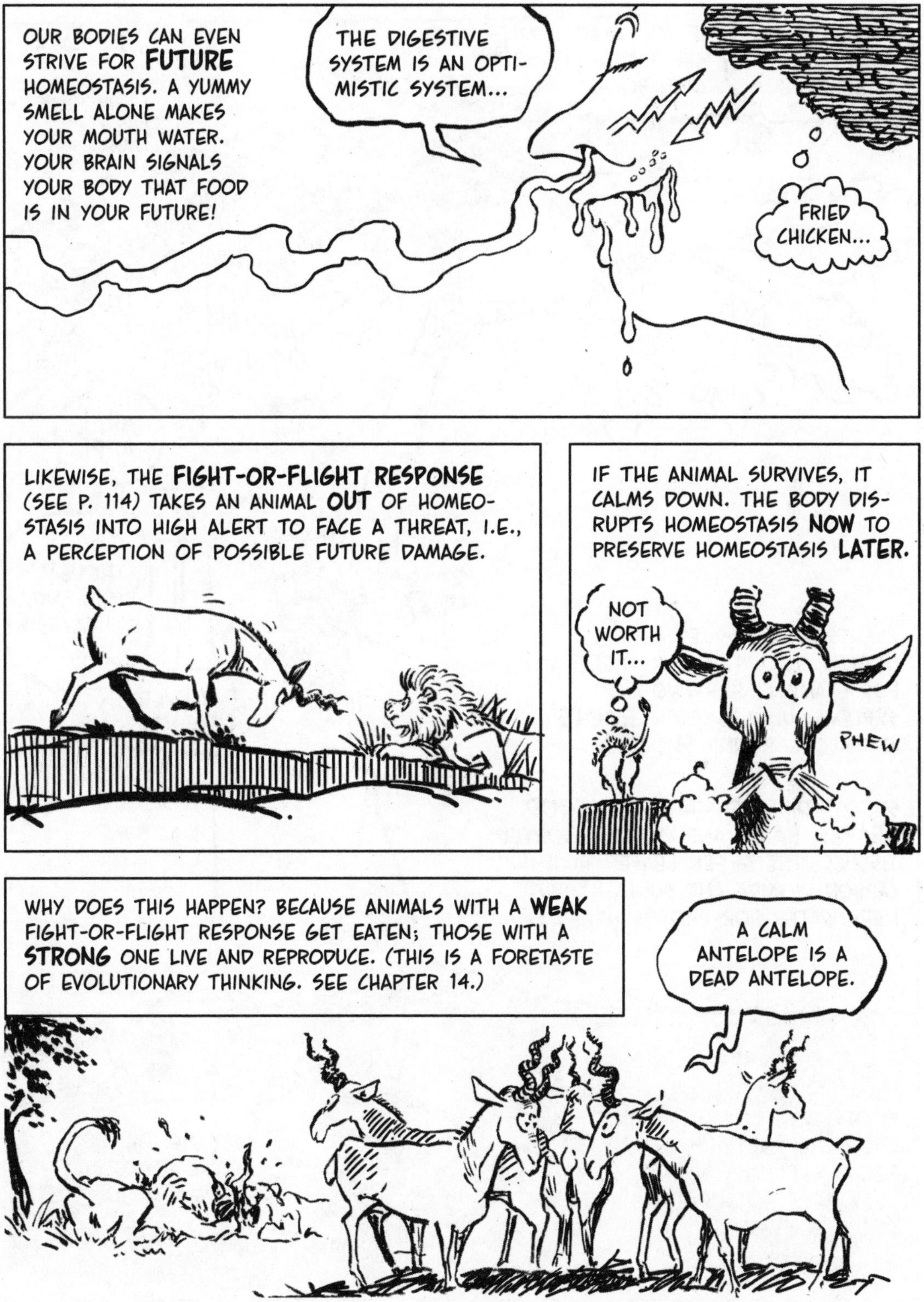
OUR BODIES CAN EVEN STRIVE FOR **FUTURE** HOMEOSTASIS. A YUMMY SMELL ALONE MAKES YOUR MOUTH WATER. YOUR BRAIN SIGNALS YOUR BODY THAT FOOD IS IN YOUR FUTURE!
THE DIGESTIVE SYSTEM IS AN OPTIMISTIC SYSTEM...
FRIED CHICKEN...
LIKEWISE, THE **FIGHT-OR-FLIGHT RESPONSE** (SEE P. 114) TAKES AN ANIMAL **OUT** OF HOMEOSTASIS INTO HIGH ALERT TO FACE A THREAT, I.E., A PERCEPTION OF POSSIBLE FUTURE DAMAGE.
IF THE ANIMAL SURVIVES, IT CALMS DOWN. THE BODY DISRUPTS HOMEOSTASIS **NOW** TO PRESERVE HOMEOSTASIS **LATER.**
NOT WORTH IT...
PHEW
WHY DOES THIS HAPPEN? BECAUSE ANIMALS WITH A **WEAK** FIGHT-OR-FLIGHT RESPONSE GET EATEN; THOSE WITH A **STRONG** ONE LIVE AND REPRODUCE. (THIS IS A FORETASTE OF EVOLUTIONARY THINKING. SEE CHAPTER 14.)
A CALM ANTELOPE IS A DEAD ANTELOPE.

WELL, THAT WAS ANIMAL-CENTRIC!
IF YOU WANT PLANT-CENTRIC, WRITE YOUR OWN BOOK.
ARF! ARF!
WHAT?
OH, I THOUGHT YOU SAID, "LEAVES BARK."
NEVER MIND.
PLANTS HAVE TISSUES AND ORGANS, TOO. THIS OAK TREE, FOR EXAMPLE, HAS TWO MAIN SYSTEMS: UNDERGROUND **ROOTS** AND ABOVEGROUND **SHOOTS.**
SHOOTS DIFFERENTIATE INTO **WOOD, LEAVES, BARK,** AND OTHER ASSORTED TISSUES. THE GREEN LEAVES TAKE IN CARBON DIOXIDE AND SUNLIGHT, BUT NEED WATER FOR PHOTOSYNTHESIS.
ROOTS TAKE IN WATER, IONS, AND OTHER NUTRIENTS. BUT ROOTS SEE NO SUNLIGHT, SO THEY MAKE NO SUGAR. HOW CAN THEY SURVIVE?
TELL ME. I'M IN THE DARK HERE.

OUR OAK, LIKE OTHER TREES, SHRUBS, AND FERNS, HAS A NETWORK OF VESSELS RUNNING ALL THE WAY FROM THE OUTERMOST SHOOTS TO THE DEEPEST ROOTS.
XYLEM
VASCULAR BUNDLES
PHLOEM
THESE PIPES, OR **VASCULAR BUNDLES,** SUPPORT TWO KINDS OF CHANNEL: **XYLEM** TOWARD THE INSIDE AND **PHLOEM** TOWARD THE OUTSIDE.

XYLEM DRAWS WATER AND DISSOLVED NUTRIENTS UP FROM ROOTS TO SHOOTS.
PHLOEM SENDS SUGAR DOWN FROM SHOOTS TO ROOTS.

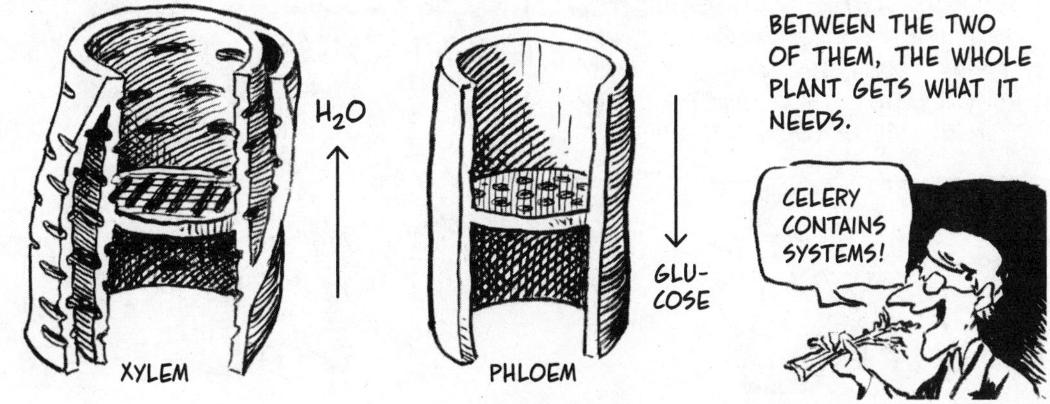
H_2O
XYLEM
GLU-COSE
PHLOEM
BETWEEN THE TWO OF THEM, THE WHOLE PLANT GETS WHAT IT NEEDS.
CELERY CONTAINS SYSTEMS!

THIS CHAPTER ON MULTICELLULARITY BARELY SCRATCHED THE SURFACE OF THE SUBJECT.
THE FULL RANGE OF BODY PLANS AND ORGAN SYSTEMS IN THE LIVING WORLD WOULD FILL MANY THICK VOLUMES.
EVEN A "SIMPLE," BRAINLESS **JELLYFISH** HAS RUDIMENTARY EYES AND A NERVOUS SYSTEM.
OW!
STINGERS, TOO!
SPEAKING OF SCRATCHING THE SURFACE, EVEN **SKIN** IS AN ORGAN, CAPABLE OF SHEDDING WATER, FIGHTING INFECTION, REGULATING BODY TEMPERATURE, PERCEIVING TOUCH, SIGNALING PAIN, AND REGENERATING ITSELF, AMONG OTHER THINGS. LIFE IS **TOO** AMAZING!
THROB THROB THROB
HMMM... CAN JELLYFISH LAUGH?

WHEW! WE'VE COVERED A LOT IN THE LAST 172 PAGES...

WE'VE TALKED ABOUT THE BASIC INGREDIENTS OF LIFE, FROM ELECTRONS AND PROTONS UP TO MACROMOLECULES...
WE'VE SHOWN HOW PHOTOSYNTHESIZERS BUILD SUGAR FROM INORGANIC RAW MATERIALS AND HOW ORGANISMS GET ENERGY BY OXIDIZING THAT SUGAR.
DOUGHNUTS
WE'VE SEEN HOW ORGANISMS COMMUNICATE, BOTH WITHIN THEMSELVES AND WITH THE OUTSIDE WORLD, TO FIND FOOD AND MAINTAIN HOMEOSTASIS.
URGENT Communiqué: Buy Doughnuts Now!
AND WE'VE SEEN CELLS BUILD PROTEINS AND RNA FROM DNA TEMPLATES.
LET'S SEE... HAVE WE LEFT ANYTHING OUT?

OH, RIGHT! I ALMOST FORGOT LIFE'S MOST AMAZING FEAT OF ALL: REPRODUCING ITSELF, GENERATION AFTER GENERATION, FOR NEARLY FOUR BILLION YEARS AND COUNTING!

Chapter 12
REPRODUCTION (Part 1)

UP TO NOW IN THIS BOOK, EVERY CELLULAR PROCESS (ASIDE FROM SOME SIGNALING) HAS TAKEN PLACE **WITHIN** AN INDIVIDUAL CELL. IN THIS CHAPTER, WE SEE HOW A CELL CAN MAKE A **NEW COPY** OF ITSELF.

ANY NEW CELL HAS TO HAVE A FULL SET OF OPERATING INSTRUCTIONS. THAT IS, A REPRODUCING CELL MUST COPY ALL ITS DNA.

IN THEIR 1953 PAPER ANNOUNCING THE DISCOVERY OF DNA'S PAIRED SPIRALS, THE AUTHORS, JAMES WATSON AND FRANCIS CRICK, WROTE A SENTENCE ALMOST AS FAMOUS AS THE PAPER ITSELF:
"IT HAS NOT ESCAPED OUR NOTICE THAT THE SPECIFIC PAIRING WE HAVE POSTULATED IMMEDIATELY SUGGESTS A POSSIBLE COPYING MECHANISM FOR THE GENETIC MATERIAL."
(IT DID ESCAPE THEIR NOTICE THAT THEY MIGHT CREDIT THEIR COLLABORATOR ROSALIND FRANKLIN, BUT THAT'S A STORY FOR THE CARTOON HISTORY OF MISOGYNY AND EGOTISM IN SCIENCE, VOL. 37.)
WE'LL BE FAMOUS!

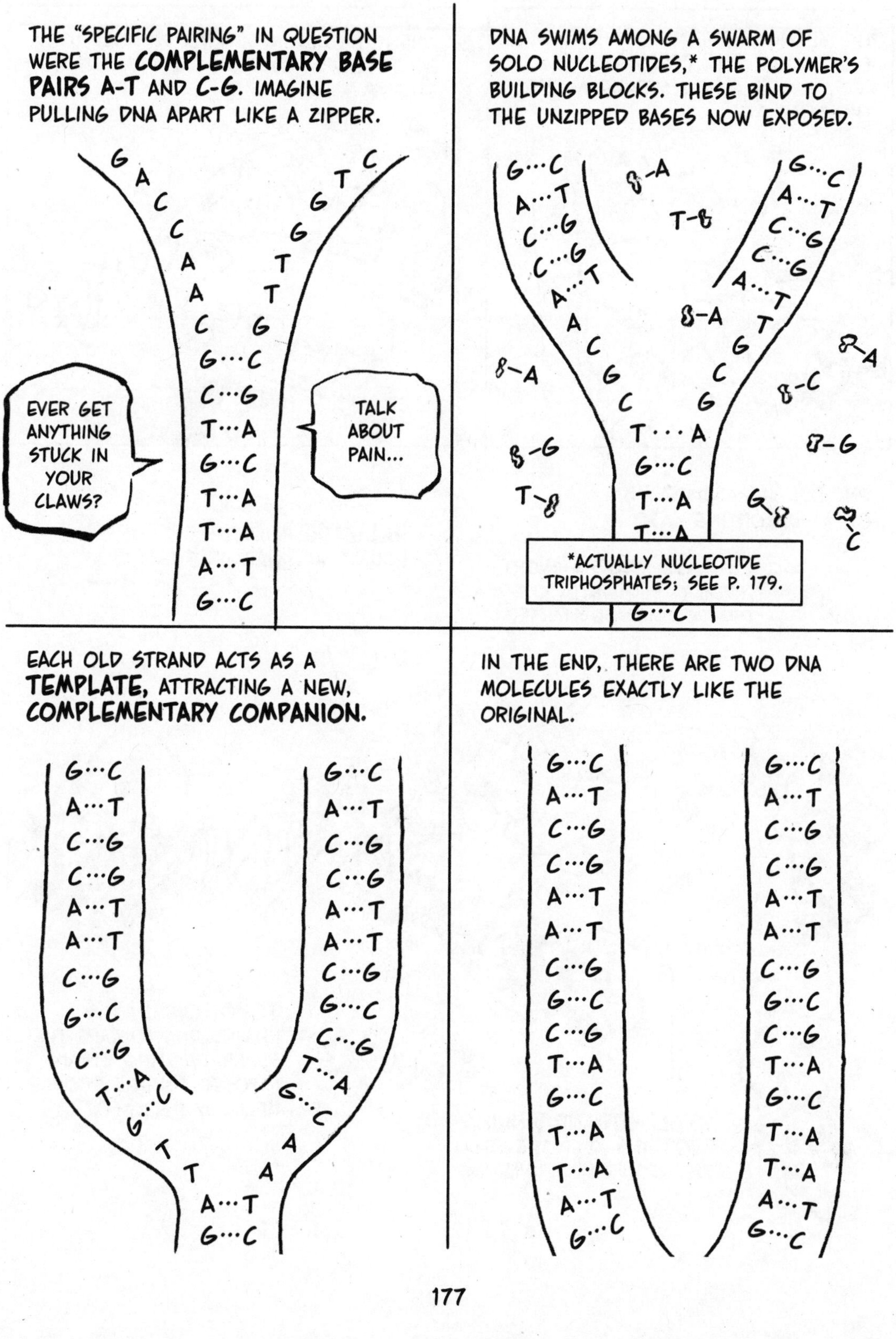
THE "SPECIFIC PAIRING" IN QUESTION WERE THE **COMPLEMENTARY BASE PAIRS A-T** AND **C-G**. IMAGINE PULLING DNA APART LIKE A ZIPPER.
EVER GET ANYTHING STUCK IN YOUR CLAWS?
TALK ABOUT PAIN...
DNA SWIMS AMONG A SWARM OF SOLO NUCLEOTIDES,* THE POLYMER'S BUILDING BLOCKS. THESE BIND TO THE UNZIPPED BASES NOW EXPOSED.
*ACTUALLY NUCLEOTIDE TRIPHOSPHATES; SEE P. 179.
EACH OLD STRAND ACTS AS A **TEMPLATE,** ATTRACTING A NEW, **COMPLEMENTARY COMPANION.**
IN THE END, THERE ARE TWO DNA MOLECULES EXACTLY LIKE THE ORIGINAL.

THE CHEMICAL HERO OF THIS DRAMA, THE ENZYME **DNA POLYMERASE** (DP), BUILDS A NEW STRAND ALONG THE TEMPLATE.
5′
3′
T
A
3′
C
G
G
C
C
5′
A

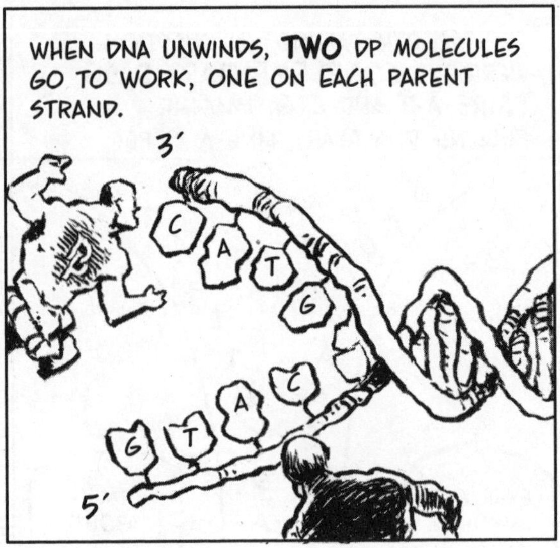
WHEN DNA UNWINDS, **TWO** DP MOLECULES GO TO WORK, ONE ON EACH PARENT STRAND.
3′
C
A
T
G
C
A
T
G
5′

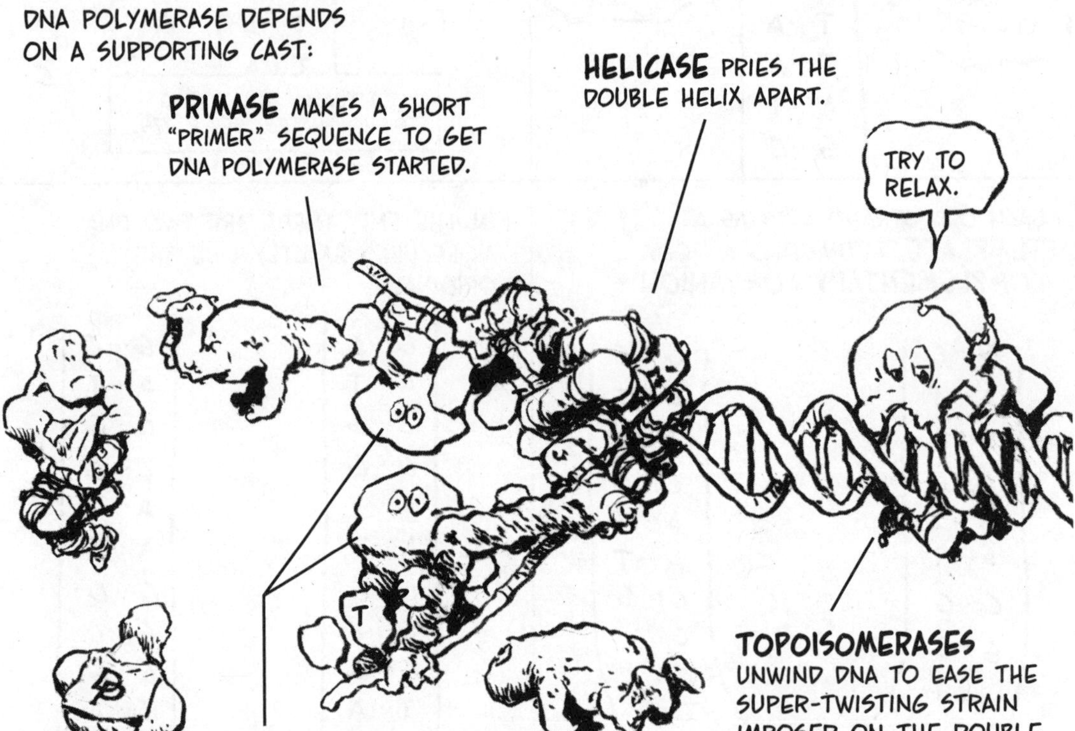
DNA POLYMERASE DEPENDS ON A SUPPORTING CAST:
PRIMASE MAKES A SHORT "PRIMER" SEQUENCE TO GET DNA POLYMERASE STARTED.
HELICASE PRIES THE DOUBLE HELIX APART.
TRY TO RELAX.
T
SINGLE-STRAND BINDING PROTEINS KEEP THE TWO STRANDS FROM REUNITING.
TOPOISOMERASES UNWIND DNA TO EASE THE SUPER-TWISTING STRAIN IMPOSED ON THE DOUBLE HELIX BY HELICASE.

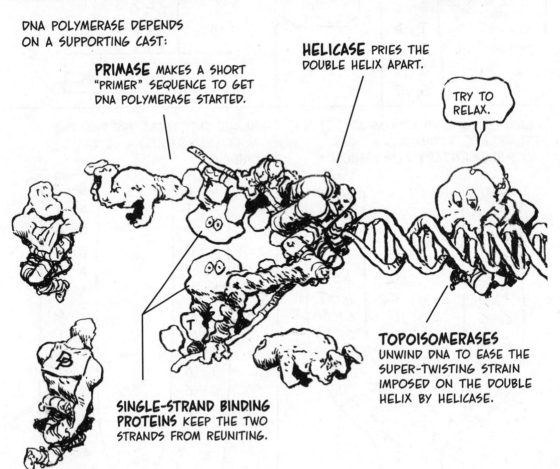

DP (LIKE RNA POLYMERASE) ALWAYS BUILDS ITS NEW STRAND RUNNING 5′→3′ ON A TEMPLATE STRAND RUNNING 3′→5′. THIS MEANS THAT ONLY ONE TEMPLATE STRAND, THE SO-CALLED **LEADING STRAND,** LETS DP **FOLLOW HELICASE** AS IT UNZIPS DNA.

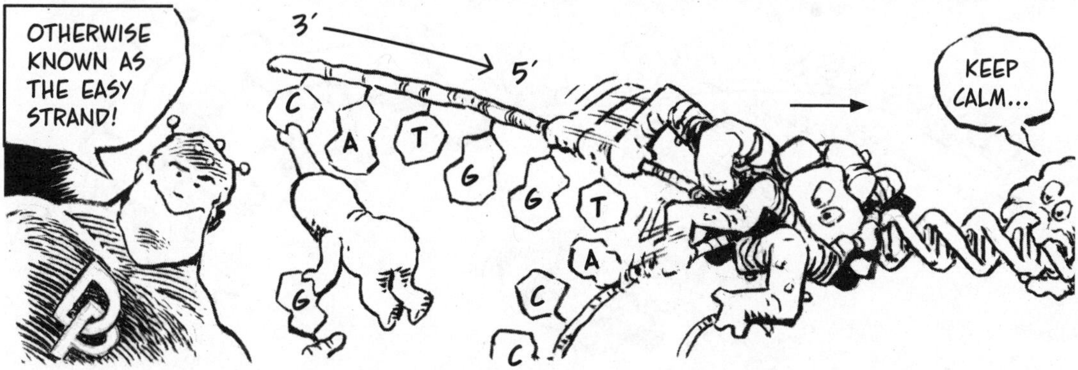

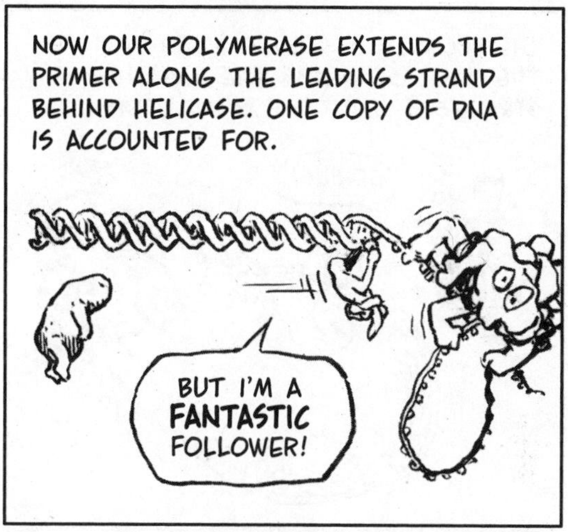
NOW OUR POLYMERASE EXTENDS THE PRIMER ALONG THE LEADING STRAND BEHIND HELICASE. ONE COPY OF DNA IS ACCOUNTED FOR.
BUT I'M A FANTASTIC FOLLOWER!

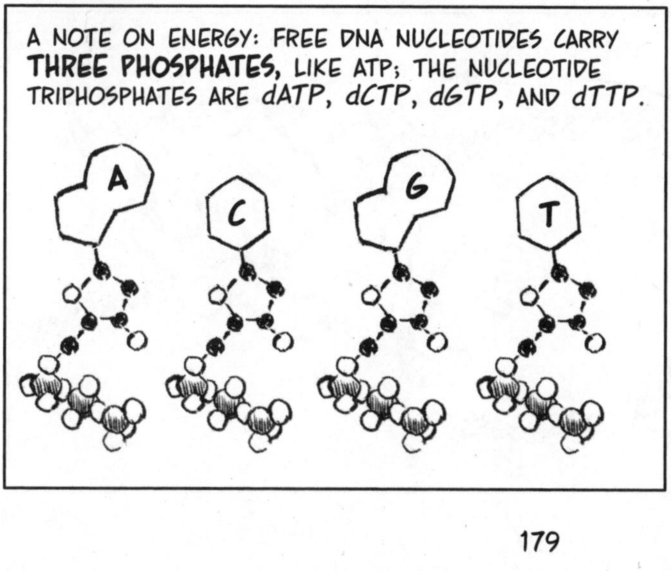
A NOTE ON ENERGY: FREE DNA NUCLEOTIDES CARRY THREE PHOSPHATES, LIKE ATP; THE NUCLEOTIDE TRIPHOSPHATES ARE dATP, dCTP, dGTP, AND dTTP.
A
C
G
T

WHEN A NUCLEOTIDE TRIPHOSPHATE FINDS ITS PLACE ON A DNA STRAND, TWO PHOSPHATES POP OFF AND DELIVER THE ENERGY THAT DRIVES REPLICATION.

REPLICATION ALONG THE OTHER, **LAGGING,** STRAND, IS MORE COMPLICATED. MOVING 3′→5′ ON THIS STRAND WOULD SEND POLYMERASE **AWAY** FROM THE ACTION!
LEADING STRAND
POLYMERASE GOES THIS WAY
5′
3′

THE PROBLEM IS SOLVED WHEN PART OF THE HELICASE GRABS THE LAGGING STRAND'S 5′ TAIL TO FORM A LOOP.
GENIUS!
5′

AS BEFORE, PRIMASE BEGINS A NEW COMPLEMENTARY STRAND, AND DP STEPS IN TO EXTEND IT.
5′

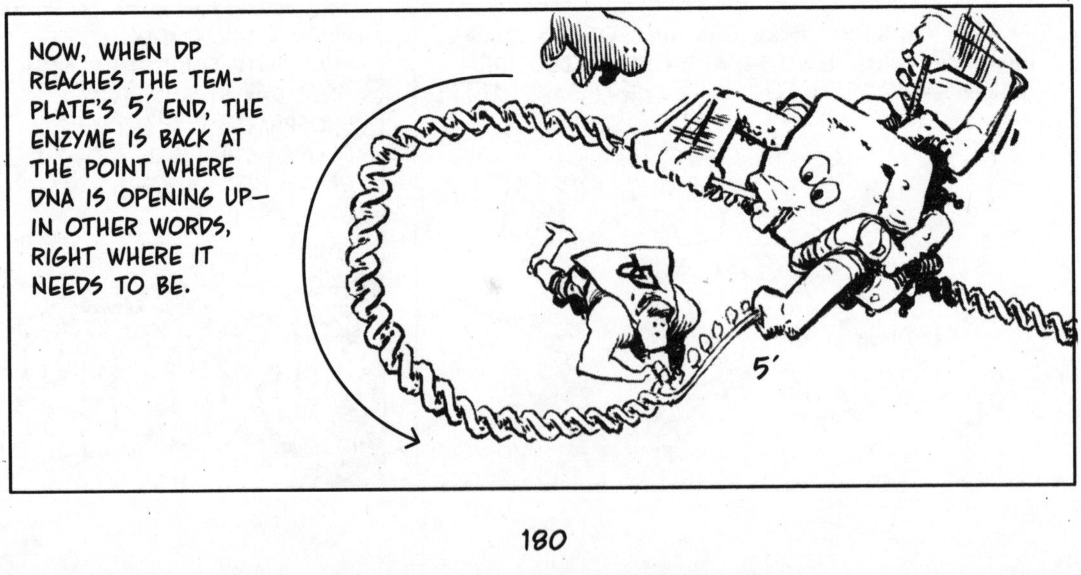
NOW, WHEN DP REACHES THE TEMPLATE'S 5′ END, THE ENZYME IS BACK AT THE POINT WHERE DNA IS OPENING UP— IN OTHER WORDS, RIGHT WHERE IT NEEDS TO BE.
5′

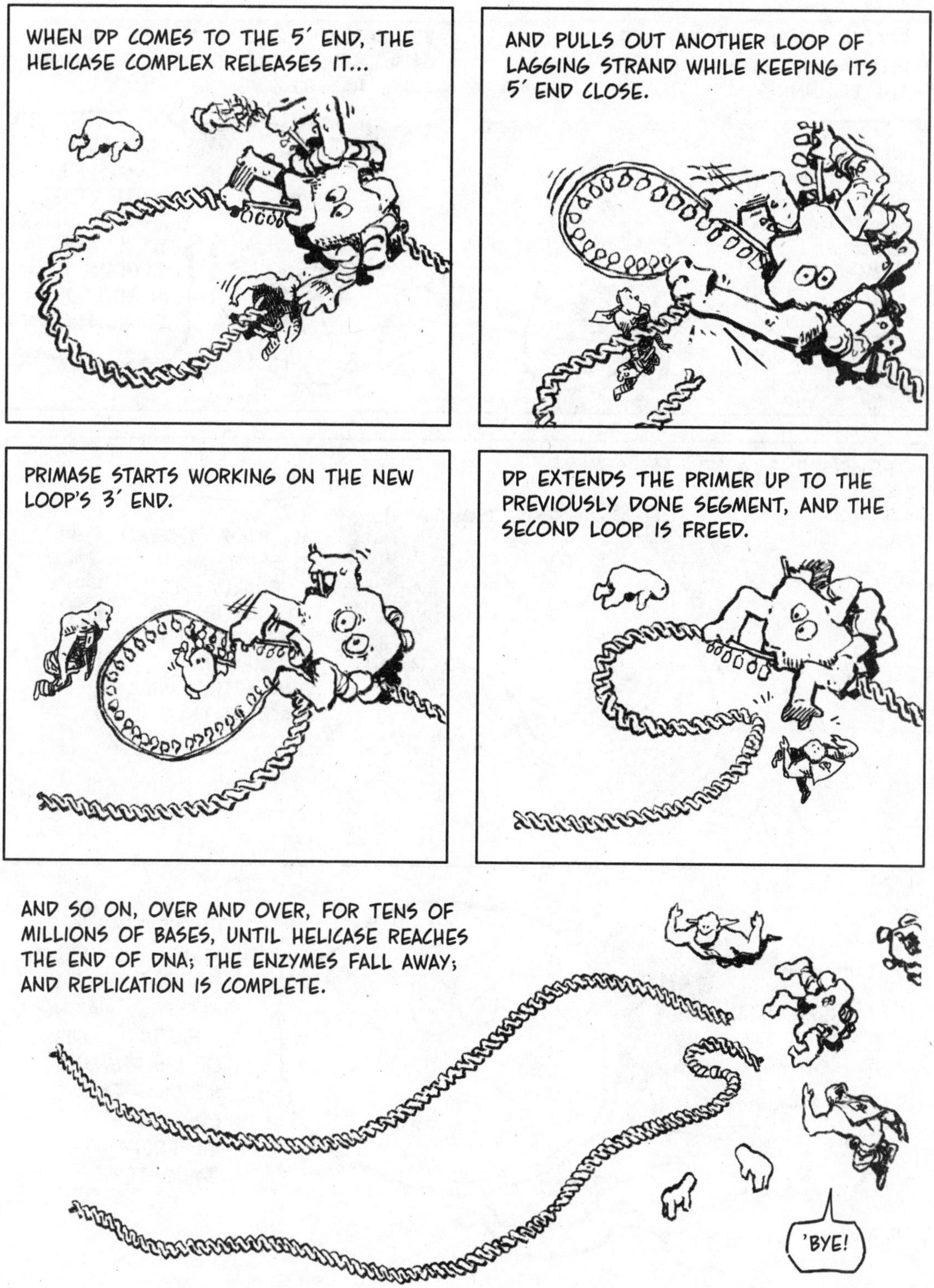
WHEN DP COMES TO THE 5′ END, THE HELICASE COMPLEX RELEASES IT...
AND PULLS OUT ANOTHER LOOP OF LAGGING STRAND WHILE KEEPING ITS 5′ END CLOSE.
PRIMASE STARTS WORKING ON THE NEW LOOP'S 3′ END.
DP EXTENDS THE PRIMER UP TO THE PREVIOUSLY DONE SEGMENT, AND THE SECOND LOOP IS FREED.
AND SO ON, OVER AND OVER, FOR TENS OF MILLIONS OF BASES, UNTIL HELICASE REACHES THE END OF DNA; THE ENZYMES FALL AWAY; AND REPLICATION IS COMPLETE.
'BYE!

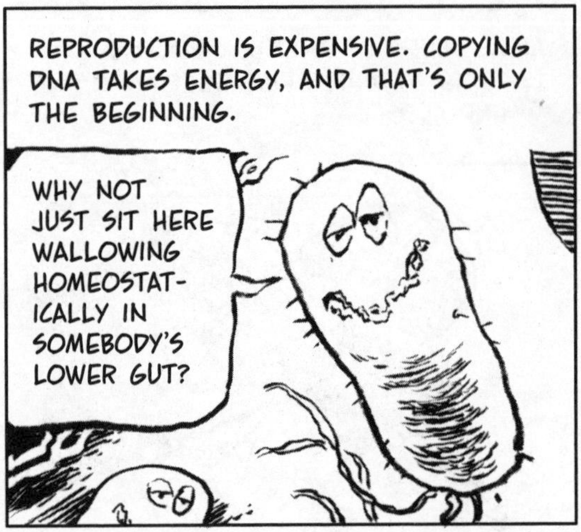
REPRODUCTION IS EXPENSIVE. COPYING DNA TAKES ENERGY, AND THAT'S ONLY THE BEGINNING.
WHY NOT JUST SIT HERE WALLOWING HOMEOSTATICALLY IN SOMEBODY'S LOWER GUT?

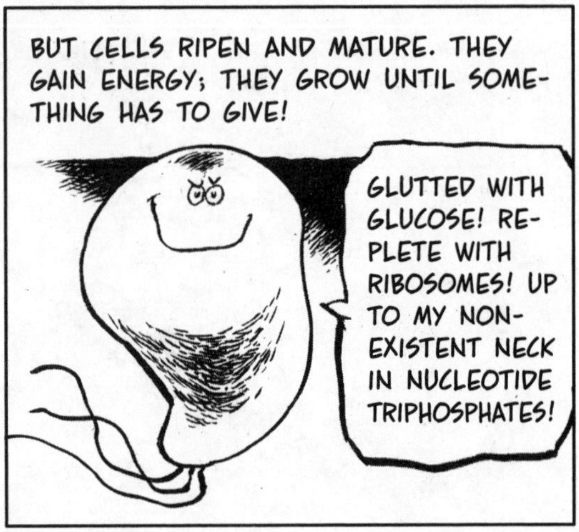
BUT CELLS RIPEN AND MATURE. THEY GAIN ENERGY; THEY GROW UNTIL SOMETHING HAS TO GIVE!
GLUTTED WITH GLUCOSE! REPLETE WITH RIBOSOMES! UP TO MY NONEXISTENT NECK IN NUCLEOTIDE TRIPHOSPHATES!

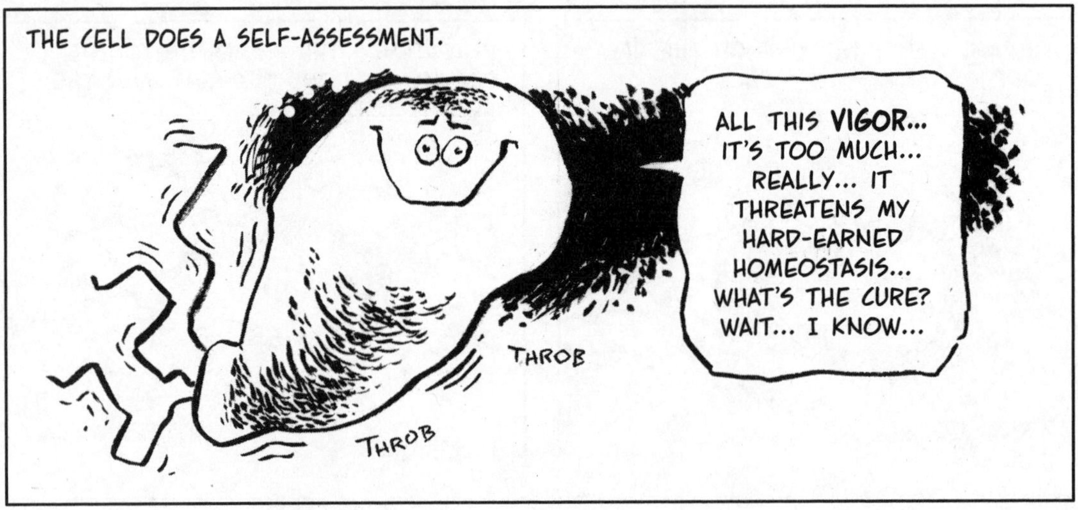
THE CELL DOES A SELF-ASSESSMENT.
ALL THIS VIGOR... IT'S TOO MUCH... REALLY... IT THREATENS MY HARD-EARNED HOMEOSTASIS... WHAT'S THE CURE? WAIT... I KNOW...
THROB
THROB

THERE NEED TO BE TWO OF ME!

IF THE TIME IS RIGHT, THE CELL **DIVIDES.** THIS REQUIRES MAKING A FAITHFUL COPY OF ITS DNA AND SENDING EACH COPY TO A DIFFERENT "DAUGHTER."

A PROKARYOTE SPLITS

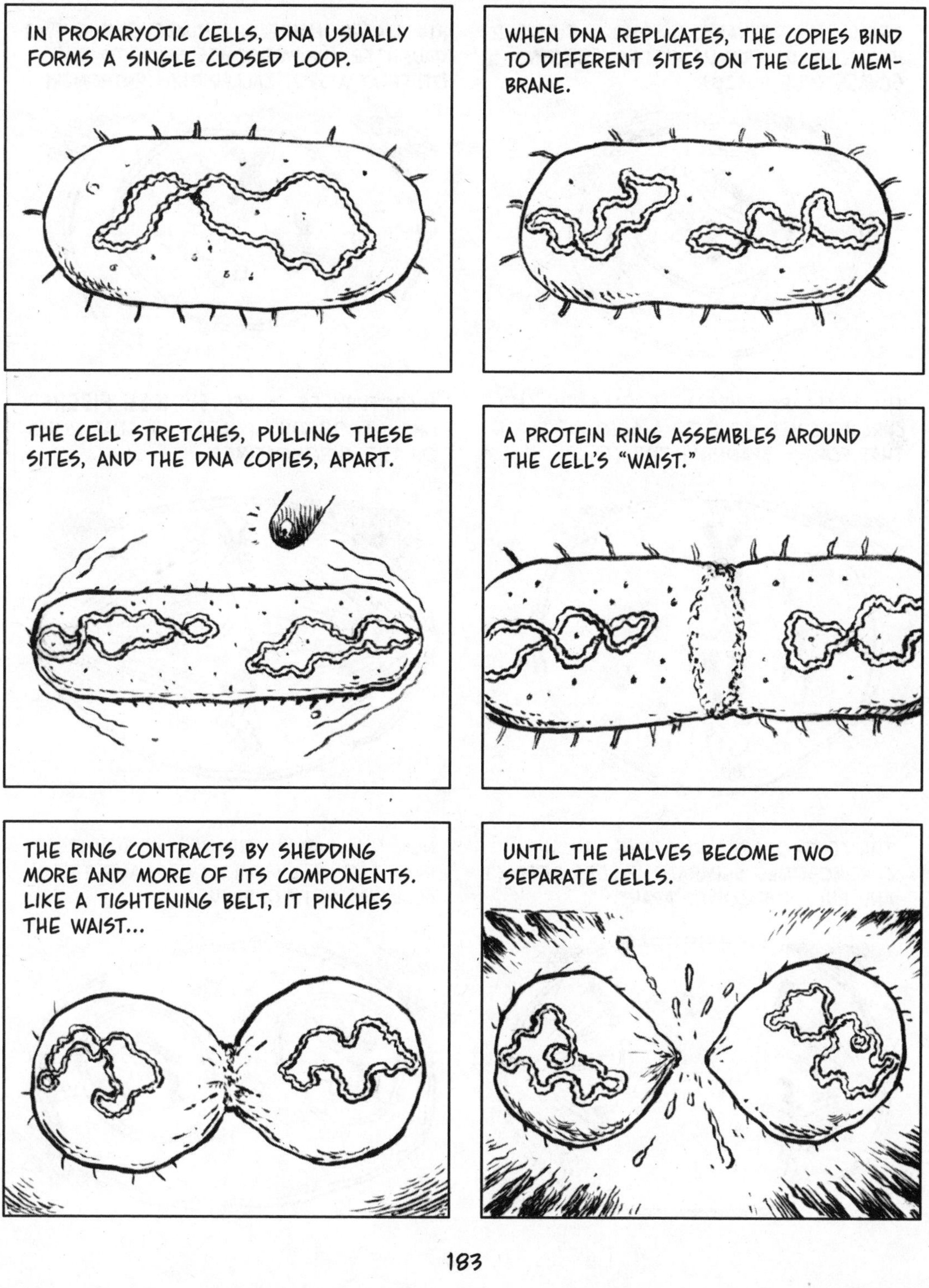
IN PROKARYOTIC CELLS, DNA USUALLY FORMS A SINGLE CLOSED LOOP.
WHEN DNA REPLICATES, THE COPIES BIND TO DIFFERENT SITES ON THE CELL MEMBRANE.
THE CELL STRETCHES, PULLING THESE SITES, AND THE DNA COPIES, APART.
A PROTEIN RING ASSEMBLES AROUND THE CELL'S "WAIST."
THE RING CONTRACTS BY SHEDDING MORE AND MORE OF ITS COMPONENTS. LIKE A TIGHTENING BELT, IT PINCHES THE WAIST...
UNTIL THE HALVES BECOME TWO SEPARATE CELLS.

MITOSIS

IS THE SEGREGATION OF DNA COPIES IN **EUKARYOTES.** ROUGHLY SPEAKING, IT MEANS "THREADING."

RECALL THAT EUKARYOTIC DNA IS BUNDLED INTO COILED STRANDS CALLED **CHROMOSOMES** (SEE P. 128).

REPLICATED INSIDE THE NUCLEUS, THE TWO DAUGHTER CHROMOSOMES STICK TO EACH OTHER AT A SPOT CALLED THE CENTROMERE.

THE NUCLEAR MEMBRANE BREAKS UP; AND CHROMOSOMES MOVE TO A SORT OF PLATE THAT FORMS, SPANNING THE CELL.

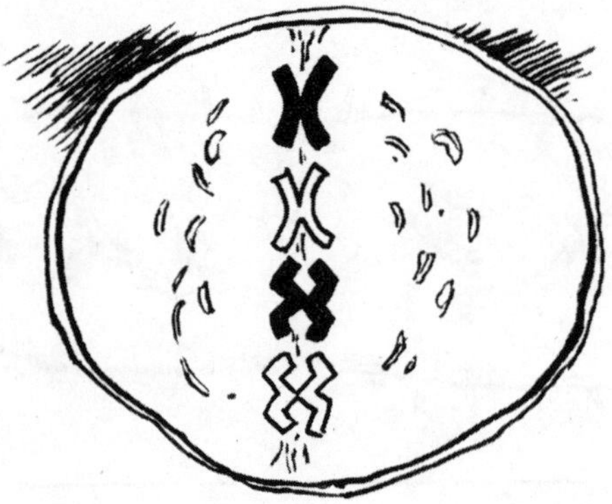

MICROTUBULES CALLED **SPINDLE FIBERS** GROW FROM TWO POLES AND GRAB HOLD OF THE CHROMOSOMES.

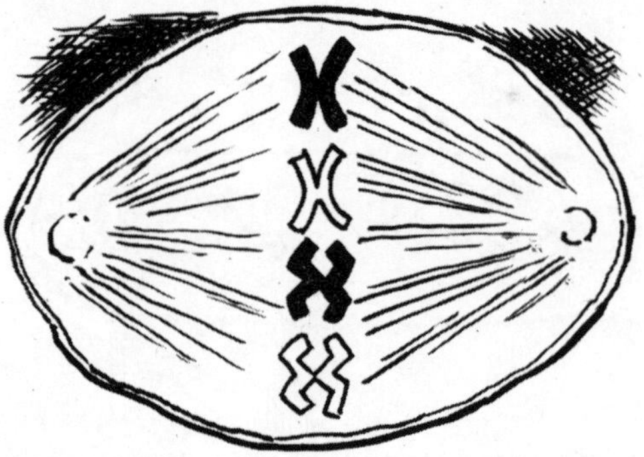

THE CENTROMERES DISSOLVE; DAUGHTER CHROMOSOMES SEPARATE; FIBERS RETRACT AND PULL THE COPIES APART.

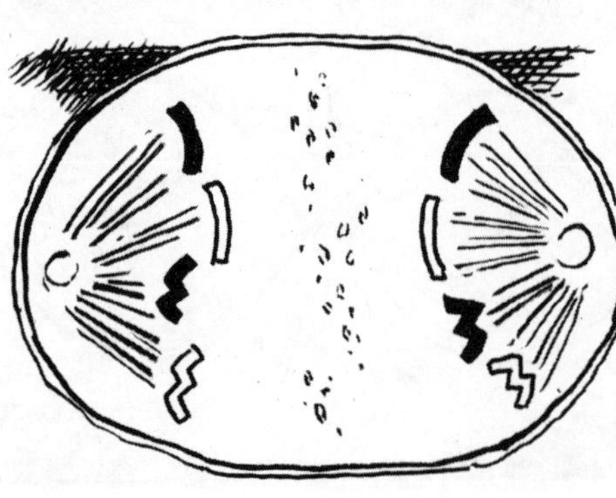

NEW NUCLEAR MEMBRANES SURROUND EACH SET OF CHROMOSOMES—AND THE CELL HAS YET TO DIVIDE!

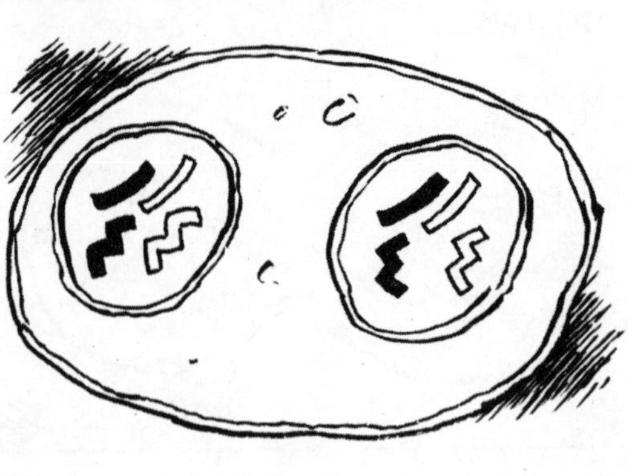

TO SURVIVE, EACH DAUGHTER CELL MUST ALSO INHERIT MITOCHONDRIA AND OTHER ESSENTIAL STUFF FROM THE PARENT.

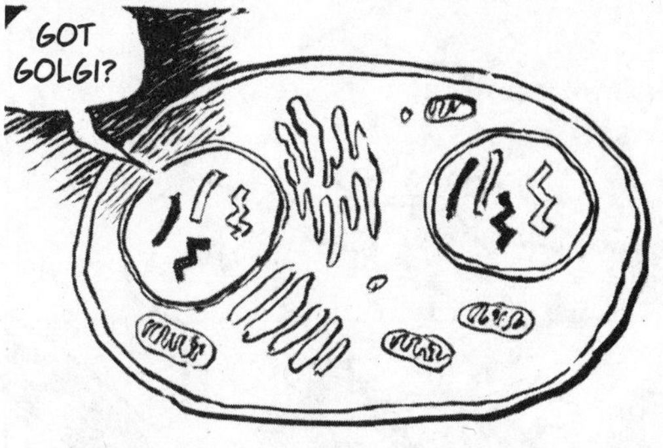
GOT GOLGI?

RANDOM DIFFUSION NEARLY ALWAYS SENDS ENOUGH ORGANELLES TO BOTH ENDS OF THE PARENT.

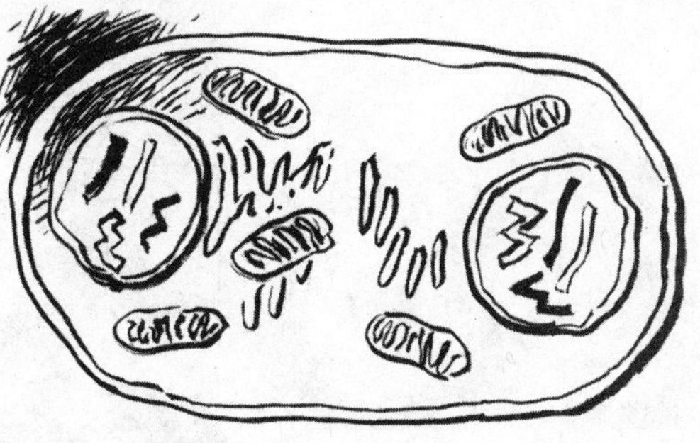

TO SHARE E.R. AND GOLGI, THE PARENT BREAKS THE STRUCTURES INTO BLOBS (VESICLES), WHICH ALSO DIFFUSE.

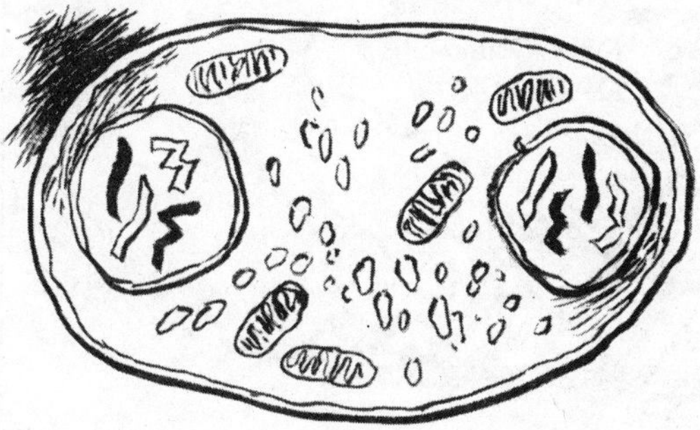

THEY REASSEMBLE AROUND BOTH NEW NUCLEI. THIS IS THE END OF MITOSIS.

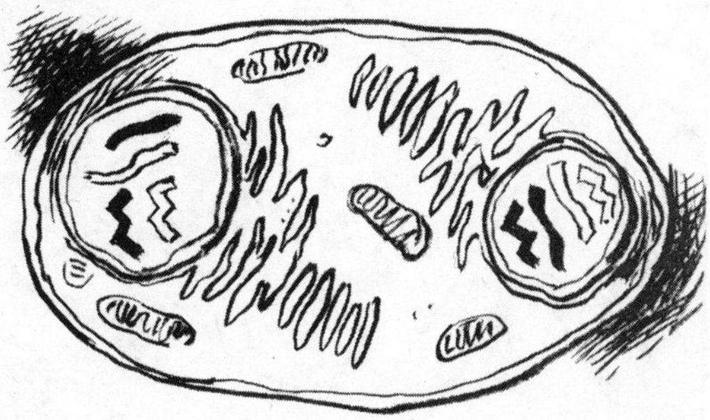

NOW DIVISION BEGINS, AS A NEW MEMBRANE GROWS ACROSS THE CELL.

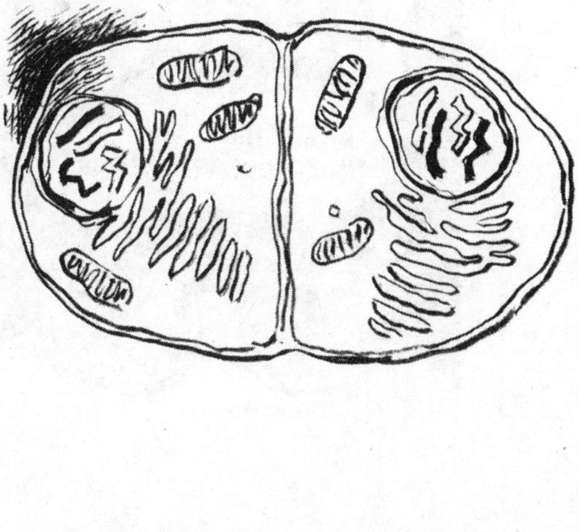

THE TWO DAUGHTERS BREAK FREE, AND THE CELL DIVIDES ("UNDERGOES CYTOKINESIS").

SIGNAL ME SOMETIME!

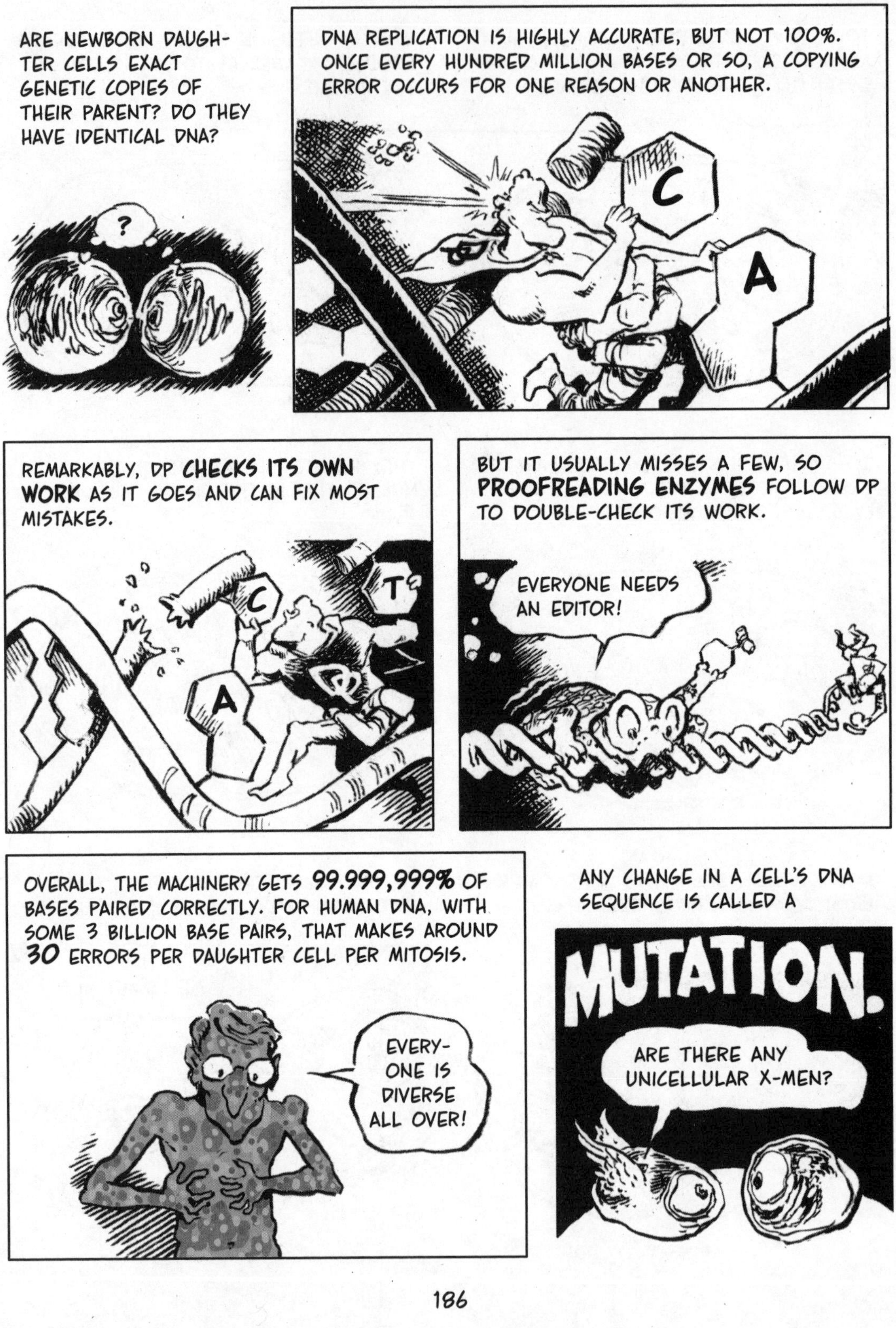
ARE NEWBORN DAUGHTER CELLS EXACT GENETIC COPIES OF THEIR PARENT? DO THEY HAVE IDENTICAL DNA?
?
DNA REPLICATION IS HIGHLY ACCURATE, BUT NOT 100%. ONCE EVERY HUNDRED MILLION BASES OR SO, A COPYING ERROR OCCURS FOR ONE REASON OR ANOTHER.
C
A
REMARKABLY, DP **CHECKS ITS OWN WORK** AS IT GOES AND CAN FIX MOST MISTAKES.
C
T
A
BUT IT USUALLY MISSES A FEW, SO **PROOFREADING ENZYMES** FOLLOW DP TO DOUBLE-CHECK ITS WORK.
EVERYONE NEEDS AN EDITOR!
OVERALL, THE MACHINERY GETS **99.999,999%** OF BASES PAIRED CORRECTLY. FOR HUMAN DNA, WITH SOME 3 BILLION BASE PAIRS, THAT MAKES AROUND **30** ERRORS PER DAUGHTER CELL PER MITOSIS.
EVERYONE IS DIVERSE ALL OVER!
ANY CHANGE IN A CELL'S DNA SEQUENCE IS CALLED A
MUTATION.
ARE THERE ANY UNICELLULAR X-MEN?

TWO KINDS OF MUTATION ARISE DURING REPLICATION.

POINT MUTATIONS HAPPEN WHEN DNA POLYMERASE PUTS THE WRONG BASE OPPOSITE THE TEMPLATE. FOR INSTANCE, IT MIGHT PUT ***G*** OPPOSITE ***T*** BY MISTAKE.
TEMPLATE
POLYMERASE AND PROOFREADERS MAY SIMPLY MISS THE MISTAKE AND LET IT PASS, LEAVING A BAD PAIR...
OR THEY MAY "CORRECT" THE MISMATCH BY REPLACING ***T*** ON THE **TEMPLATE** STRAND INSTEAD OF ***G*** ON THE NEW ONE.
THE NEW DNA STILL HAS COMPLEMENTARY BASE PAIRS, BUT ITS SEQUENCE DIFFERS FROM THE ORIGINAL.

POLYMERASE MAY ALSO ERR BY **ADDING** OR **REMOVING** A BASE IN THE NEW STRAND, AND AGAIN THE TEMPLATE STRAND IS "REPAIRED." THESE **INSERTION** AND **DELETION** ERRORS CHANGE DNA'S LENGTH AS WELL AS ITS SEQUENCE.

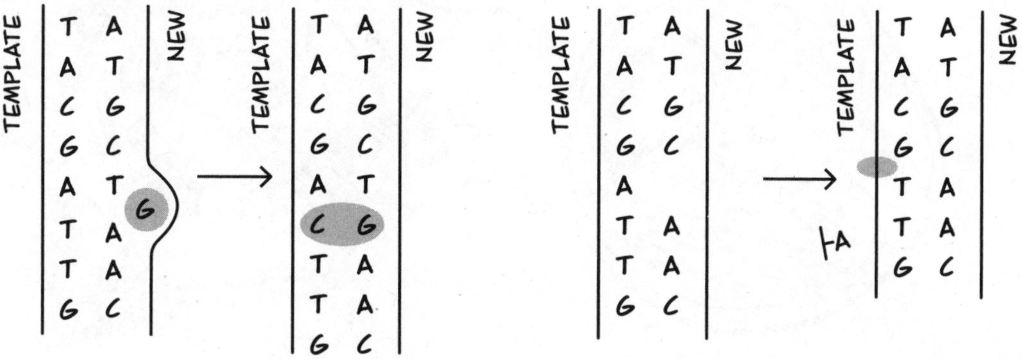

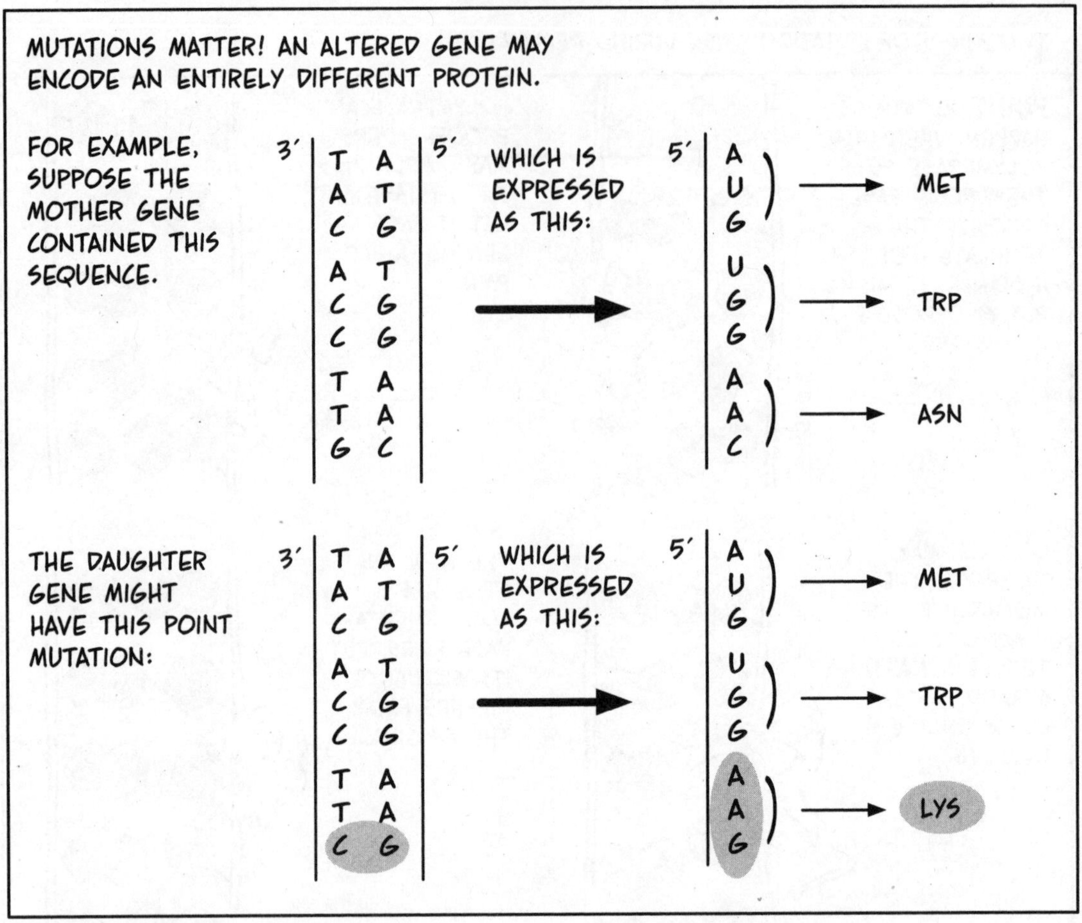
MUTATIONS MATTER! AN ALTERED GENE MAY ENCODE AN ENTIRELY DIFFERENT PROTEIN.
FOR EXAMPLE, SUPPOSE THE MOTHER GENE CONTAINED THIS SEQUENCE.
3′ TACACCTTG 5′
ATGTGGAAC
WHICH IS EXPRESSED AS THIS:
5′ AUG UGG AAC
MET
TRP
ASN
THE DAUGHTER GENE MIGHT HAVE THIS POINT MUTATION:
3′ TACACCTTC 5′
ATGTGGAAG
WHICH IS EXPRESSED AS THIS:
5′ AUG UGG AAG
MET
TRP
LYS

A CHANGE OF EVEN ONE AMINO ACID CAN AFFECT HOW A PROTEIN FOLDS UP. IN THE DAUGHTER CELL, THE MUTANT PROTEIN MAY FUNCTION QUITE DIFFERENTLY, USUALLY FOR THE WORSE, BUT DIFFERENTLY, ANYWAY.

WOW!

NOW SUPPOSE OUR ORIGINAL SEQUENCE FROM THE PREVIOUS PAGE UNDERGOES A DIFFERENT POINT MUTATION:

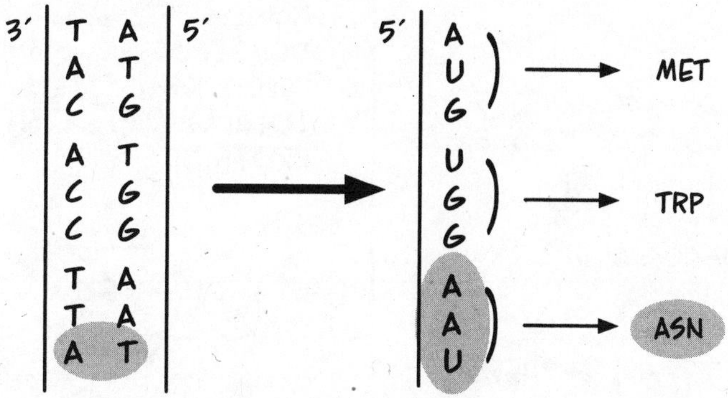

THIS CHANGE HAS **NO EFFECT** ON THE PROTEIN, BECAUSE THE GENETIC CODE IS **REDUNDANT.** **AAC** AND **AAU** BOTH ENCODE THE **SAME AMINO ACID,** ASPARGINE.

THIS IS A **SILENT** MUTATION. THE GENE CHANGES, BUT THE PROTEIN STAYS THE SAME.

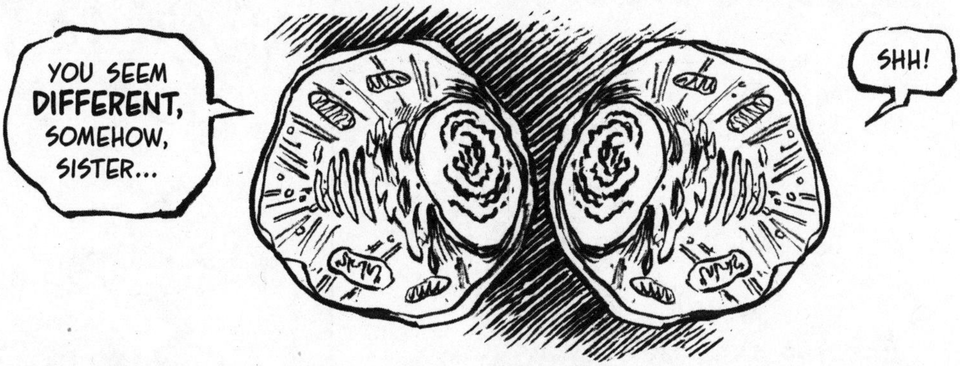

SILENT MUTATIONS CAN ACCUMULATE OVER MANY GENERATIONS WITHOUT HAVING MUCH EFFECT ON A POPULATION.

INSERTIONS AND **DELETIONS** CAN DO MUCH MORE DAMAGE. FOR INSTANCE, SAY WE START WITH THIS SEQUENCE:

3′ TAC ACC TTG CCA ACC GTT AAG... 5′
5′ ATG TGG AAC GGT TGG CAA TTC... 3′

↓ ↓ ↓ ↓ ↓ ↓ ↓

5′ AUG UGG AAC GGU UGG CAA UUC... 3′

↓ ↓ ↓ ↓ ↓ ↓ ↓

MET-TRP-ASN-GLY-TRP-GLN-PHE...

AND SUPPOSE THE MUTATION DELETES THE NINTH PAIR, ***G-C***, WITH THIS RESULT:

3′ TAC ACC TT CCA ACC GTT AAG... 5′
5′ ATG TGG AA GGT TGG CAA TTC... 3′

↓ ↓ ↓ ↓ ↓

5′ AUG UGG AAG GUU GGC AAU UC... 3′

↓ ↓ ↓ ↓ ↓

MET-TRP-LYS-VAL-GLY-ASN- ...

EVERY AMINO ACID DOWNSTREAM FROM **TRP** IS POTENTIALLY CHANGED! THE MUTATION **SHIFTS** THE "READING FRAME."

CA NYO URE ADTHI S?

FRAME-SHIFT MUTATIONS (INSERTIONS AND DELETIONS) CREATE A **TOTALLY DIFFERENT PROTEIN.**

DAUGHTER?

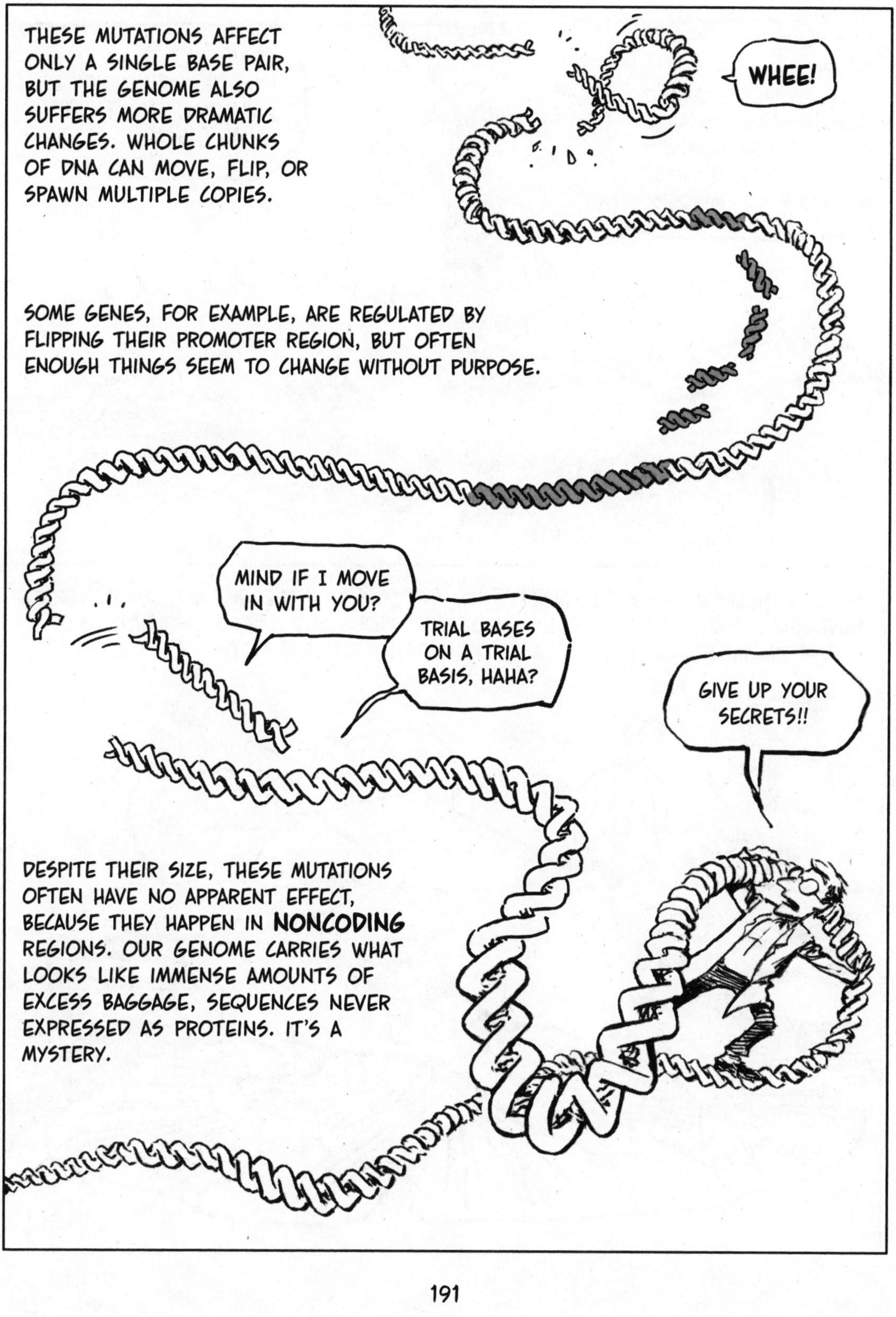
THESE MUTATIONS AFFECT ONLY A SINGLE BASE PAIR, BUT THE GENOME ALSO SUFFERS MORE DRAMATIC CHANGES. WHOLE CHUNKS OF DNA CAN MOVE, FLIP, OR SPAWN MULTIPLE COPIES.
WHEE!
SOME GENES, FOR EXAMPLE, ARE REGULATED BY FLIPPING THEIR PROMOTER REGION, BUT OFTEN ENOUGH THINGS SEEM TO CHANGE WITHOUT PURPOSE.
MIND IF I MOVE IN WITH YOU?
TRIAL BASES ON A TRIAL BASIS, HAHA?
GIVE UP YOUR SECRETS!!
DESPITE THEIR SIZE, THESE MUTATIONS OFTEN HAVE NO APPARENT EFFECT, BECAUSE THEY HAPPEN IN NONCODING REGIONS. OUR GENOME CARRIES WHAT LOOKS LIKE IMMENSE AMOUNTS OF EXCESS BAGGAGE, SEQUENCES NEVER EXPRESSED AS PROTEINS. IT'S A MYSTERY.

DESPITE MUTATIONS, CELL DIVISION WORKS. HUMANS, FOR INSTANCE, CAN LIVE FOR MANY DECADES WITH THEIR 30 MISTAKES PER MITOSIS.
STILL, MUTATIONS DO PILE UP WITH REPEATED DIVISIONS, AND THAT IS NOT GOOD!

SO LIFE INVENTED AN ALTERNATIVE TO MITOSIS, A DIFFERENT WAY TO GENERATE NEW CELLS, A WAY THAT SPREADS GOOD MUTATIONS AND WEEDS OUT BAD ONES. YOU'VE PROBABLY HEARD OF IT. AND THOUGHT ABOUT IT. A LOT.

Chapter 13
REPRODUCTION (Part 2)

SEX AND SUCH

MITOSIS IS REALLY, TRULY REPRODUCTION. IT WORKS LIKE A COPY MACHINE. A CELL REPLICATES ITS DNA AND BEGETS A DUPLICATE OF ITSELF. A GERANIUM CUTTING IN WATER WILL SPROUT ROOTS AND GROW BY MITOSIS ALONE INTO A PLANT GENETICALLY IDENTICAL TO ITS PARENT, A **CLONE**.

YET OFTEN ENOUGH, "REPRODUCTION" MEANS SOMETHING ENTIRELY DIFFERENT: **TWO** PARENTS JOIN TO BEGET OFFSPRING THAT DIFFER FROM BOTH OF THEM. THIS "REPRODUCTION"—**SEXUAL REPRODUCTION**—IS **NOT** STRICTLY **COPYING!**

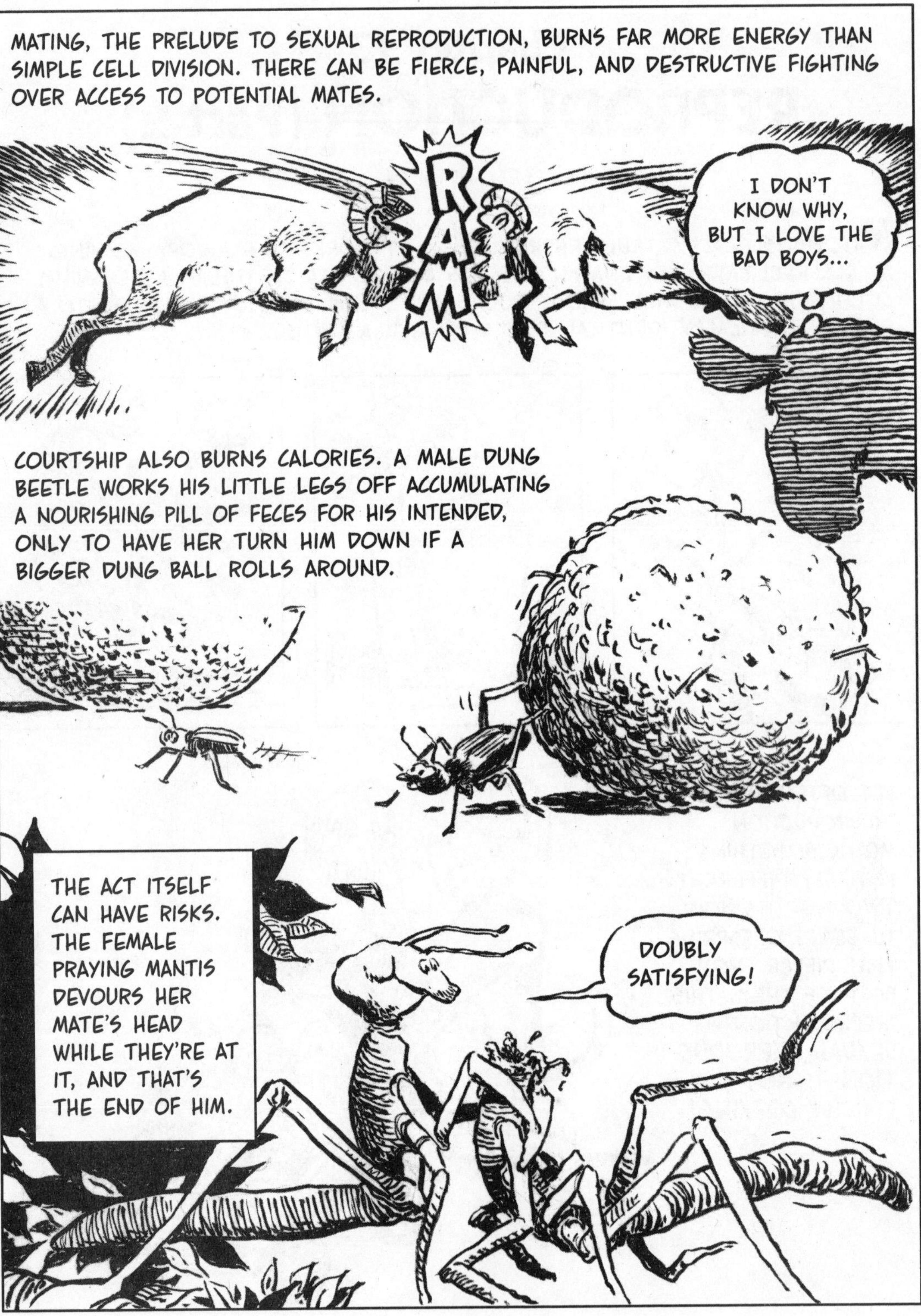
MATING, THE PRELUDE TO SEXUAL REPRODUCTION, BURNS FAR MORE ENERGY THAN SIMPLE CELL DIVISION. THERE CAN BE FIERCE, PAINFUL, AND DESTRUCTIVE FIGHTING OVER ACCESS TO POTENTIAL MATES.
RAM
I DON'T KNOW WHY, BUT I LOVE THE BAD BOYS...
COURTSHIP ALSO BURNS CALORIES. A MALE DUNG BEETLE WORKS HIS LITTLE LEGS OFF ACCUMULATING A NOURISHING PILL OF FECES FOR HIS INTENDED, ONLY TO HAVE HER TURN HIM DOWN IF A BIGGER DUNG BALL ROLLS AROUND.
THE ACT ITSELF CAN HAVE RISKS. THE FEMALE PRAYING MANTIS DEVOURS HER MATE'S HEAD WHILE THEY'RE AT IT, AND THAT'S THE END OF HIM.
DOUBLY SATISFYING!

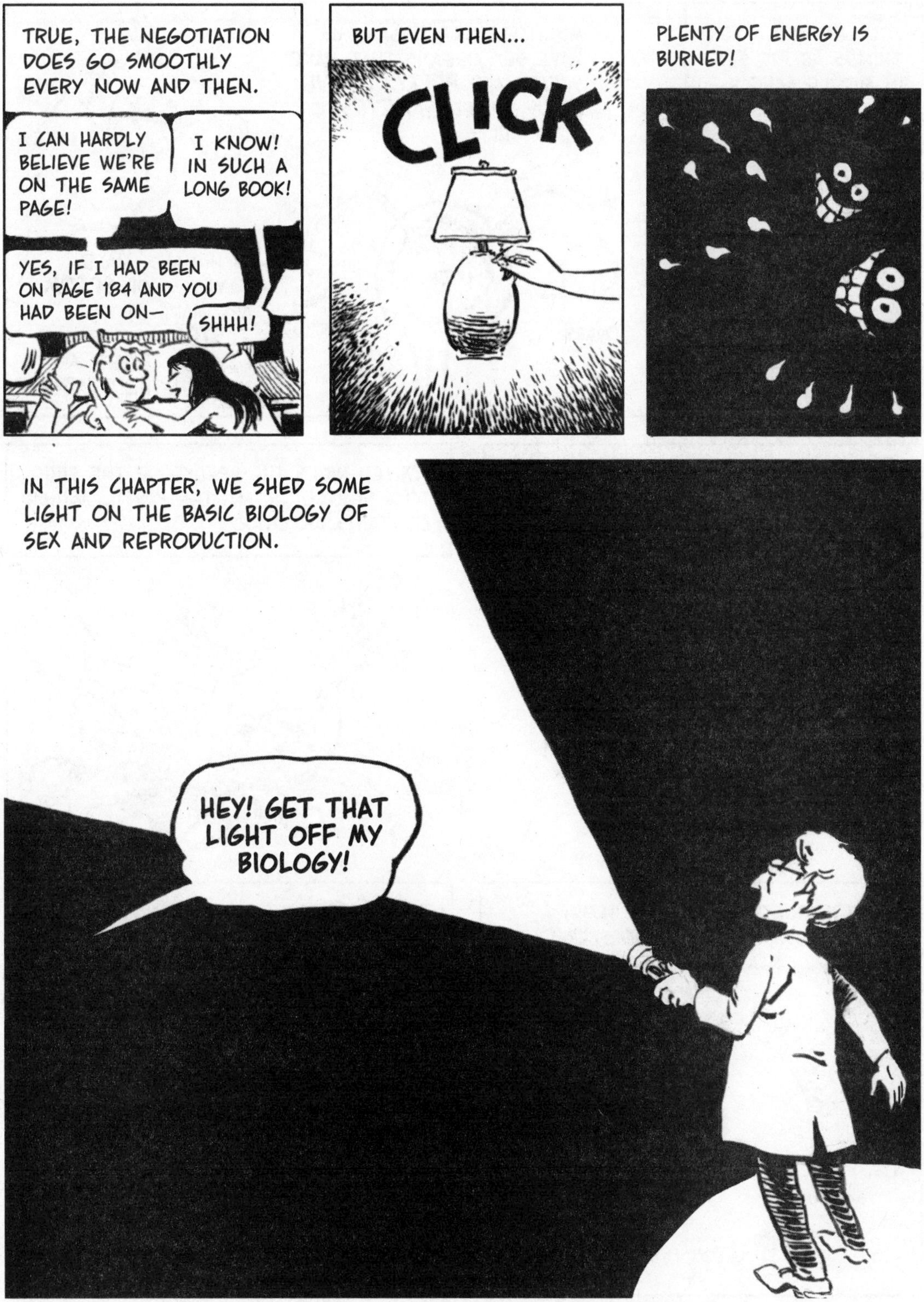
TRUE, THE NEGOTIATION DOES GO SMOOTHLY EVERY NOW AND THEN.
I CAN HARDLY BELIEVE WE'RE ON THE SAME PAGE!
I KNOW! IN SUCH A LONG BOOK!
YES, IF I HAD BEEN ON PAGE 184 AND YOU HAD BEEN ON—
SHHH!
BUT EVEN THEN...
CLICK
PLENTY OF ENERGY IS BURNED!
IN THIS CHAPTER, WE SHED SOME LIGHT ON THE BASIC BIOLOGY OF SEX AND REPRODUCTION.
HEY! GET THAT LIGHT OFF MY BIOLOGY!

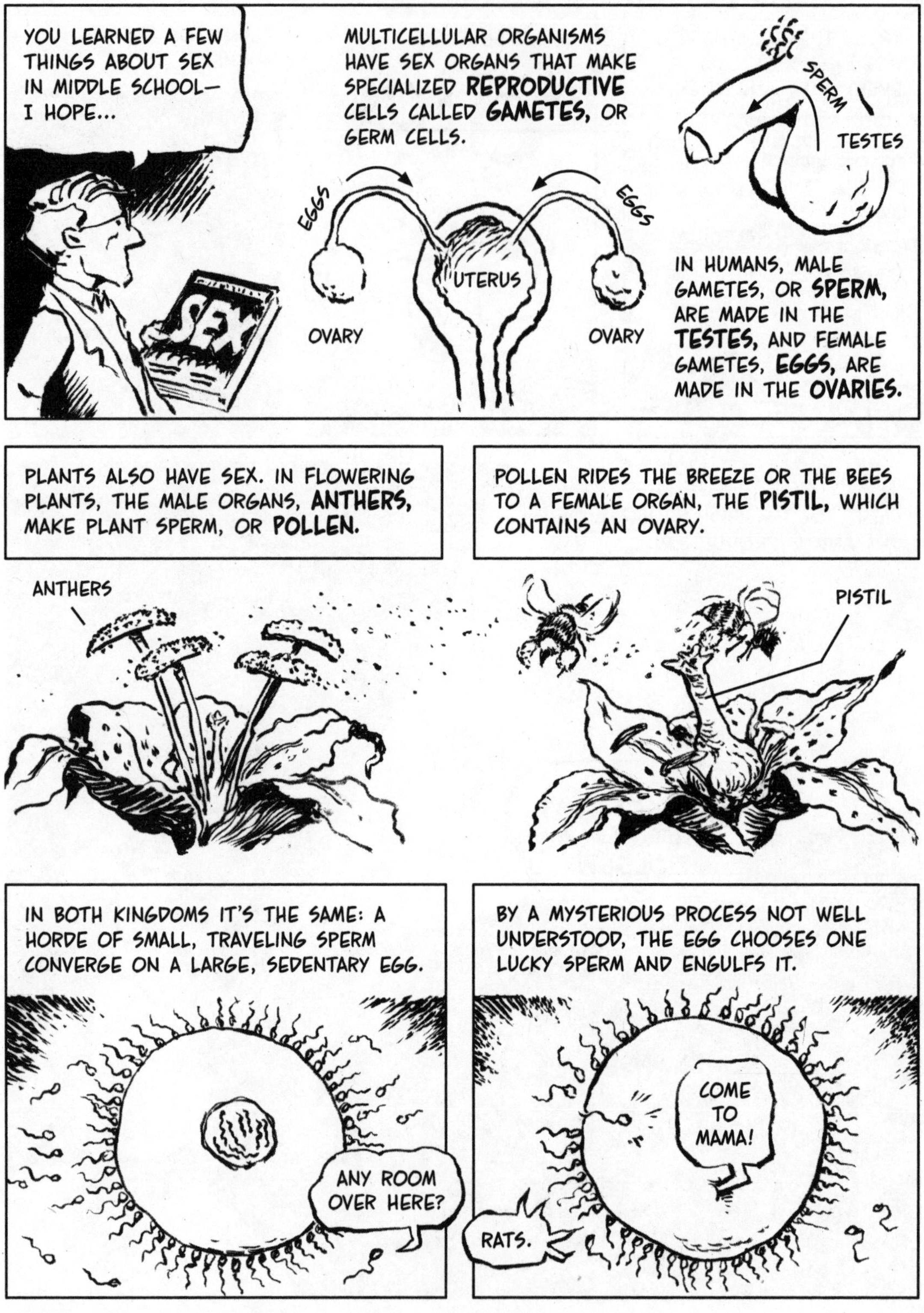
YOU LEARNED A FEW THINGS ABOUT SEX IN MIDDLE SCHOOL— I HOPE...
SEX
MULTICELLULAR ORGANISMS HAVE SEX ORGANS THAT MAKE SPECIALIZED **REPRODUCTIVE** CELLS CALLED **GAMETES,** OR GERM CELLS.
EGGS
EGGS
UTERUS
OVARY
OVARY
SPERM
TESTES
IN HUMANS, MALE GAMETES, OR **SPERM,** ARE MADE IN THE **TESTES,** AND FEMALE GAMETES, **EGGS,** ARE MADE IN THE **OVARIES.**
PLANTS ALSO HAVE SEX. IN FLOWERING PLANTS, THE MALE ORGANS, **ANTHERS,** MAKE PLANT SPERM, OR **POLLEN.**
POLLEN RIDES THE BREEZE OR THE BEES TO A FEMALE ORGAN, THE **PISTIL,** WHICH CONTAINS AN OVARY.
ANTHERS
PISTIL
IN BOTH KINGDOMS IT'S THE SAME: A HORDE OF SMALL, TRAVELING SPERM CONVERGE ON A LARGE, SEDENTARY EGG.
ANY ROOM OVER HERE?
BY A MYSTERIOUS PROCESS NOT WELL UNDERSTOOD, THE EGG CHOOSES ONE LUCKY SPERM AND ENGULFS IT.
COME TO MAMA!
RATS.

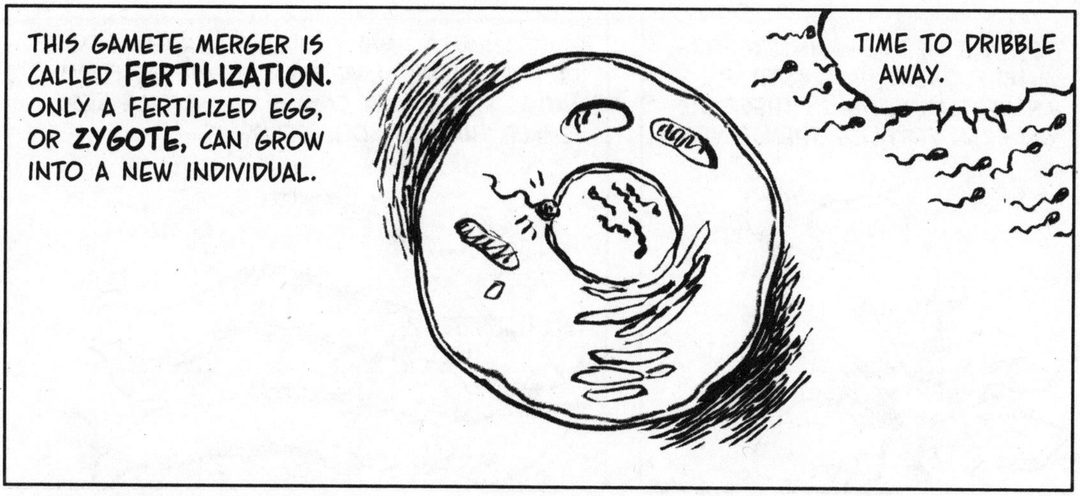
THIS GAMETE MERGER IS CALLED **FERTILIZATION.** ONLY A FERTILIZED EGG, OR **ZYGOTE,** CAN GROW INTO A NEW INDIVIDUAL.
TIME TO DRIBBLE AWAY.

IN FLOWERING PLANTS, ZYGOTES SWELL INTO SEEDS, AND THE SEEDS FALL TO EARTH, WHERE THEY SPROUT.

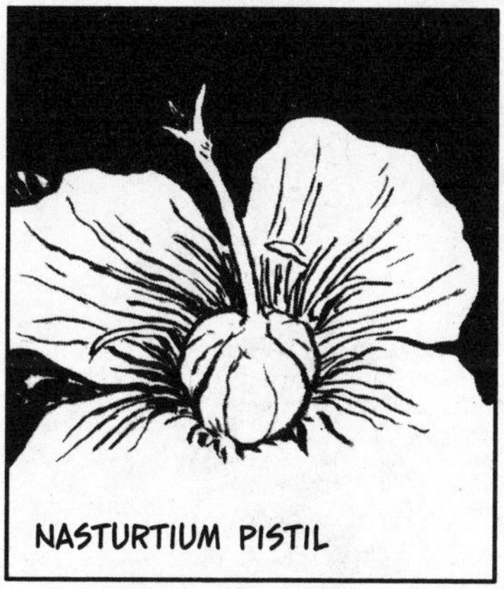
NASTURTIUM PISTIL

WITHERED FLOWER SHEDS SEED.

NEW NASTURTIUM GROWS.

IN ANIMALS, THE FERTILIZED EGG **GESTATES,** DIVIDING MANY TIMES, UNTIL IT HATCHES A NEW INDIVIDUAL. GESTATION CAN BE INSIDE OR OUTSIDE THE MOTHER, DEPENDING ON THE TYPE OF ANIMAL.

TO MAKE LIFE MORE INTERESTING, MANY FLOWERS HAVE BOTH ANTHERS AND STIGMA; THESE GUY-GALS CAN FERTILIZE THEMSELVES.
HUH!

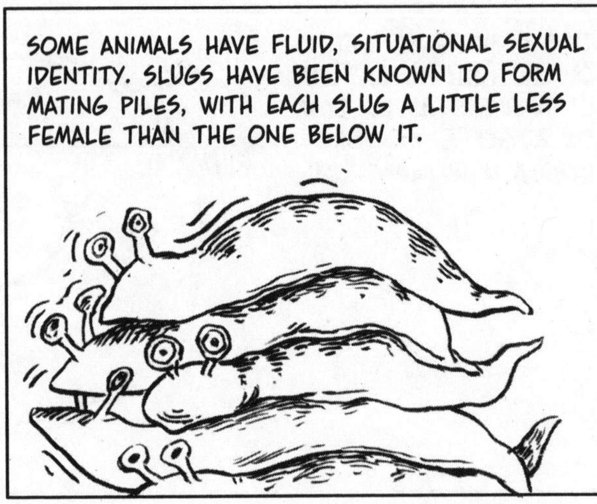
SOME ANIMALS HAVE FLUID, SITUATIONAL SEXUAL IDENTITY. SLUGS HAVE BEEN KNOWN TO FORM MATING PILES, WITH EACH SLUG A LITTLE LESS FEMALE THAN THE ONE BELOW IT.

BY AND LARGE, MOST FAMILIAR ANIMALS ARE EITHER/OR, MALE OR FEMALE, WITH JUST ONE SET OF SEX ORGANS (AND THE TWO SEXES MAY LOOK DIFFERENT).

YEAH. SHOULD I CONSIDER SURGERY?
ALL OF WHICH RAISES SOME QUESTIONS.

WHY DOES NATURE DIVIDE THINGS INTO SEXES, ONLY TO RECOMBINE THEM? WHY DO OFFSPRING LOOK LIKE THEIR PARENTS? HOW ARE TRAITS INHERITED FROM TWO DIFFERENT INDIVIDUALS? FOR EONS, PEOPLE HAVE MUSED, THEORIZED, DOGMATIZED, LEGISLATED, OBSESSED, AND DROOLED ABOUT SEX.
OBVIOUSLY, THE SPERM CARRIES THE SEED OF THE FUTURE BEING, WHILE THE FEMALE SERVES AS A PASSIVE RECEPTACLE TO NURTURE THE PRECIOUS MALE GIFT—WHICH I SAY IN THE MOST UNBIASED WAY POSSIBLE.
CLEARLY, EVERY SPERM CELL CARRIES A LITTLE REPLICA OF THE FUTURE OFFSPRING, A HOMUNCULUS WITH ITS OWN MINIATURE SPERM EACH WITH ITS OWN MINIATURE HOMUNCULI, AND SO ON, HOMUNCULI ALL THE WAY DOWN!
MANSPLAINING THROUGH THE AGES!
WELL, WE'RE NOT ALWAYS WRONG.

LUCKILY FOR SCIENCE, ONE ZYGOTE GREW UP TO BE

GREGOR MENDEL

(1822–1884). THIS AUSTRIAN MONK'S VOWS PRECLUDED HIS FERTILIZING OTHER HUMANS, BUT HE BRED A LARGE GARDEN OF PEA PLANTS—AND KEPT METICULOUS RECORDS.

TENDING HUNDREDS OF PLANTS IN THE BRNO MONASTERY'S LARGE PEA PATCH, MENDEL NOTICED SOMETHING STRIKING: EACH PLANT WAS EITHER TALL OR SHORT, BUT NEVER IN BETWEEN; EACH PLANT HAD EITHER PURPLE OR WHITE FLOWERS, BUT NEVER BOTH OR A BLEND; AND SO ON FOR A NUMBER OF OTHER TRAITS.

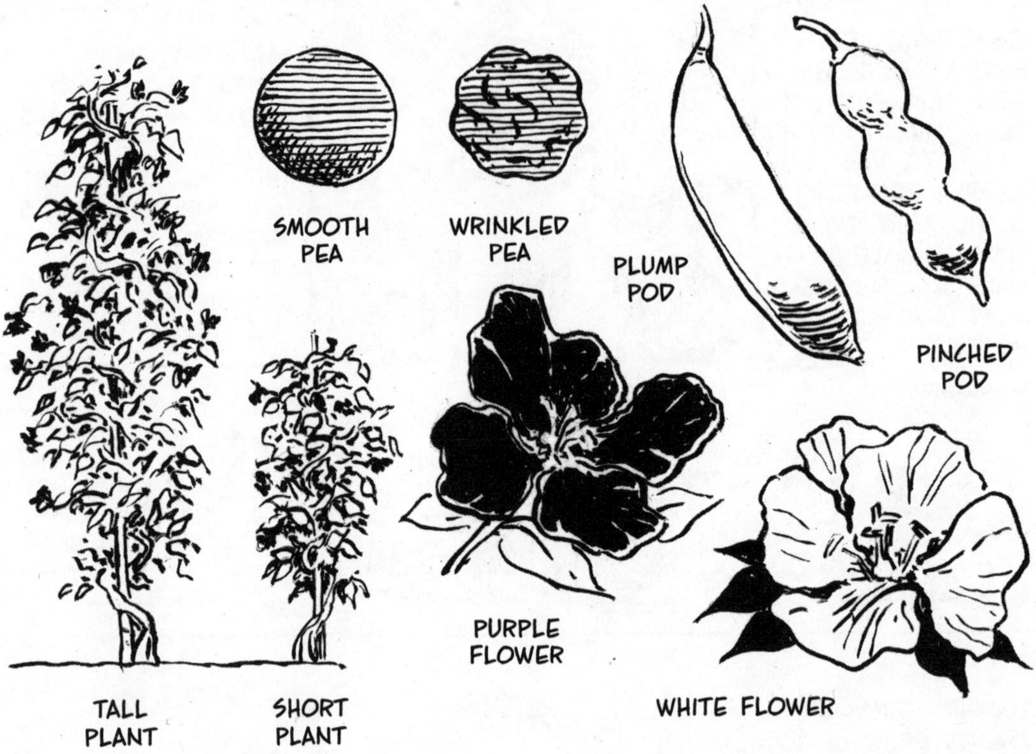
SMOOTH PEA
WRINKLED PEA
PLUMP POD
PINCHED POD
PURPLE FLOWER
WHITE FLOWER
TALL PLANT
SHORT PLANT

MENDEL CAREFULLY MATED, OR CROSSED, ONE PLANT WITH ANOTHER BY PLACING POLLEN FROM ONE FLOWER'S ANTHER ON ANOTHER'S PISTIL.
IT'S FUN TO PLAY GOD!

HE SNIPPED OFF THE "MOM'S" ANTHERS TO PREVENT SELF-FERTILIZATION AND BAGGED THESE FLOWERS TO PREVENT ANY OTHER POLLINATION.
DON'T TELL THE ABBOT!

AFTER MANY CROSSINGS, MENDEL FOUND THAT CERTAIN GROUPS OF PLANTS ALWAYS BRED **TRUE TO TYPE.**

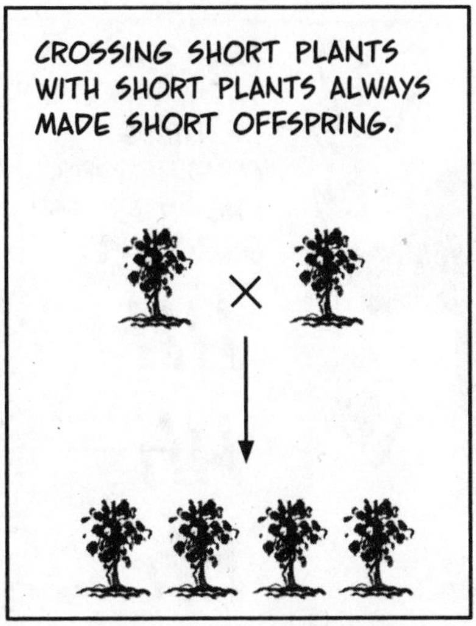
CROSSING SHORT PLANTS WITH SHORT PLANTS ALWAYS MADE SHORT OFFSPRING.
×

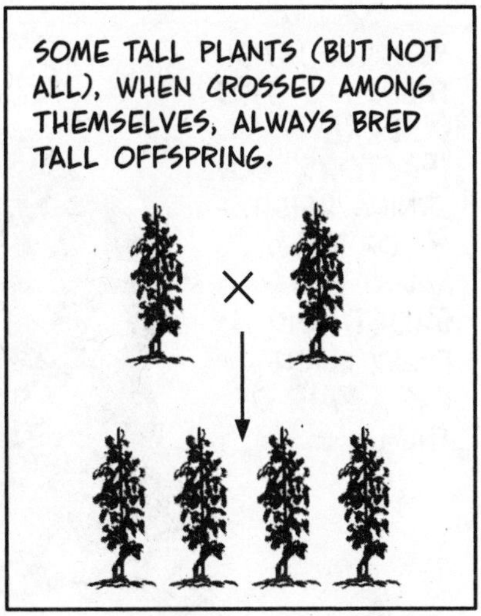
SOME TALL PLANTS (BUT NOT ALL), WHEN CROSSED AMONG THEMSELVES, ALWAYS BRED TALL OFFSPRING.
×

NOW LET THE EXPERIMENTS BEGIN: MENDEL CROSSES TRUE-BREEDING TALLS WITH SHORTS AND DISCOVERS DOMINANCE. TALLNESS IS DOMINANT OVER SHORTNESS. ALL OFFSPRING ARE TALL!
IN NOMINI DOMINI!
×

NEXT, HE CROSSES THE **HYBRID** OFFSPRING WITH **EACH OTHER.** THE SEEDS SPROUT; THE PLANTS GROW; AND THE OFFSPRING ARE AROUND **1/4** SHORT AND **3/4** TALL.

I SENSE INVISIBLE FORCES AT WORK HERE...
×

MENDEL'S INSIGHT #1: THERE ARE **TWO DIFFERENT "FACTORS"** GOVERNING HEIGHT—*H* FOR **TALL** AND *h* FOR **SHORT**—AND EVERY PLANT HAS A **PAIR** OF THEM.

EACH PLANT, THEN, HAS ONE OF THESE COMBINATIONS, KNOWN AS ITS **GENOTYPE:**

HH

Hh

hh

PEAS WORK IN MYSTERIOUS WAYS...

HH AND ***Hh*** ARE BOTH **TALL.** WE SAY THAT ***H*** IS A **DOMINANT TRAIT** AND ***h*** A **RECESSIVE** ONE: A SINGLE TALL FACTOR ***H*** IS ENOUGH TO MAKE A TALL PLANT. A TALL PLANT MAY HAVE GENOTYPE ***HH*** OR ***Hh.*** A SHORT PLANT'S GENOTYPE MUST BE ***hh.*** THE PLANT'S APPEARANCE (TALL OR SHORT) IS CALLED ITS **PHENOTYPE.**

WHO DECIDES THE PHENOTYPES AROUND HERE?
YOU, MASTER!

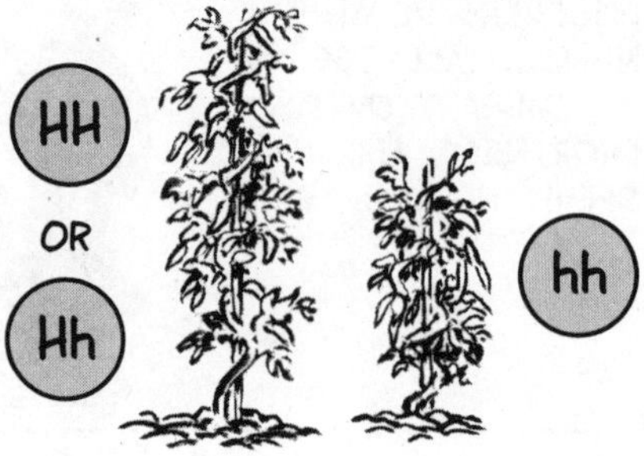
HH
OR
Hh
hh

INSIGHT #2, MENDEL'S FIRST LAW, THE **LAW OF SEGREGATION:** EACH **GAMETE** (SPERM OR EGG) GETS ONLY **ONE** FACTOR FROM THE PARENTAL PAIR. THIS DIVISION HAPPENS RANDOMLY, SO HALF THE GAMETES GET ONE FACTOR AND HALF GET THE OTHER.

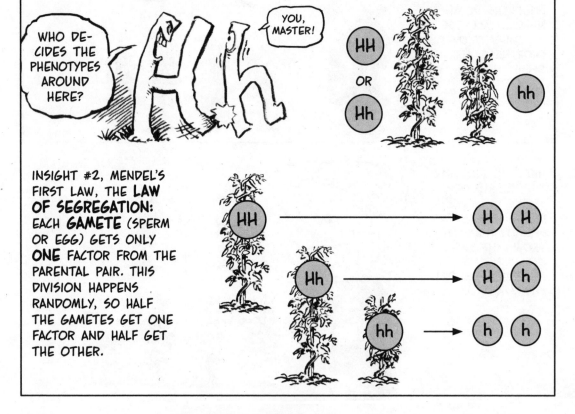

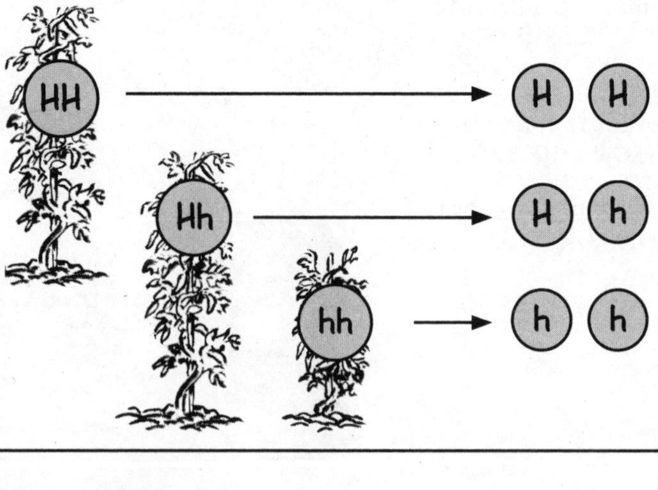
HH
H H
Hh
H h
hh
h h

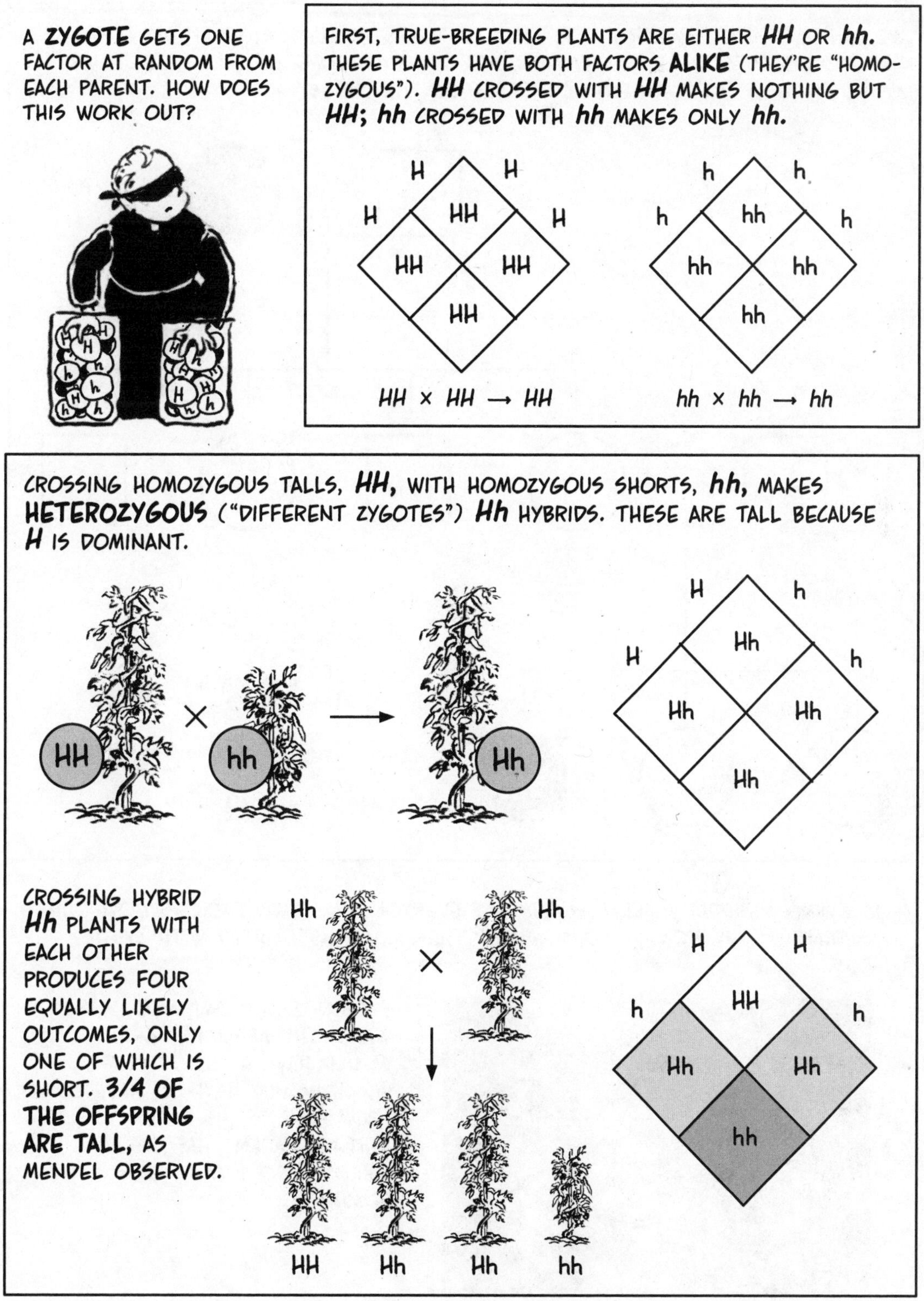

A **ZYGOTE** GETS ONE FACTOR AT RANDOM FROM EACH PARENT. HOW DOES THIS WORK OUT?

FIRST, TRUE-BREEDING PLANTS ARE EITHER **HH** OR **hh**. THESE PLANTS HAVE BOTH FACTORS **ALIKE** (THEY'RE "HOMOZYGOUS"). **HH** CROSSED WITH **HH** MAKES NOTHING BUT **HH**; **hh** CROSSED WITH **hh** MAKES ONLY **hh**.

HH × HH → HH   hh × hh → hh

CROSSING HOMOZYGOUS TALLS, **HH,** WITH HOMOZYGOUS SHORTS, **hh,** MAKES **HETEROZYGOUS** ("DIFFERENT ZYGOTES") **Hh** HYBRIDS. THESE ARE TALL BECAUSE **H** IS DOMINANT.

CROSSING HYBRID **Hh** PLANTS WITH EACH OTHER PRODUCES FOUR EQUALLY LIKELY OUTCOMES, ONLY ONE OF WHICH IS SHORT. **3/4 OF THE OFFSPRING ARE TALL,** AS MENDEL OBSERVED.

WE CAN SEE SOME SIMILAR INHERITANCE PATTERNS IN HUMANS, WHEN WE LOOK AT GENETIC DISEASES. HERE IS A FAMILY TREE IN WHICH SOME OF THE PEOPLE HAVE CYSTIC FIBROSIS (C.F.), A GENETIC DISEASE OF THE LUNGS.

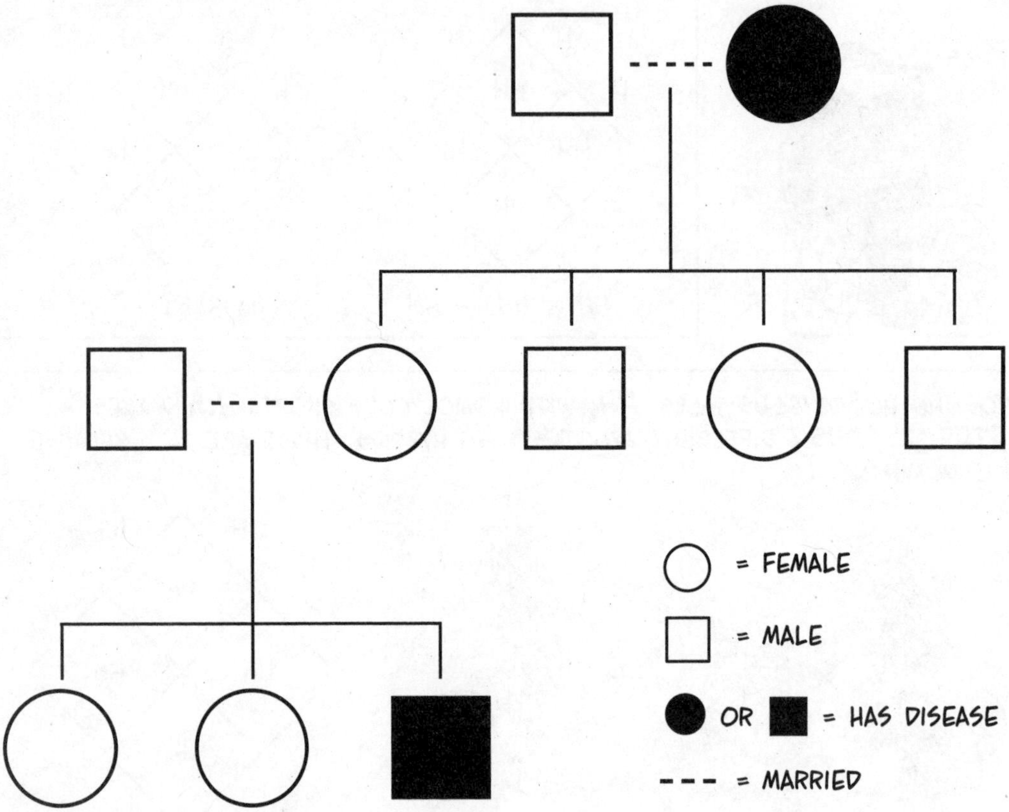

FOLLOWING MENDEL, WE'LL ASSUME THAT EVERYONE HAS TWO FACTORS GOVERNING THIS TRAIT: ***F*** THE NORMAL FORM, AND ***f*** THE FACTOR ASSOCIATED WITH CYSTIC FIBROSIS.

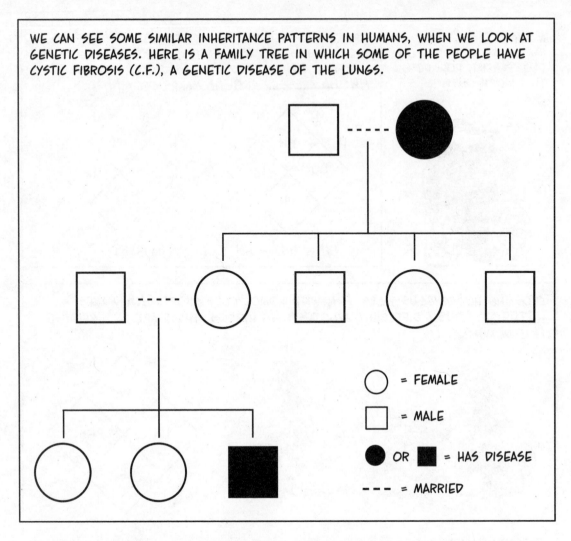

f CANNOT BE DOMINANT. IF IT WERE, THE AFFLICTED GRANDSON WOULD HAVE AT LEAST ONE ***f***, WHICH WOULD HAVE TO HAVE COME FROM ONE OF HIS PARENTS. BUT NEITHER PARENT HAS CYSTIC FIBROSIS, SO ***f*** CAN'T BE DOMINANT.

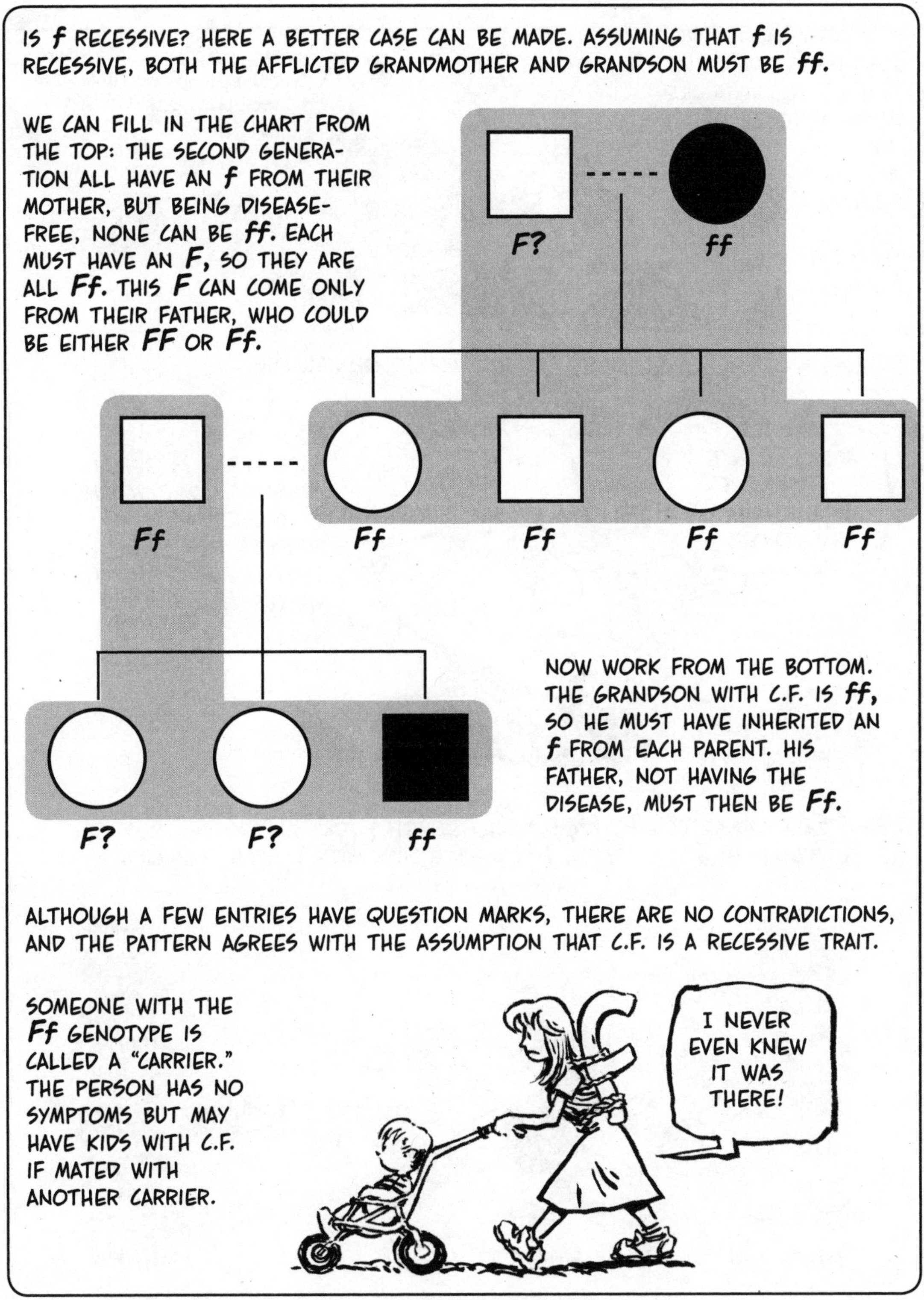
IS f RECESSIVE? HERE A BETTER CASE CAN BE MADE. ASSUMING THAT f IS RECESSIVE, BOTH THE AFFLICTED GRANDMOTHER AND GRANDSON MUST BE ff.
WE CAN FILL IN THE CHART FROM THE TOP: THE SECOND GENERATION ALL HAVE AN f FROM THEIR MOTHER, BUT BEING DISEASE-FREE, NONE CAN BE ff. EACH MUST HAVE AN F, SO THEY ARE ALL Ff. THIS F CAN COME ONLY FROM THEIR FATHER, WHO COULD BE EITHER FF OR Ff.
F?
ff
Ff
Ff
Ff
Ff
Ff
F?
F?
ff
NOW WORK FROM THE BOTTOM. THE GRANDSON WITH C.F. IS ff, SO HE MUST HAVE INHERITED AN f FROM EACH PARENT. HIS FATHER, NOT HAVING THE DISEASE, MUST THEN BE Ff.
ALTHOUGH A FEW ENTRIES HAVE QUESTION MARKS, THERE ARE NO CONTRADICTIONS, AND THE PATTERN AGREES WITH THE ASSUMPTION THAT C.F. IS A RECESSIVE TRAIT.
SOMEONE WITH THE Ff GENOTYPE IS CALLED A "CARRIER." THE PERSON HAS NO SYMPTOMS BUT MAY HAVE KIDS WITH C.F. IF MATED WITH ANOTHER CARRIER.
I NEVER EVEN KNEW IT WAS THERE!

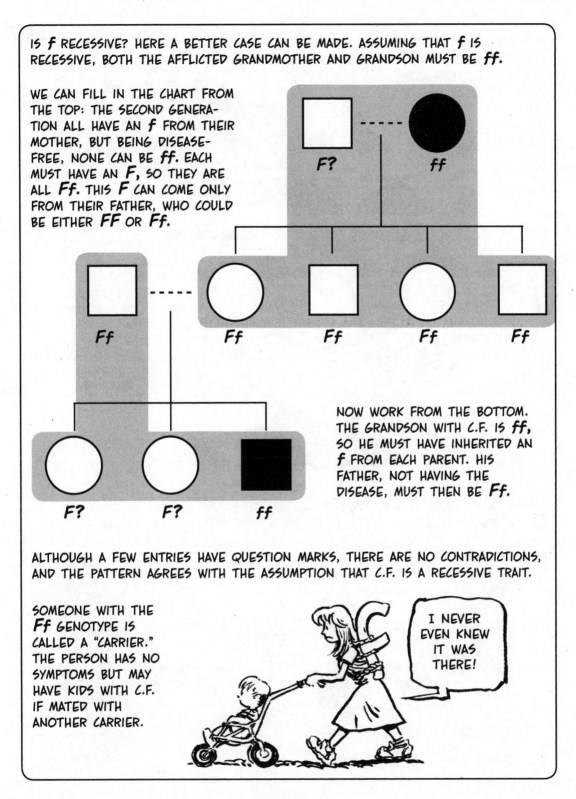

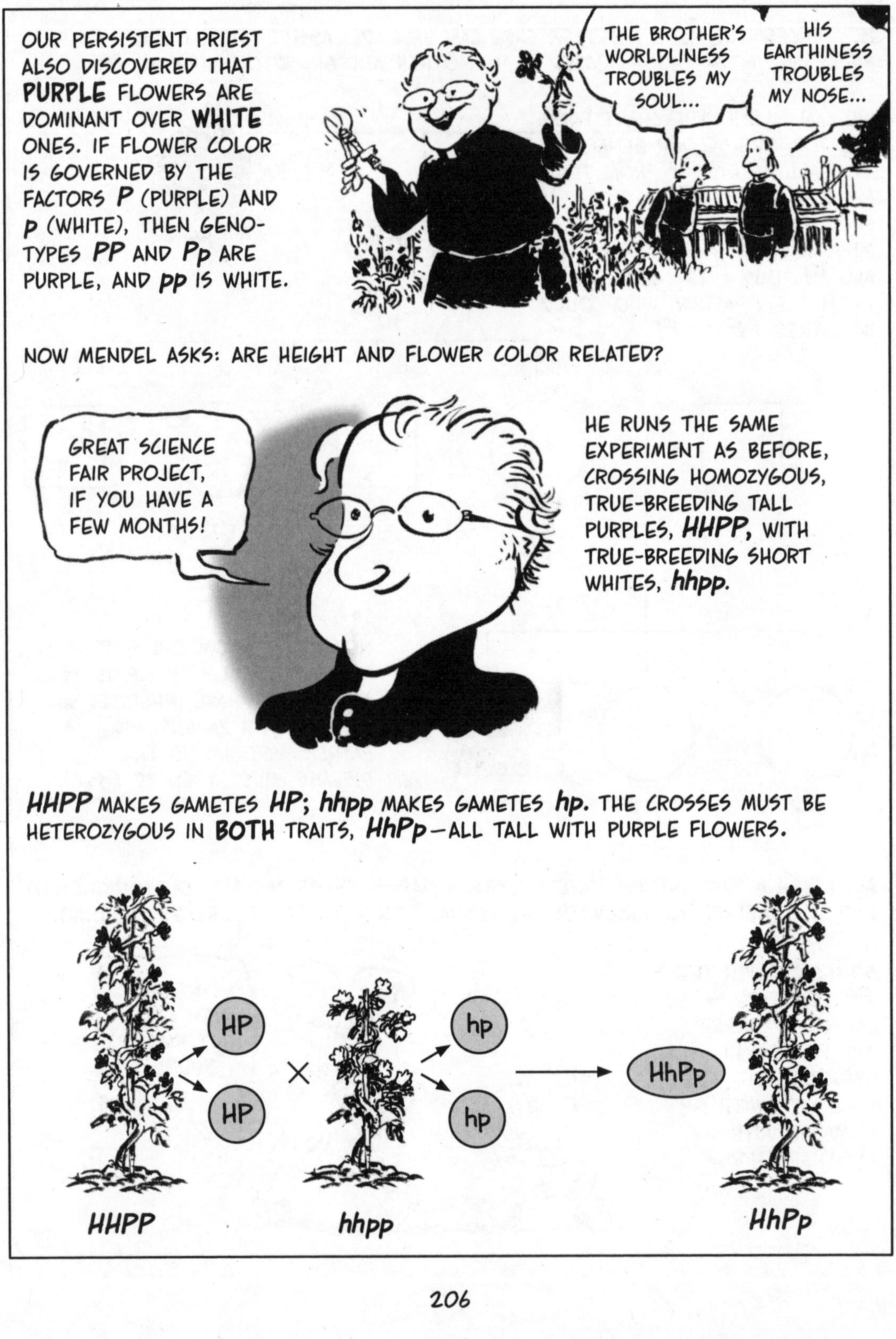
OUR PERSISTENT PRIEST ALSO DISCOVERED THAT PURPLE FLOWERS ARE DOMINANT OVER WHITE ONES. IF FLOWER COLOR IS GOVERNED BY THE FACTORS P (PURPLE) AND p (WHITE), THEN GENOTYPES PP AND Pp ARE PURPLE, AND pp IS WHITE.
THE BROTHER'S WORLDLINESS TROUBLES MY SOUL...
HIS EARTHINESS TROUBLES MY NOSE...
NOW MENDEL ASKS: ARE HEIGHT AND FLOWER COLOR RELATED?
GREAT SCIENCE FAIR PROJECT, IF YOU HAVE A FEW MONTHS!
HE RUNS THE SAME EXPERIMENT AS BEFORE, CROSSING HOMOZYGOUS, TRUE-BREEDING TALL PURPLES, HHPP, WITH TRUE-BREEDING SHORT WHITES, hhpp.
HHPP MAKES GAMETES HP; hhpp MAKES GAMETES hp. THE CROSSES MUST BE HETEROZYGOUS IN BOTH TRAITS, HhPp—ALL TALL WITH PURPLE FLOWERS.
HP
HP
×
hp
hp
HhPp
HHPP
hhpp
HhPp

CROSSING THE HYBRIDS, **HhPp x HhPp,** MENDEL GOT THESE OFFSPRING: ONE SHORT WHITE FOR EACH **3** SHORT PURPLES, **3** TALL WHITES, AND **9** TALL PURPLES.

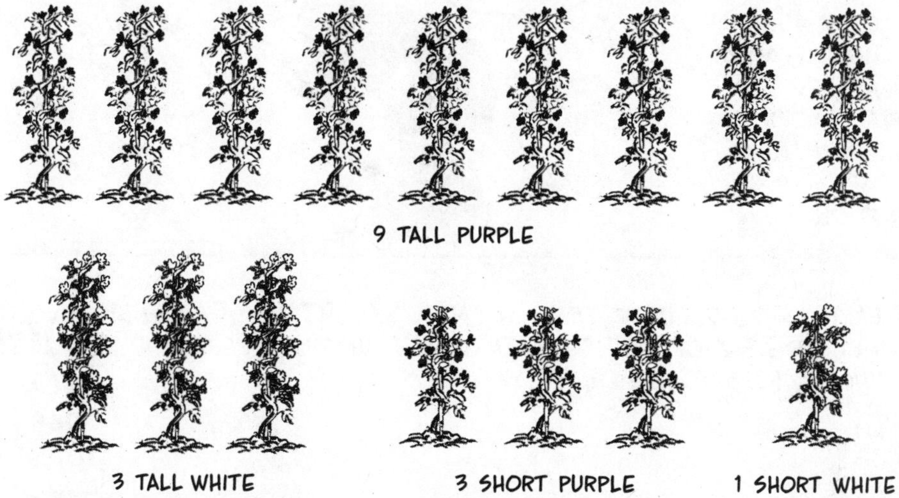

WHY? BECAUSE ***HhPp*** PARENTS MAKE GAMETES ***HP, Hp, hP,*** AND ***hp*** IN **EQUAL AMOUNTS.** COMBINING THEM RANDOMLY GENERATES **16** EQUALLY LIKELY OUTCOMES.

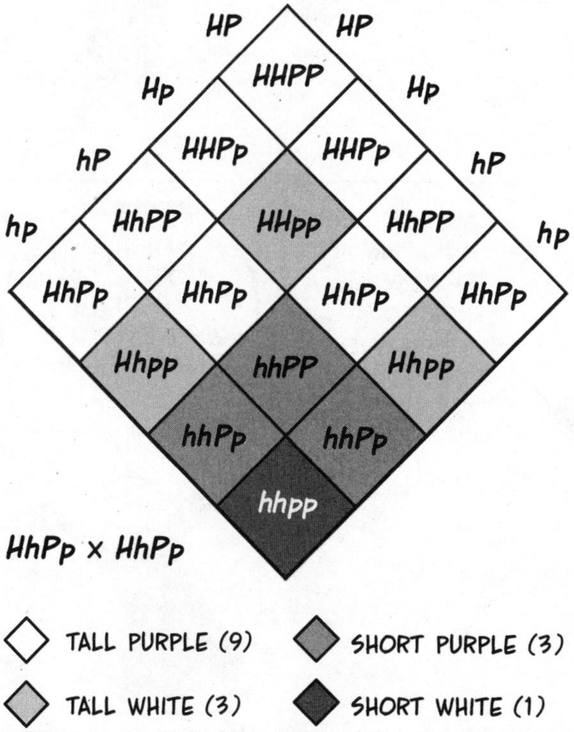

MENDEL CALLED THIS THE **PRINCIPLE OF INDEPENDENT ASSORTMENT:** A GAMETE GETS ***H*** OR ***h*** **INDEPENDENTLY** OF WHETHER IT GETS ***P*** OR ***p.*** THE TWO TRAITS HAVE **NO EFFECT** ON EACH OTHER'S INHERITANCE.

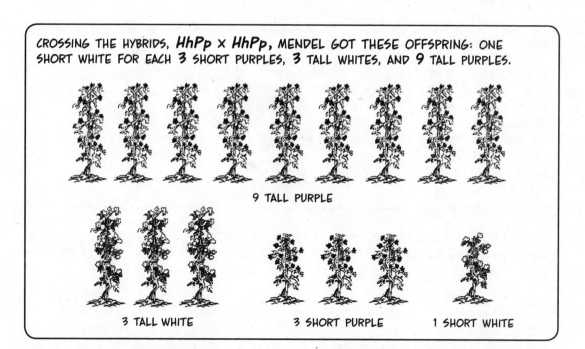

YOUR SECOND LAW!
DO YOU THINK YOU COULD CALL IT A "COMMANDMENT"?

MENDEL'S PEA-BREEDING WAS A TRIUMPH OF OUTSIDE-IN BIOLOGY. WORKING WITH WHOLE PLANTS, HE UNEARTHED INVISIBLE "FACTORS" OF HEREDITY AND THEIR PATTERNS OF COMBINATION.
I KNOW YOU'RE IN THERE!

THESE FACTORS, HE DEDUCED, COME IN PAIRS, **EXCEPT IN SEX CELLS.** GAMETES HAVE ONLY **ONE** FACTOR APIECE. FERTILIZATION, UNITING SPERM AND EGG, PUTS THEIR TWO FACTORS TOGETHER.

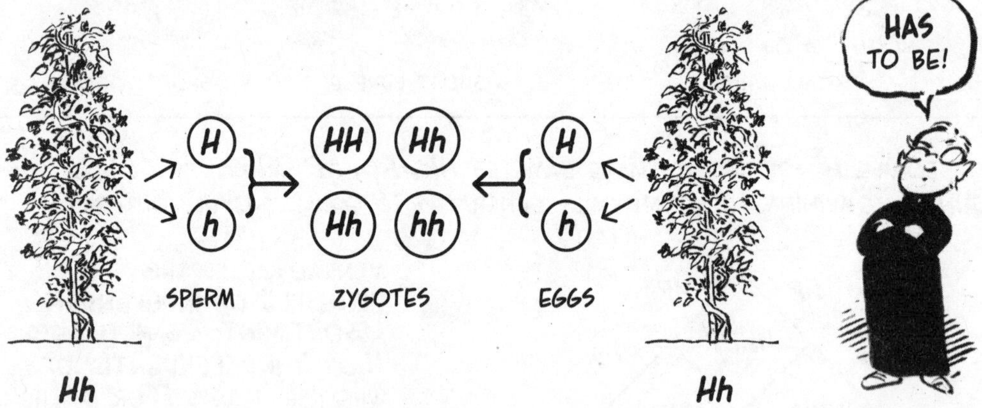

TRULY, A TRIUMPH—EXCEPT THAT MENDEL HAD NO IDEA WHAT THESE "FACTORS" MIGHT ACTUALLY BE.
EASY FOR YOU TO SAY!
CAN OUR INSIDE-OUT APPROACH TO BIOLOGY SHED SOME LIGHT ON THIS QUESTION? OF COURSE IT CAN.
MANNNN... ANYONE CAN DO BETTER WITH 150 MORE YEARS UNDER THEIR BELT!

THE HEREDITARY PARTICLES ARE **GENES,** MADE OF DNA. GENES ENCODE PROTEINS THAT GIVE ORGANISMS THEIR SPECIFIC FEATURES (THEIR PHENOTYPE).

GENES

I DID NOT KNOW THAT.

PROTEINS

PHENOTYPIC TRAITS

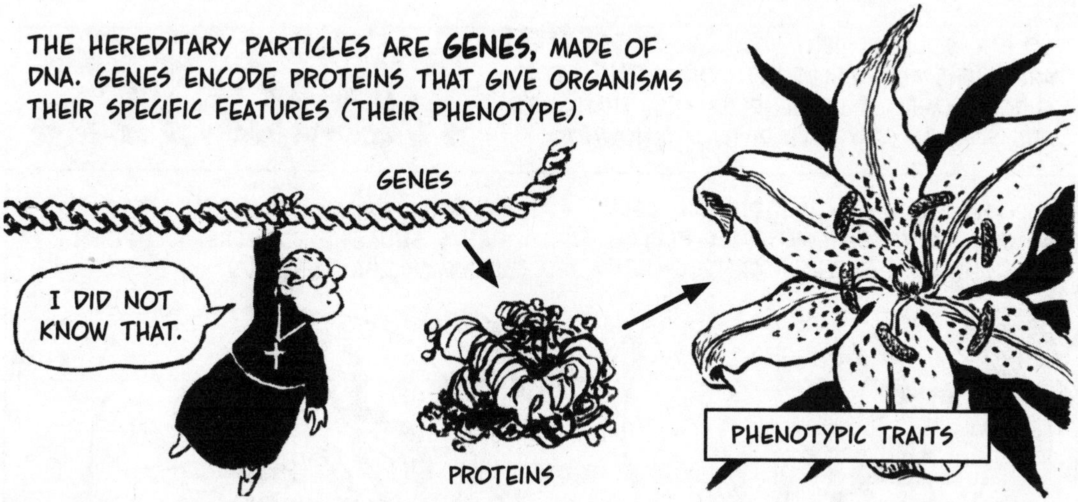

IN A EUKARYOTIC CELL NUCLEUS, RECALL, DNA IS RANGED ACROSS SEVERAL **CHROMO- SOMES.** IT SO HAPPENS THAT PEAS, PEOPLE, AND MOST OTHER EUKARYOTES HAVE A **DOUBLE SET** OF CHROMOSOMES (SEE P. 128). A PEA PLANT CELL HAS 14 OF THEM, TWO SETS OF SEVEN. A CELL WITH TWO SETS IS CALLED **DIPLOID.**

I DID NOT KNOW THAT.

TWO CORRESPONDING CHROMO- SOMES FROM DIFFERENT SETS ARE CALLED A **HOMOLOGOUS PAIR.** THEY HAVE EXACTLY THE SAME GENES IN THE SAME ORDER AND SPACING.

I DID NOT KNOW THAT.

THAT ONLY MAKES YOU MORE IM- PRESSIVE, GREGGO!

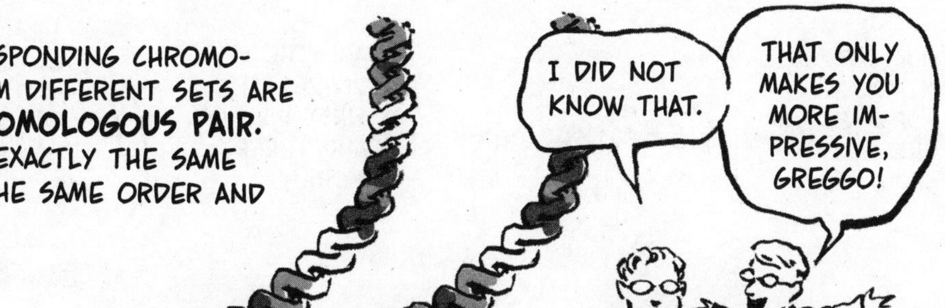

SO FAR, SO GOOD—BUT WHAT ABOUT **GAMETES?** ARE THERE ALSO CELLS WITH ONLY **ONE** SET OF CHROMOSOMES, À LA MENDEL? **YES,** THANKS TO THE SPECIAL, SEXY CELL DIVISION KNOWN AS

MEIOSIS.

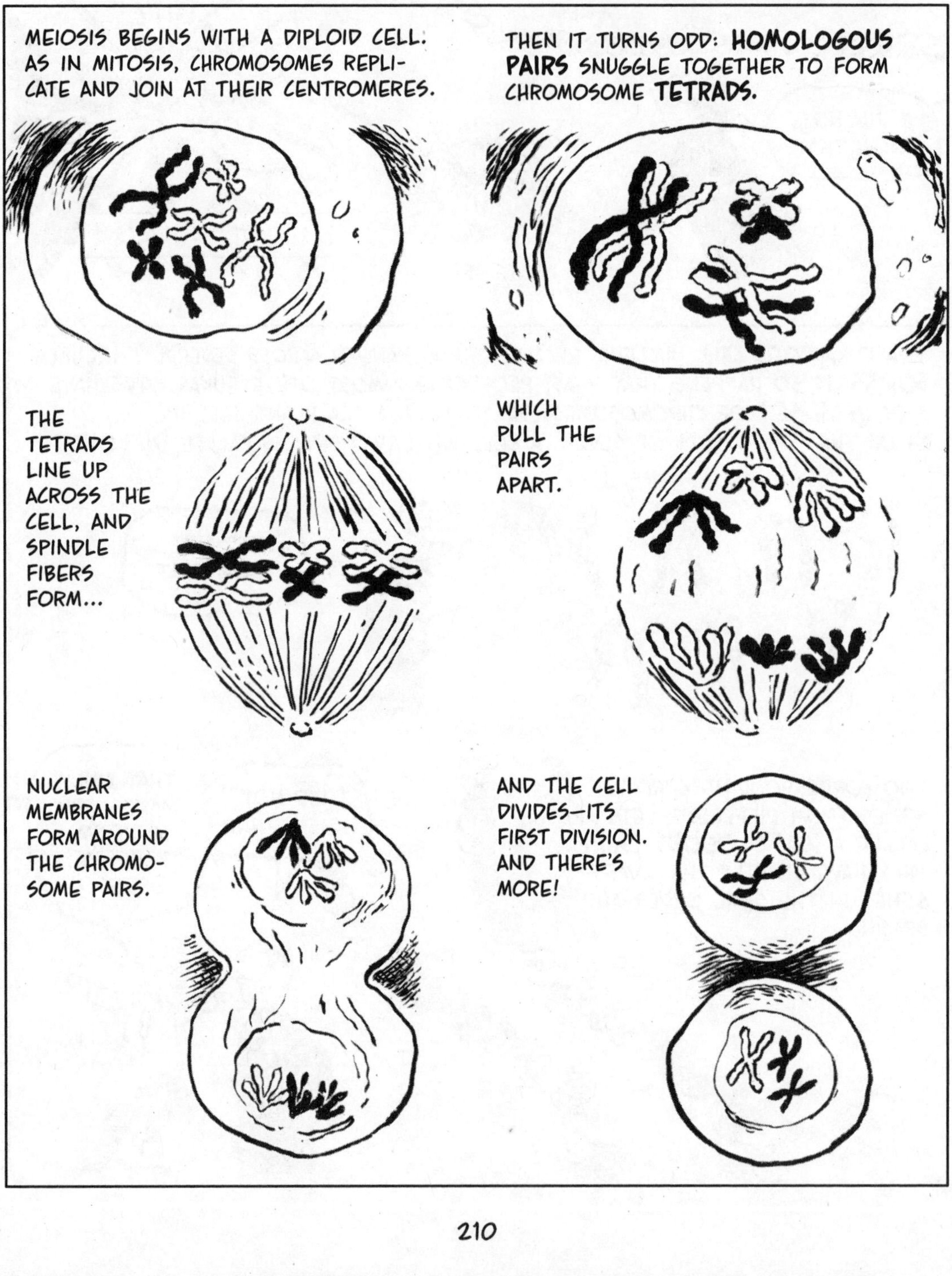
MEIOSIS BEGINS WITH A DIPLOID CELL. AS IN MITOSIS, CHROMOSOMES REPLICATE AND JOIN AT THEIR CENTROMERES.
THEN IT TURNS ODD: HOMOLOGOUS PAIRS SNUGGLE TOGETHER TO FORM CHROMOSOME TETRADS.
THE TETRADS LINE UP ACROSS THE CELL, AND SPINDLE FIBERS FORM...
WHICH PULL THE PAIRS APART.
NUCLEAR MEMBRANES FORM AROUND THE CHROMOSOME PAIRS.
AND THE CELL DIVIDES—ITS FIRST DIVISION. AND THERE'S MORE!

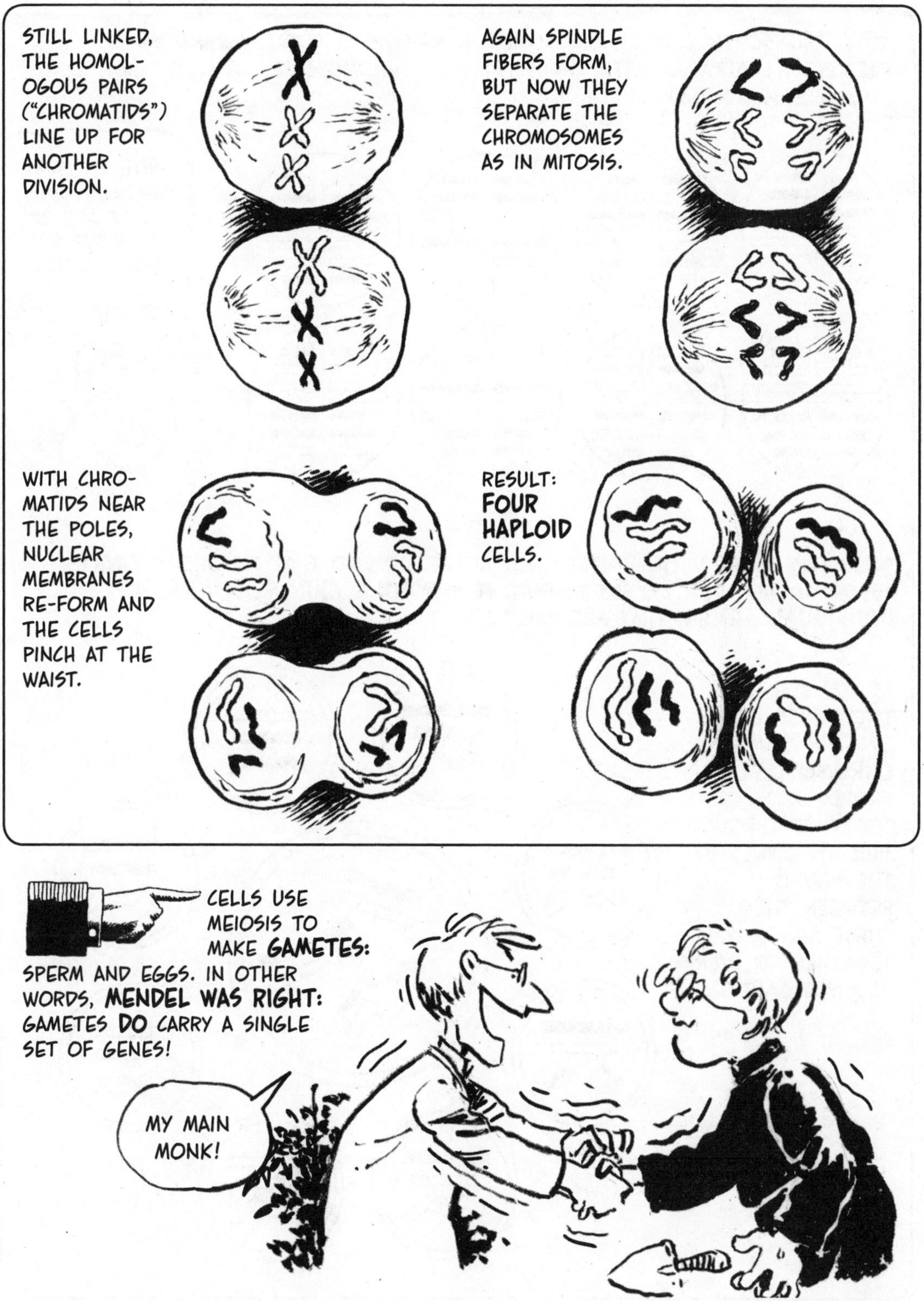
STILL LINKED, THE HOMOLOGOUS PAIRS ("CHROMATIDS") LINE UP FOR ANOTHER DIVISION.
AGAIN SPINDLE FIBERS FORM, BUT NOW THEY SEPARATE THE CHROMOSOMES AS IN MITOSIS.
WITH CHROMATIDS NEAR THE POLES, NUCLEAR MEMBRANES RE-FORM AND THE CELLS PINCH AT THE WAIST.
RESULT: FOUR HAPLOID CELLS.
CELLS USE MEIOSIS TO MAKE GAMETES: SPERM AND EGGS. IN OTHER WORDS, MENDEL WAS RIGHT: GAMETES DO CARRY A SINGLE SET OF GENES!
MY MAIN MONK!

NOTE: THE SORTING OF CHROMOSOMES IN MEIOSIS IS TOTALLY **RANDOM.** IN THE CELL ILLUSTRATED, WITH THREE PAIRS OF CHROMOSOMES, EACH OF EIGHT OUTCOMES IS EQUALLY LIKELY.

WITH 23 PAIRS (AS IN HUMANS), THERE ARE 2^{23} OR 8,388,608 POSSIBILITIES!

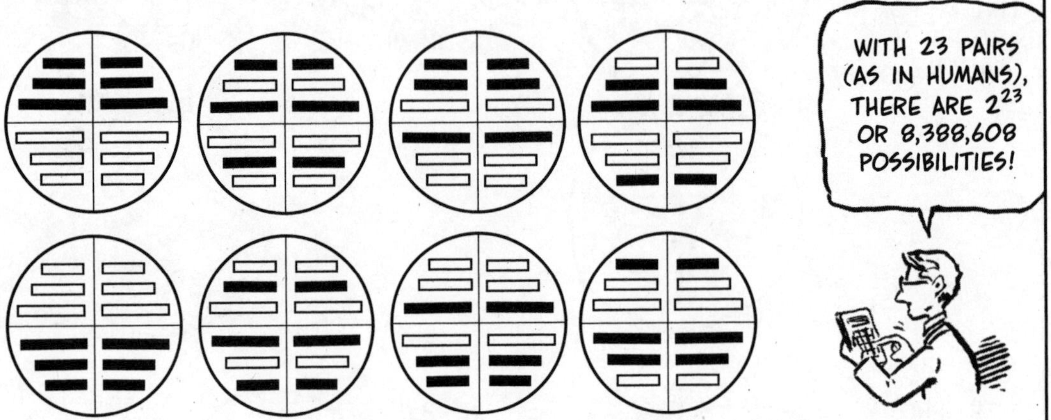

DOES THIS AGREE WITH MENDEL? NOT QUITE. HE SAID THAT DIFFERENT **TRAITS** ASSORT INDEPENDENTLY, BUT IN FACT, IT IS **WHOLE CHROMOSOMES, NOT INDIVIDUAL GENES,** THAT ARE SHUFFLED.

AS YOU SEE HERE, TWO GENES **LYING ON THE SAME CHROMOSOME** SHOULD STAY TOGETHER THROUGH MEIOSIS. JOINED BY THE STRAND BETWEEN THEM, THESE GENES GO TOGETHER TO EVERY POSSIBLE GAMETE.

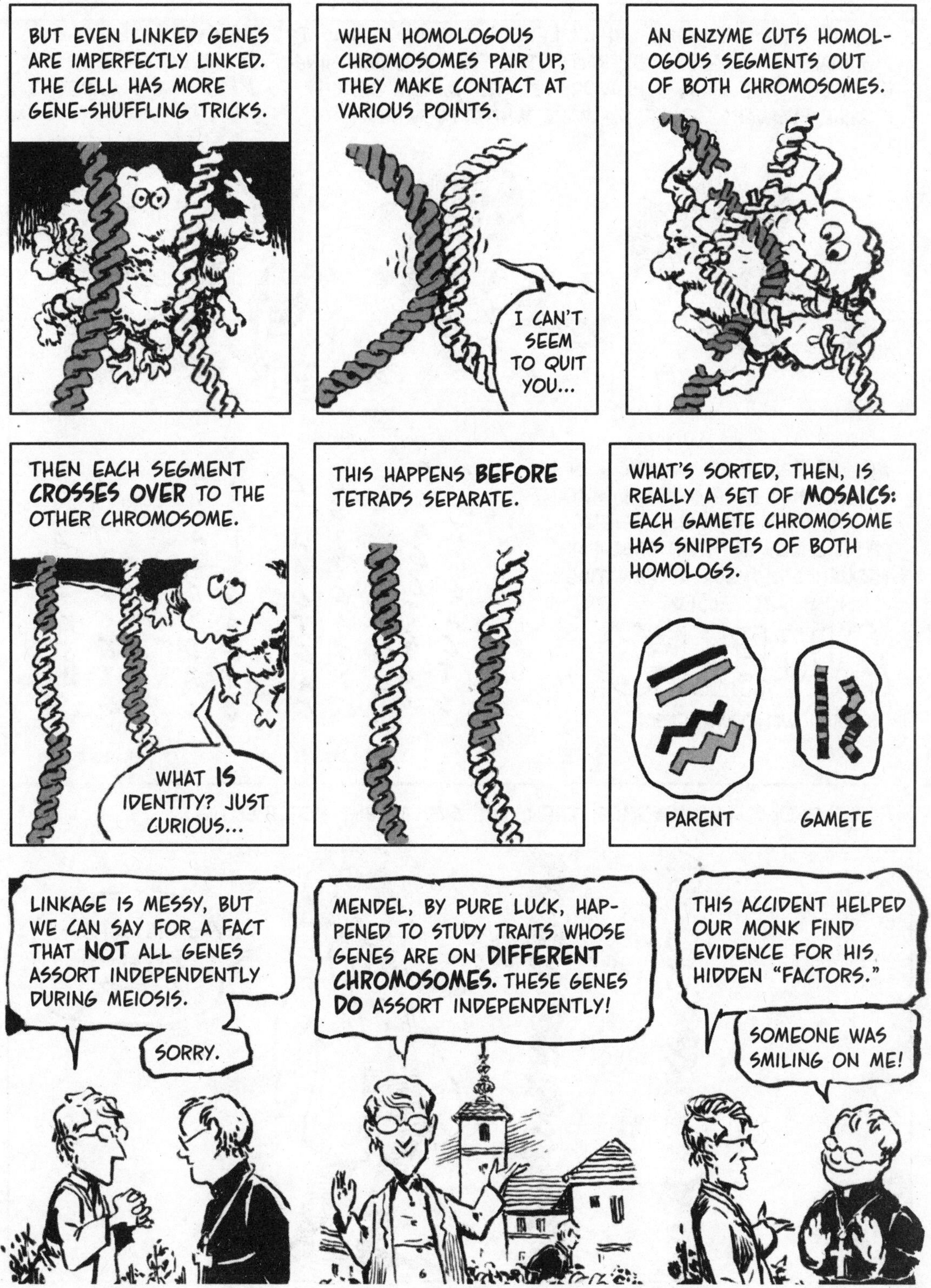
BUT EVEN LINKED GENES ARE IMPERFECTLY LINKED. THE CELL HAS MORE GENE-SHUFFLING TRICKS.
WHEN HOMOLOGOUS CHROMOSOMES PAIR UP, THEY MAKE CONTACT AT VARIOUS POINTS.
I CAN'T SEEM TO QUIT YOU...
AN ENZYME CUTS HOMOLOGOUS SEGMENTS OUT OF BOTH CHROMOSOMES.
THEN EACH SEGMENT **CROSSES OVER** TO THE OTHER CHROMOSOME.
WHAT **IS** IDENTITY? JUST CURIOUS...
THIS HAPPENS **BEFORE** TETRADS SEPARATE.
WHAT'S SORTED, THEN, IS REALLY A SET OF **MOSAICS**: EACH GAMETE CHROMOSOME HAS SNIPPETS OF BOTH HOMOLOGS.
PARENT
GAMETE
LINKAGE IS MESSY, BUT WE CAN SAY FOR A FACT THAT **NOT** ALL GENES ASSORT INDEPENDENTLY DURING MEIOSIS.
SORRY.
MENDEL, BY PURE LUCK, HAPPENED TO STUDY TRAITS WHOSE GENES ARE ON **DIFFERENT CHROMOSOMES.** THESE GENES **DO** ASSORT INDEPENDENTLY!
THIS ACCIDENT HELPED OUR MONK FIND EVIDENCE FOR HIS HIDDEN "FACTORS."
SOMEONE WAS SMILING ON ME!

MENDEL'S FIRST FINDING, WHICH LED TO THE REST, WAS THE DISCOVERY OF DOMINANT AND RECESSIVE TRAITS. THE FACTOR GOVERNING FLOWER COLOR, FOR EXAMPLE, CAME IN TWO FLAVORS, *P* AND *p*. THE GENOTYPES *PP* AND *Pp* MADE PURPLE FLOWERS, WHILE *pp* MADE WHITE FLOWERS.

PP Pp PP

WE SEE THE SAME VARIATION IN DNA. A GENE MAY HAVE SEVERAL MUTANT FORMS, WITH (USUALLY) SLIGHT DIFFERENCES IN THEIR BASE-PAIR SEQUENCES. THESE ALTERNATIVE VERSIONS ARE CALLED

ALLELES

(RHYMES WITH "WHEELS").

```
G C    G C    G C    G C
G C    G C    G C    A T
T C    T C    T C    T A
C A    G A    G A    T A
G G    A C    A C    A T
A C    C T    C T    C G
C T    T G    T G    G A
T G    G A    G A    T C
G C    C    A C    G
```

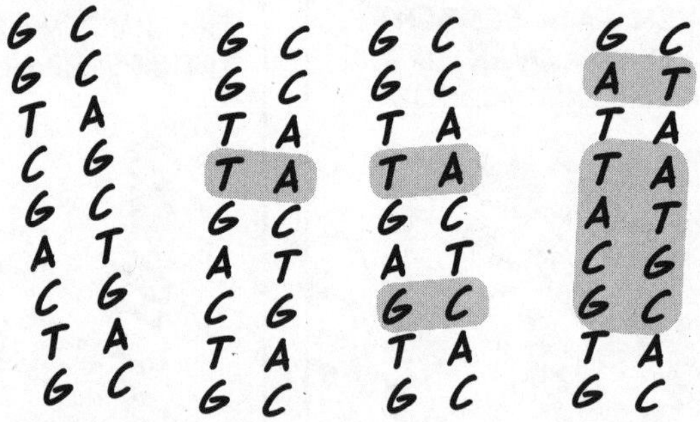

HOMOLOGOUS CHROMOSOMES, THEN, ARE SIMILAR BUT NOT IDENTICAL.

SAME GENES, DIFFERENT ALLELES!!!

HOW CAN ONE ALLELE BE DOMINANT OVER ANOTHER? HERE IS ONE WAY:

SUPPOSE ALLELE ***P*** ENCODES A PROTEIN INVOLVED IN MAKING PURPLE PIGMENT.

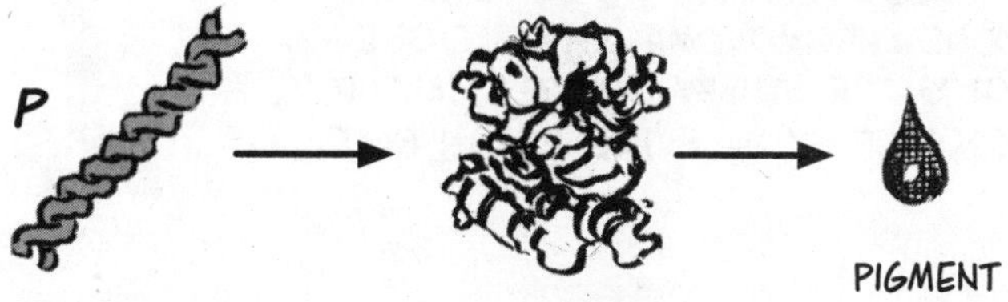

SUPPOSE ALLELE ***p*** HAS A MUTATION THAT DISABLES ITS PROTEIN PRODUCT FROM MAKING THAT PIGMENT.

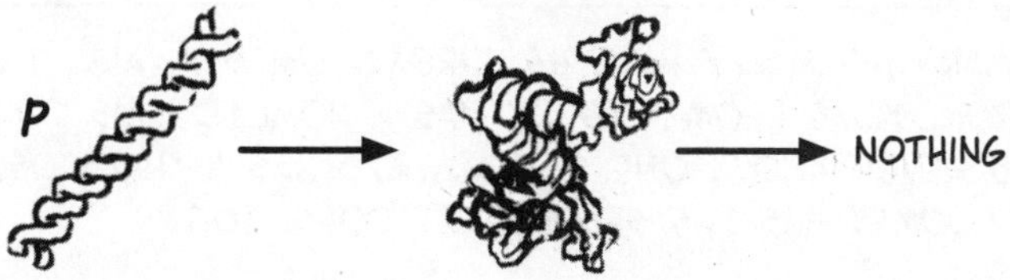

THEN ANY PEA PLANT WITH ALLELE ***P*** WILL MAKE PIGMENT, AND ITS FLOWERS WILL BE PURPLE. ONLY THE DOUBLE RECESSIVE ***pp*** MAKES NO PIGMENT AND SO HAS WHITE FLOWERS.

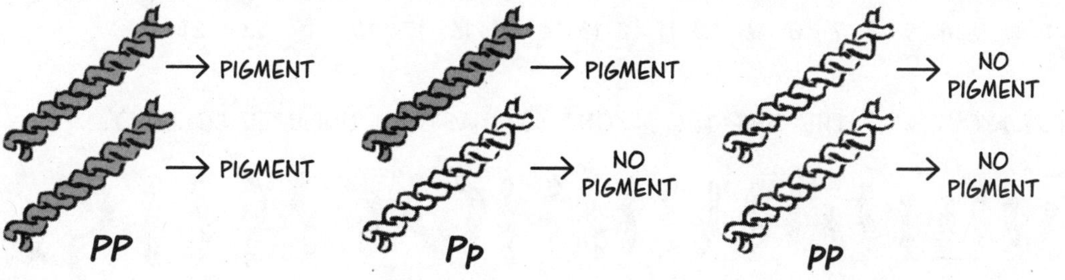

GARDENERS BREED DOUBLE RECESSIVES FOR FUN AND PROFIT. HUMANS SEEM TO LIKE MUTANTS, OR SOME MUTANTS, ANYWAY.

WHILE MOST CHROMOSOMES HAVE A HOMOLOGOUS MATE, WE SOMETIMES FIND A LONE CHROMOSOME WITHOUT ONE. SUCH IS THE STUBBY ODDBALL CALLED Y, POSSESSED BY HALF THE HUMAN RACE.

HUMANS NORMALLY HAVE 46 CHROMOSOMES IN ALL, TWO SETS OF 23. EVERY CHROMOSOME FROM 1 TO 22 HAS A HOMOLOGOUS TWIN, BUT THE 23RD, WELL, YOU NEVER KNOW. ONE SET **ALWAYS** HAS A CHROMOSOME CALLED **X** FOR #23, AND SOMETIMES THE SECOND SET DOES, TOO.

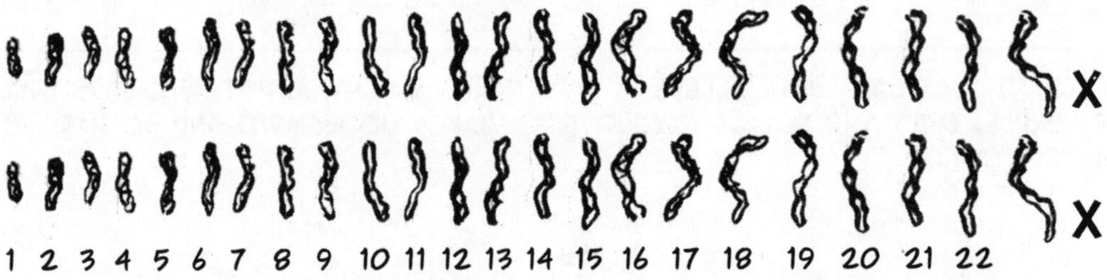

BUT ABOUT HALF THE TIME THE SECOND SET HAS THAT UNPAIRED LONER, Y.

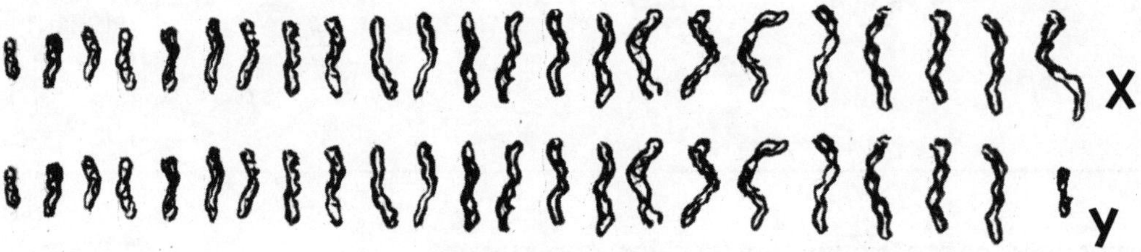

THERE'S A NAME FOR PEOPLE WITH THE **XX** GENOTYPE:

FEMALES.

PEOPLE WITH **XY** ARE CALLED

MALES.

YOU CAN PROBABLY THINK OF OTHER NAMES, TOO.

WITH THIS SYSTEM, EVERY EGG IS **X**, WHILE HALF OF SPERM ARE **X** AND HALF ARE **Y**.

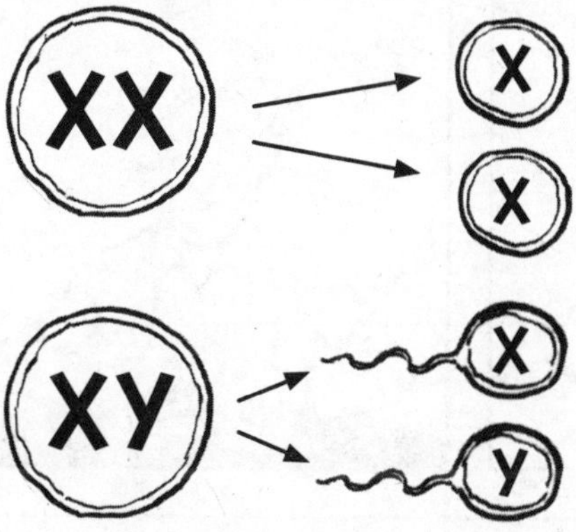

RESULT: ROUGHLY HALF OF ALL FERTILIZED EGGS ARE **XX**; HALF ARE **XY**. BOYS AND GIRLS ARE BORN IN EQUAL NUMBERS, MORE OR LESS.

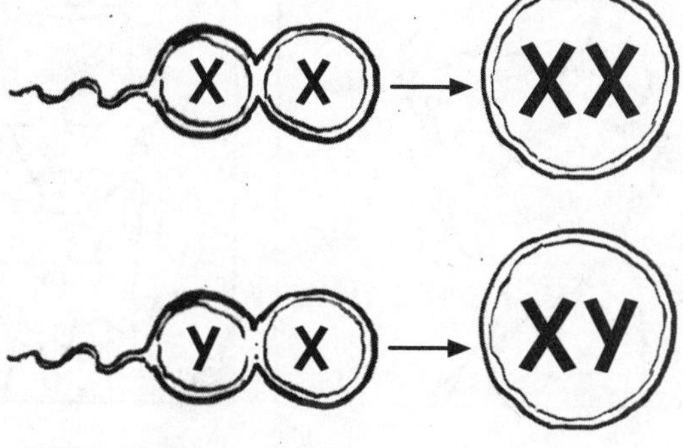

THAT DOUBLE **X**, BY THE WAY, GIVES AN ADVANTAGE TO WOMEN: A BACK-UP COPY OF THE **X** CHROMOSOME, WHICH CAN COVER FOR FLAWS IN THE OTHER COPY.

MALES, LACKING THE EXTRA **X**, ARE VULNERABLE TO MUTATIONS ON THE **X** CHROMOSOME; **HEMOPHILIA** AND **COLOR BLINDNESS** ARE TWO CONDITIONS CAUSED BY RECESSIVE ALLELES ON THE **X** CHROMOSOME.

NATURE HAS MANY WAYS TO CREATE SEX DIFFERENCES. THE *AGAMA AGAMA* LIZARD, FOR INSTANCE, HATCHES FEMALES FROM COOLER EGGS AND MALES FROM WARMER ONES. THE SEXES HAVE THE SAME GENES; TEMPERATURE AFFECTS THEIR EXPRESSION.

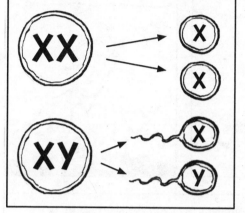

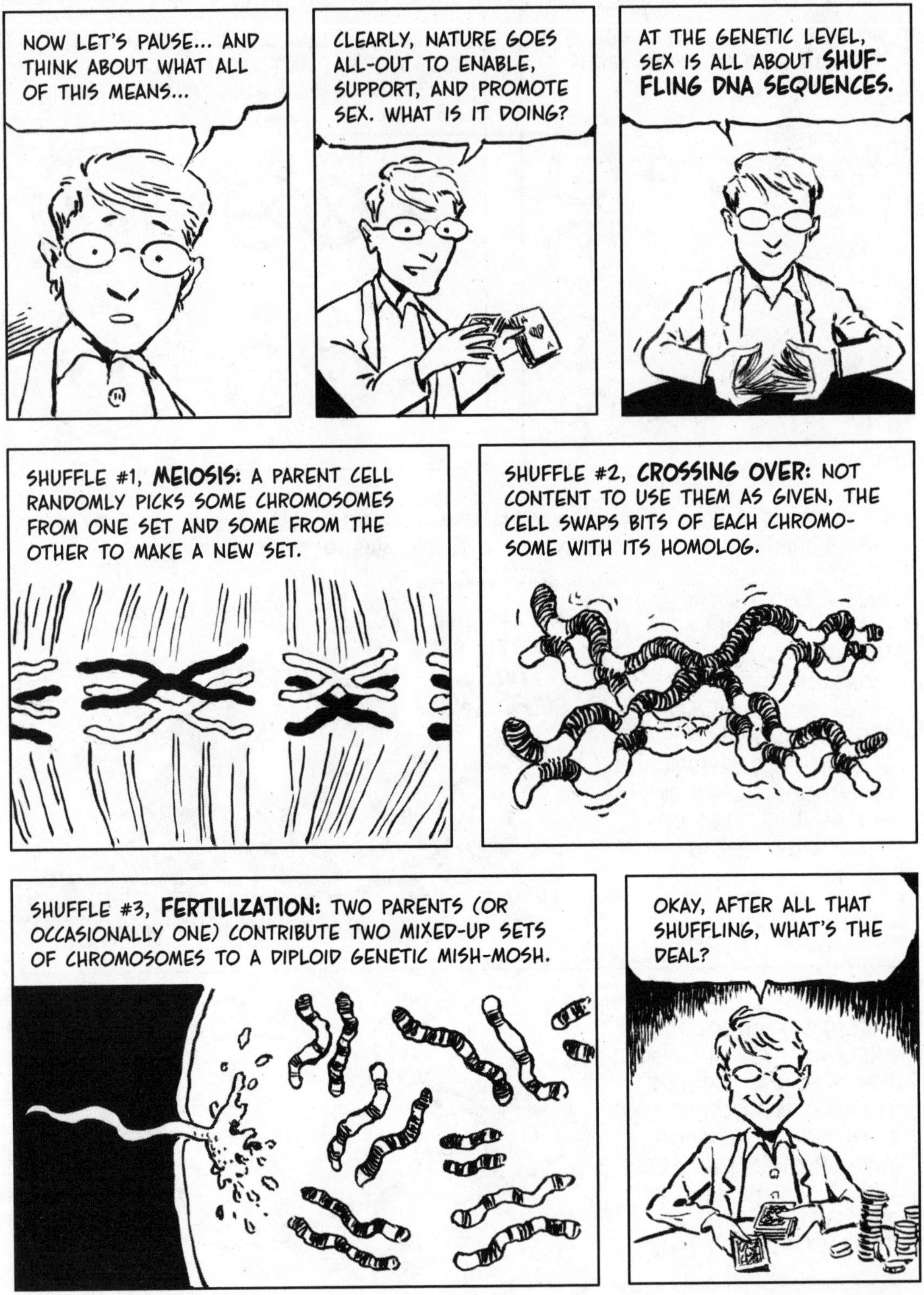
NOW LET'S PAUSE... AND THINK ABOUT WHAT ALL OF THIS MEANS...
CLEARLY, NATURE GOES ALL-OUT TO ENABLE, SUPPORT, AND PROMOTE SEX. WHAT IS IT DOING?
AT THE GENETIC LEVEL, SEX IS ALL ABOUT **SHUFFLING DNA SEQUENCES.**
SHUFFLE #1, **MEIOSIS:** A PARENT CELL RANDOMLY PICKS SOME CHROMOSOMES FROM ONE SET AND SOME FROM THE OTHER TO MAKE A NEW SET.
SHUFFLE #2, **CROSSING OVER:** NOT CONTENT TO USE THEM AS GIVEN, THE CELL SWAPS BITS OF EACH CHROMOSOME WITH ITS HOMOLOG.
SHUFFLE #3, **FERTILIZATION:** TWO PARENTS (OR OCCASIONALLY ONE) CONTRIBUTE TWO MIXED-UP SETS OF CHROMOSOMES TO A DIPLOID GENETIC MISH-MOSH.
OKAY, AFTER ALL THAT SHUFFLING, WHAT'S THE DEAL?

SEX SIMPLY **MUST** BE IMPORTANT. WE SEE SIGNS OF GENE SWAPPING AND SHARING THROUGHOUT THE LIVING WORLD, EVEN IN SOME DECIDEDLY UNSEXY CREATURES.

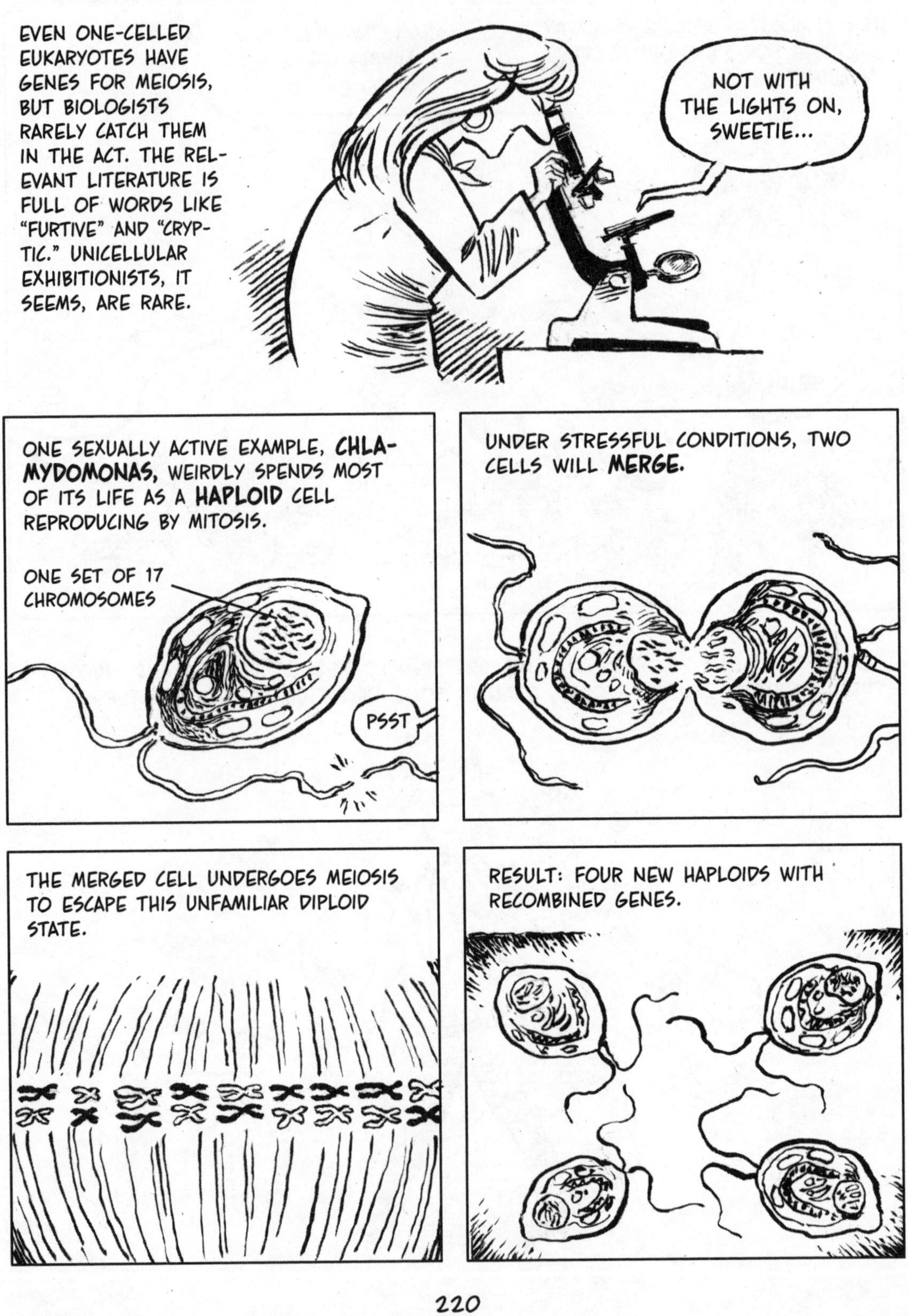
EVEN ONE-CELLED EUKARYOTES HAVE GENES FOR MEIOSIS, BUT BIOLOGISTS RARELY CATCH THEM IN THE ACT. THE RELEVANT LITERATURE IS FULL OF WORDS LIKE "FURTIVE" AND "CRYPTIC." UNICELLULAR EXHIBITIONISTS, IT SEEMS, ARE RARE.
NOT WITH THE LIGHTS ON, SWEETIE...
ONE SEXUALLY ACTIVE EXAMPLE, **CHLAMYDOMONAS,** WEIRDLY SPENDS MOST OF ITS LIFE AS A **HAPLOID** CELL REPRODUCING BY MITOSIS.
ONE SET OF 17 CHROMOSOMES
PSST
UNDER STRESSFUL CONDITIONS, TWO CELLS WILL **MERGE.**
THE MERGED CELL UNDERGOES MEIOSIS TO ESCAPE THIS UNFAMILIAR DIPLOID STATE.
RESULT: FOUR NEW HAPLOIDS WITH RECOMBINED GENES.

AND PROKARYOTES? NO PROKARYOTE DOES MEIOSIS; PROKARYOTES NEVER HAVE SEX; AND YET PROKARYOTES ALSO FIDDLE WITH THEIR GENES.

SOMETIMES A CELL SIMPLY GRABS STRAY BITS OF DNA FLOATING BY.

SOMETIMES A **VIRUS*** CARRIES DNA FROM ONE BACTERIUM TO ANOTHER.

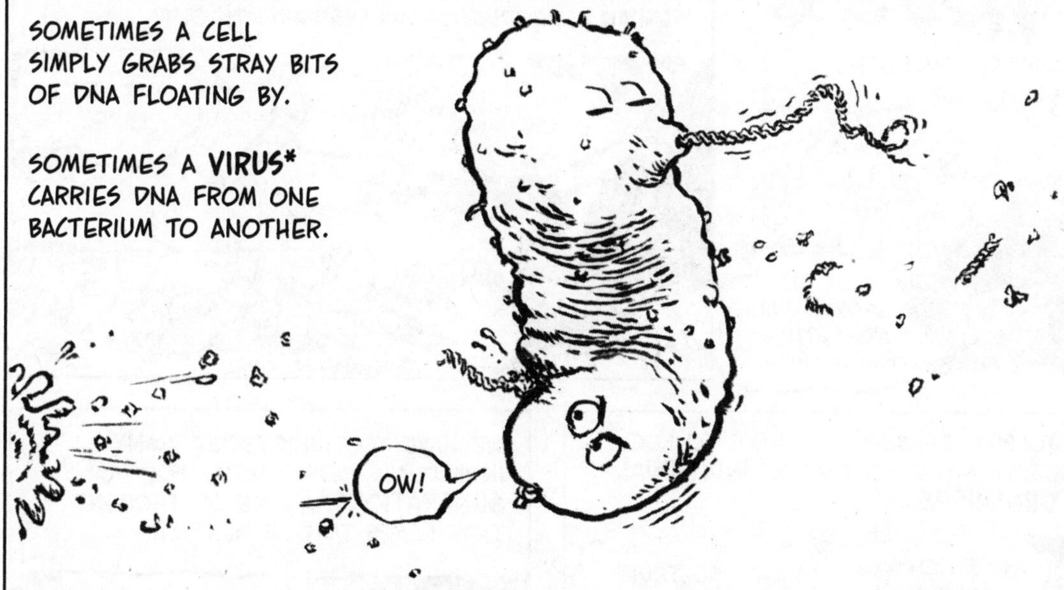

*A **VIRUS** IS A NONLIVING PACKET OF GENETIC MATERIAL WRAPPED IN A PROTEIN COAT. WITHOUT A METABOLISM OF ITS OWN, A VIRUS CAN INFECT A HOST CELL AND HIJACK ITS CHEMISTRY TO PRODUCE MORE VIRUSES, WHICH BURST THE HOST AND GO ON TO INFECT NEW HOSTS.

AND SOMETIMES, TWO PROKARYOTES ENGAGE IN **CONJUGATION:** THEY SNUGGLE UP, BUILD A TUNNEL BETWEEN THEMSELVES, AND PASS DNA BACK AND FORTH.

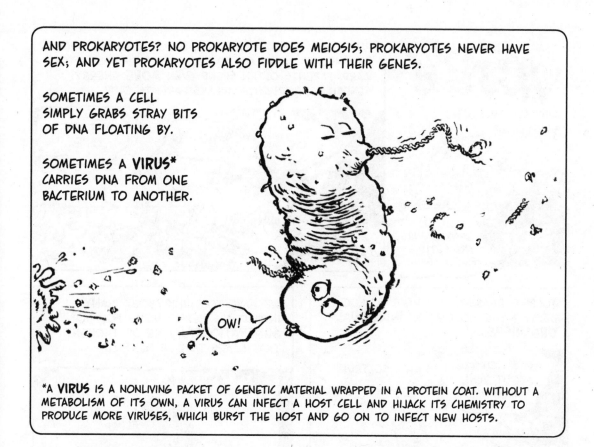

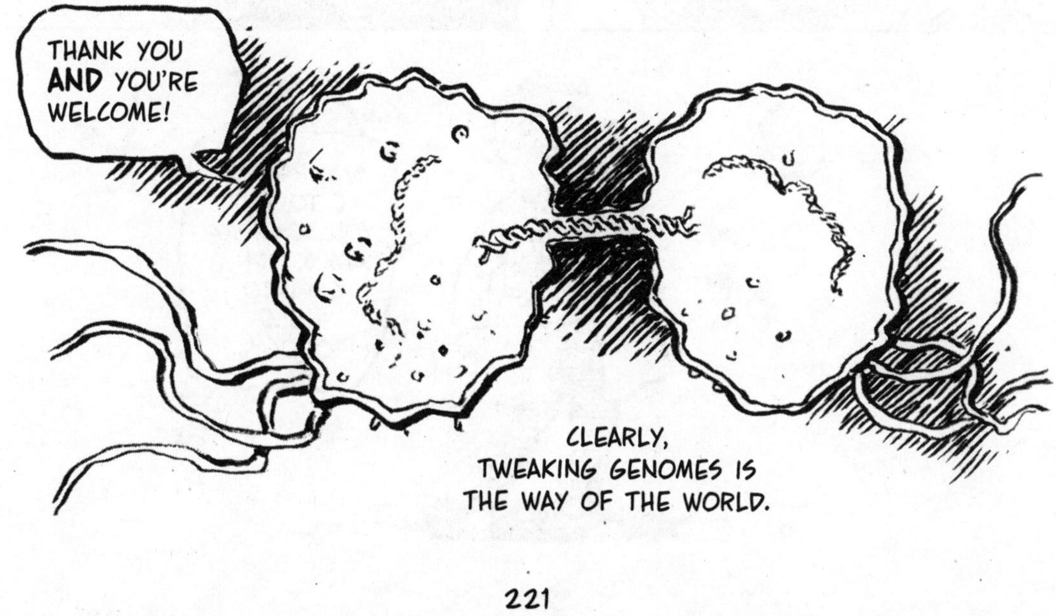

THANK YOU **AND** YOU'RE WELCOME!
CLEARLY, TWEAKING GENOMES IS THE WAY OF THE WORLD.

AGAIN,
WHY?
WHY ALL THIS GENE-SHUFFLING?
RUN! THE BRONTOSAURS ARE GETTING FRISKY AGAIN!
PARENTS, WHO BEAR THE COST, GET NO BENEFIT. THEIR RECOMBINED GENES GO TO THE OFFSPRING, AND AFTERWARD, PARENTS OFTEN SPEND EVEN MORE ENERGY. SEXUAL REPRODUCTION WEARS PARENTS OUT!
AND YET I CARE...
TO FIND REASONS, WE HAVE TO LOOK BEYOND THE BIOLOGY OF INDIVIDUAL ORGANISMS...
IT'S NOT ALL ABOUT YOU, MAAHHM!
SIGH... SO TRUE...
WE HAVE TO THINK ABOUT MANY INDIVIDUALS REPRODUCING FOR MANY GENERATIONS, A LINE OF THOUGHT THAT LEADS TO OUR NEXT CH—
PLURT
PLURP
SAY, REMIND ME TO TELL YOU SOMEDAY WHY I DIDN'T GO INTO FIELD BIOLOGY...

Chapter 14
EVOLUTION

EVERY SEXUAL BEING HAS TO ASK: SEX WITH WHOM? OR RATHER, **REPRODUCE** WITH WHOM? DESPITE THEIR CHEMICAL AND CELLULAR SIMILARITIES, A MOUSE CAN NO MORE MATE WITH A MUSK OX THAN THE MAN IN THE MOON.

I REMAIN OPTIMISTIC.

IF TWO INDIVIDUALS **CAN** MATE AND PRODUCE FERTILE OFFSPRING, WE SAY THEY BELONG TO THE SAME

SPECIES.

MOUSE MATES WITH MOUSE, MUSK OX WITH MUSK OX. A SPECIES IS AN INTERBREEDING POPULATION OF INDIVIDUALS.

NO ONE KNOWS HOW MANY SPECIES LIVE ON EARTH; ONE ESTIMATE TOPS 8 MILLION, AND THAT'S JUST EUKARYOTES.* THEY COME IN A CRAZY VARIETY OF SHAPES, SIZES, COLORS, LIFE SPANS, HABITS, AND DIETARY PREFERENCES.
HOW DID THIS RIOT OF LIFE COME TO BE? WHAT GENERATED AND SHAPED SUCH A PROFUSION OF FORMS FROM THE SAME BASIC MATERIALS AND CHEMISTRY? THOSE ARE THE QUESTIONS WE ADDRESS IN THIS CHAPTER.

*WHAT'S A PROKARYOTIC SPECIES? GOOD QUESTION—AND NOT ALWAYS EASY TO ANSWER!

FOR MANY YEARS, SCIENTISTS LEFT THESE QUESTIONS ALONE; IT WAS EASIER TO ACCEPT THE SCRIPTURAL VIEW THAT LIFE HAS ALWAYS BEEN THE SAME, SINCE THE BEGINNING—OR AT LEAST SINCE THE FLOOD.

STILL, THERE WERE REASONS FOR DOUBT. **FOSSILS,** ANCIENT BODY PARTS SET IN STONE LONG AGO, SHOWED PLANTS AND ANIMALS LIKE NOTHING ALIVE. SOME SPECIES, IT SEEMED, HAD DIED OUT.

IT WAS DROWNED IN THE DELUGE.
FISH CAN DROWN?
Xiphactinus

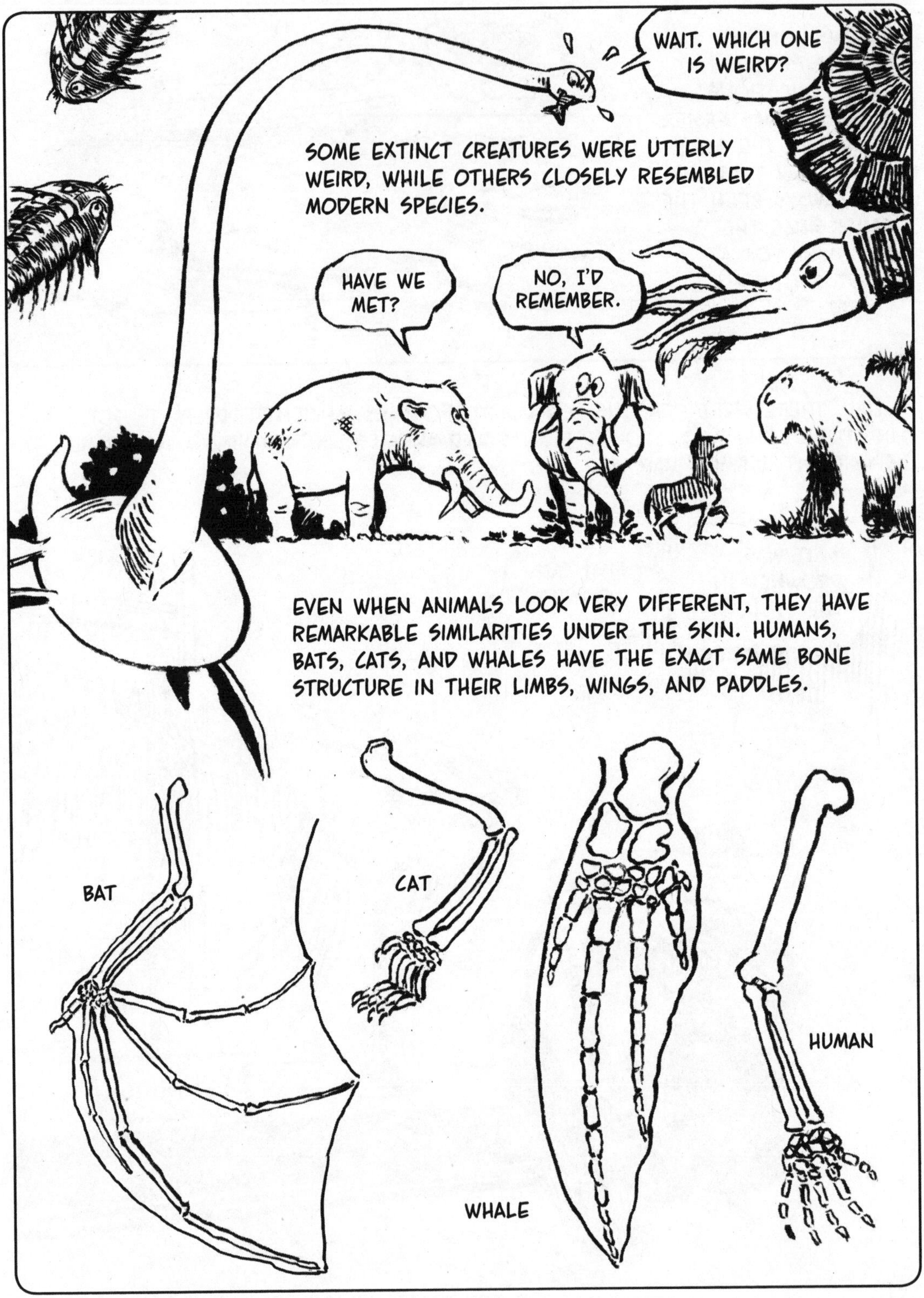
WAIT. WHICH ONE IS WEIRD?
SOME EXTINCT CREATURES WERE UTTERLY WEIRD, WHILE OTHERS CLOSELY RESEMBLED MODERN SPECIES.
HAVE WE MET?
NO, I'D REMEMBER.
EVEN WHEN ANIMALS LOOK VERY DIFFERENT, THEY HAVE REMARKABLE SIMILARITIES UNDER THE SKIN. HUMANS, BATS, CATS, AND WHALES HAVE THE EXACT SAME BONE STRUCTURE IN THEIR LIMBS, WINGS, AND PADDLES.
BAT
CAT
WHALE
HUMAN

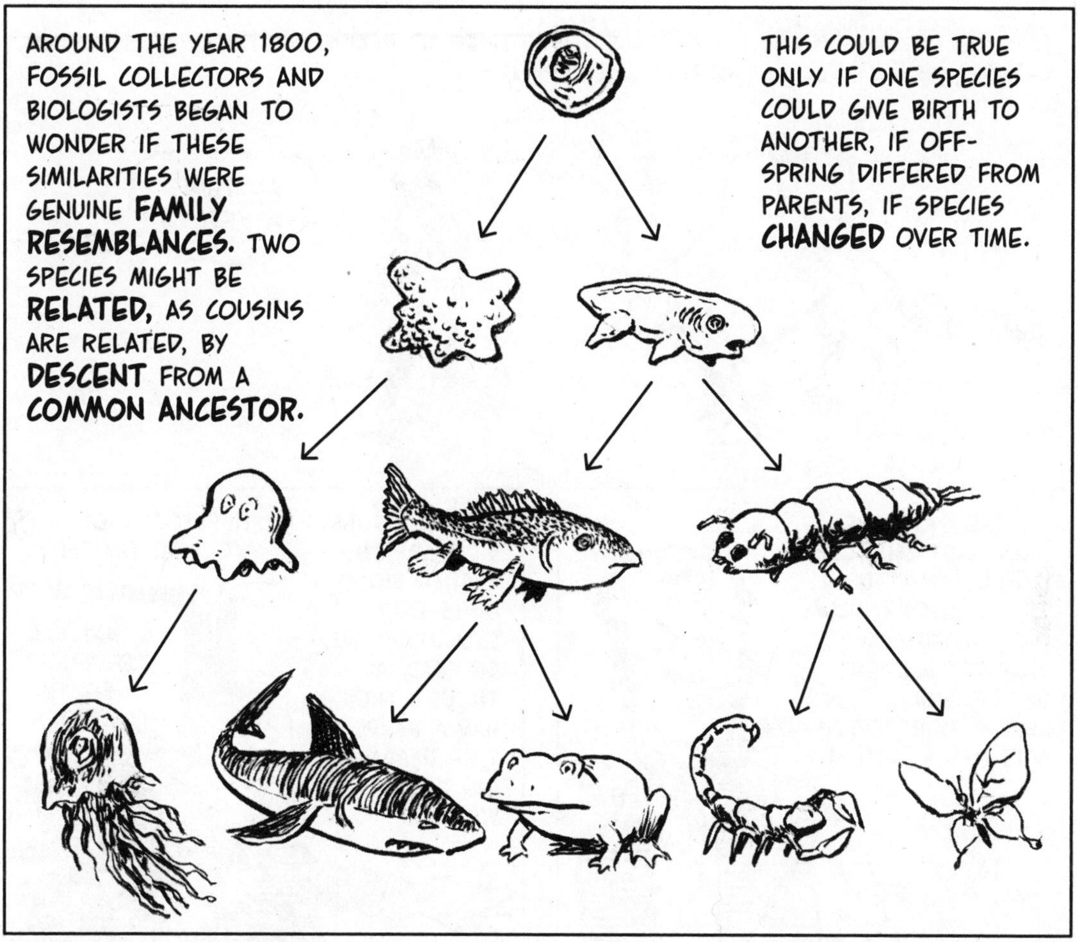
AROUND THE YEAR 1800, FOSSIL COLLECTORS AND BIOLOGISTS BEGAN TO WONDER IF THESE SIMILARITIES WERE GENUINE **FAMILY RESEMBLANCES.** TWO SPECIES MIGHT BE **RELATED,** AS COUSINS ARE RELATED, BY **DESCENT** FROM A **COMMON ANCESTOR.**
THIS COULD BE TRUE ONLY IF ONE SPECIES COULD GIVE BIRTH TO ANOTHER, IF OFF-SPRING DIFFERED FROM PARENTS, IF SPECIES **CHANGED** OVER TIME.

NO ONE HAD EVER SEEN SUCH A THING HAPPEN, BUT THEORIES WERE FLOATED—AND PUNCTURED.

IF AN ANTELOPE CRANES ITS NECK, IT'LL BIRTH A GIRAFFE!
IF I WRAP MY HEAD AROUND YOUR IDEA, WILL MY CHILDREN HAVE SKULLS SHAPED LIKE DONUTS?
Lamarck
Cuvier

FOR THE NEXT 50 YEARS, LEADING SCIENTISTS TRIED TO DECIDE WHETHER SPECIES COULD CHANGE, OR **EVOLVE,** AS THE FRENCH LIKED TO SAY.

ANYTHING YET?

ON THE "NO" SIDE, GEOLOGIST **CHARLES LYELL** ADMITTED THAT SPECIES COULD DIE OUT, BUT SAID THAT REPLACEMENT SPECIES WOULD APPEAR BY REPEATED ACTS OF **CREATION.**

ON THE OTHER SIDE, VARIOUS FRENCH BIOLO-GISTS FELT THAT EVOLUTION MUST BE REAL, AL-THOUGH THEY HAD A HARD TIME EXPLAIN-ING HOW IT WORKED.
INHERITANCE OF ACQUIRED TRAITS!
A "NATURAL TENDENCY" TO FIT THE ENVIRONMENT.

IN FACT, LYELL'S GEOLOGY IS CONSISTENT WITH EVO-LUTION. HE SAW RADICAL ENVIRONMENTAL CHANGE, AS THE EARTH'S CRUST RISES, FALLS, AND FOLDS OVER MILLIONS OF YEARS.

HOW COULD LIFE **NOT** HAVE RESPONDED?

IT MUST HAVE. I CAN FEEL IT IN MY BONES...

ENTER **CHARLES DARWIN,** A FRESH CAMBRIDGE GRADUATE WITH A PASSION FOR NATURE. IN 1831, HE BEGAN FIVE YEARS OF TROPICAL TRAVEL IN SEARCH OF SPECIMENS AND IDEAS.
AS DARWIN'S IDEAS TOOK SHAPE, HE REALIZED THAT HE WAS ONTO SOMETHING NEW, RADICAL, AND GUARANTEED TO RAISE HACKLES (OTHER PEOPLE'S HACKLES, THAT IS).
HACKLES
WHEN HACKLES GO UP, ANIMALS **BITE**... AND **PEOPLE** ARE ANIMALS...
SO HE FIRMED UP HIS THESIS WITH TWENTY YEARS OF THINKING, WRITING, AND EXPERIMENTS. WHY ARE NO FROGS FOUND ON OCEANIC ISLANDS? BECAUSE NONE WERE CREATED THERE? OR BECAUSE NONE COULD MIGRATE FROM THE MAINLAND?
SALT WATER KILLS FROG EGGS.
THEN, IN 1858, CAME A LETTER FROM ONE **ALFRED WALLACE,** WHO HAD INDEPENDENTLY THOUGHT DARWIN'S THOUGHTS.
"MY DEAR DARWIN..." MM-HM... MM-HMM... **AAAKKK!**
C. D. SPRANG INTO ACTION.
I AM THE **FITTEST!!**
WITHIN A YEAR, HE HAD PUT TOGETHER THE MOST INFLUENTIAL BIOLOGY BOOK EVER WRITTEN, *ON THE ORIGIN OF SPECIES.*
HACKLES BE HANGED!
MEET th
AUTHO

DARWIN BEGINS BY TELLING HOW PEOPLE CREATE NEW VARIETIES THROUGH
SELECTIVE BREEDING.
COME, FIDO!
SUPPOSE THAT A DOG BREEDER WANTS TO GENERATE A SHORT-SNOUTED BREED. FROM A LITTER SHE SELECTS ONE OR TWO WITH THE SHORTEST SNOUTS.
YOU.
THESE ARE RAISED AND CROSSED WITH THE SHORTEST-SNOUTED OFFSPRING OF OTHER PARENTS.
THEIR OFFSPRING WILL ALSO HAVE A RANGE OF SNOUTS, BUT SHORTER OVERALL. AGAIN BREED THE SHORTEST.
REPEATED SEVERAL TIMES, SNOUTS GET SHORTER AND SHORTER...
UNTIL EVENTUALLY, YOU HAVE A PUG, A PEKINESE, OR SOMETHING ELSE WITH A SMUSHED-IN FACE.
GREAT-GRANDMA?

BY MATING DOGS WITH DESIRED FEATURES OVER SEVERAL GENERATIONS, PEOPLE HAVE MADE SHORT, TALL, SHAGGY, BALD, STOUT, AND SKINNY BREEDS IN JUST ABOUT EVERY COLOR BUT PURPLE.

FARMERS AND STOCKBREEDERS BRED (AND STILL BREED) DOMESTICATED PLANTS AND ANIMALS THE SAME WAY: ALLOW ONLY THE "BEST" INDIVIDUALS TO REPRODUCE.

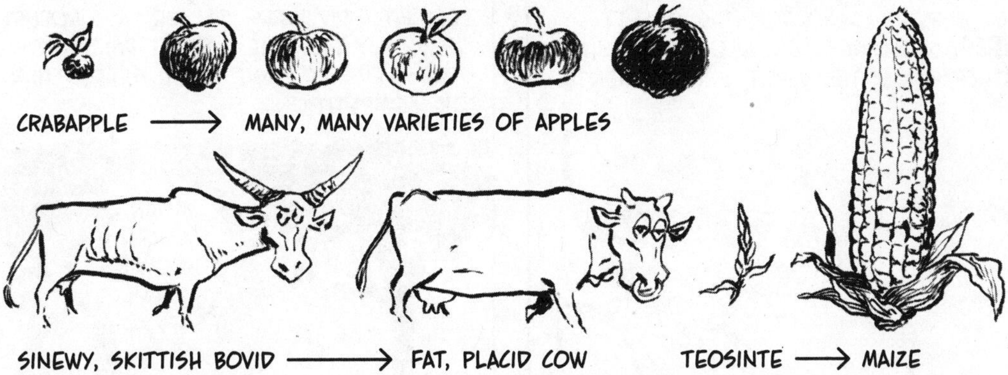

BREEDERS TAKE ADVANTAGE OF **VARIATIONS** AMONG OFFSPRING: NOT ALL PUPS, LAMBS, CALVES, APPLE TREES, OR CORN PLANTS ARE EXACTLY ALIKE.

NOW THE QUESTION IS: DOES **NATURE** DO THE SAME? AND IF SO, WHAT IS NATURE'S WAY OF PICKING THE "BEST"? THE ANSWER IS GRIM.

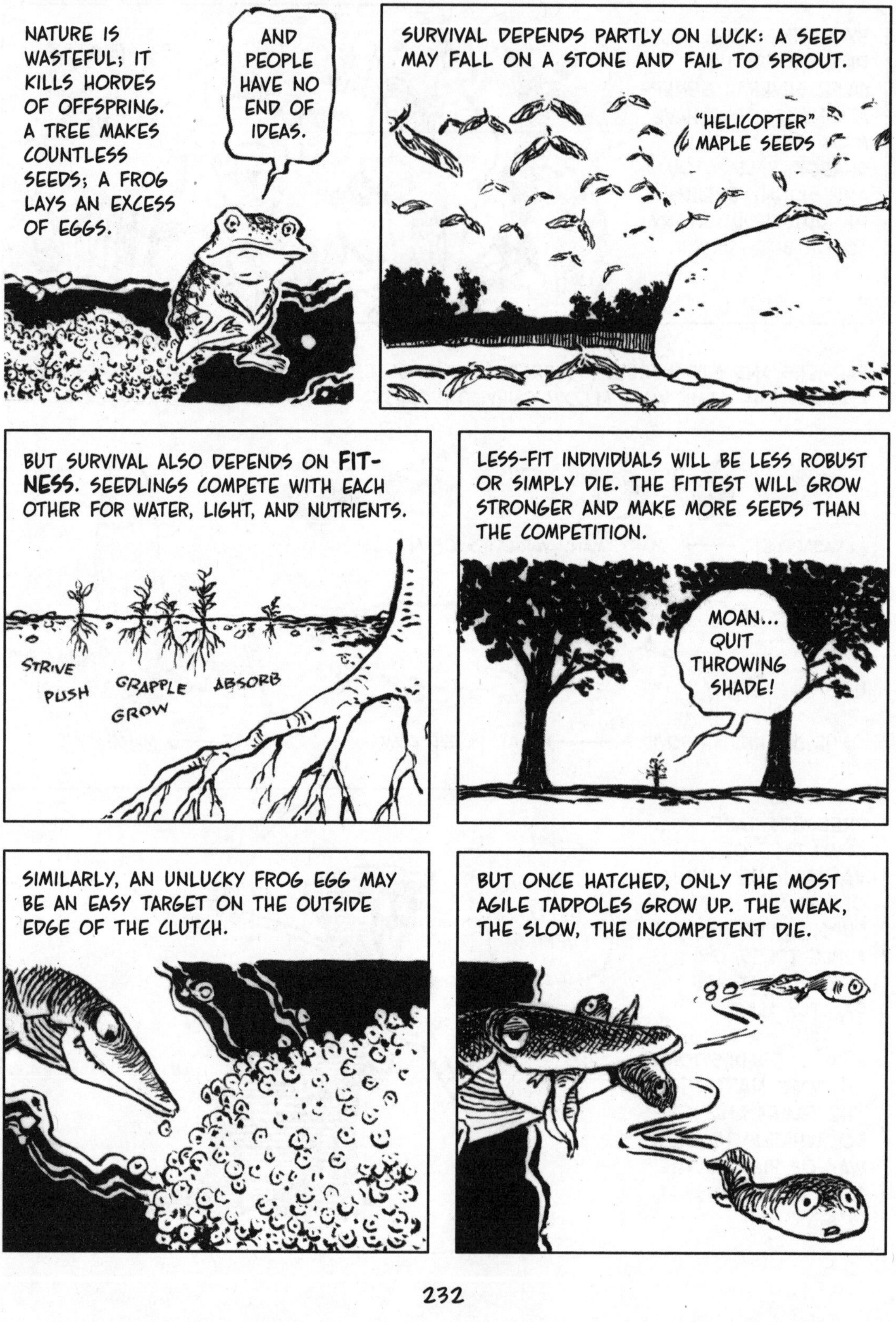
NATURE IS WASTEFUL; IT KILLS HORDES OF OFFSPRING. A TREE MAKES COUNTLESS SEEDS; A FROG LAYS AN EXCESS OF EGGS.
AND PEOPLE HAVE NO END OF IDEAS.
SURVIVAL DEPENDS PARTLY ON LUCK: A SEED MAY FALL ON A STONE AND FAIL TO SPROUT.
"HELICOPTER" MAPLE SEEDS
BUT SURVIVAL ALSO DEPENDS ON FITNESS. SEEDLINGS COMPETE WITH EACH OTHER FOR WATER, LIGHT, AND NUTRIENTS.
STRIVE
PUSH
GRAPPLE
GROW
ABSORB
LESS-FIT INDIVIDUALS WILL BE LESS ROBUST OR SIMPLY DIE. THE FITTEST WILL GROW STRONGER AND MAKE MORE SEEDS THAN THE COMPETITION.
MOAN... QUIT THROWING SHADE!
SIMILARLY, AN UNLUCKY FROG EGG MAY BE AN EASY TARGET ON THE OUTSIDE EDGE OF THE CLUTCH.
BUT ONCE HATCHED, ONLY THE MOST AGILE TADPOLES GROW UP. THE WEAK, THE SLOW, THE INCOMPETENT DIE.

NATURE, SAYS DARWIN, IS A LIFE-AND-DEATH COMPETITION FOR FOOD, SPACE, AND SAFETY. IN EACH GENERATION, THE **FITTEST** INDIVIDUALS **SURVIVE AND REPRODUCE MORE OFTEN** THAN THEIR LESS-FIT PEERS. DARWIN CALLS THIS

NATURAL SELECTION

OR "SURVIVAL OF THE FITTEST."

SORRY, I DON'T MAKE THE RULES, I ONLY REPORT 'EM!

HE SEES NO GUIDING HAND, NO CREATION, NO PLAN, NO DESIGN OR DESIGNER, ONLY ENVIRONMENTAL PRESSURE (SELECTION) AND BLIND CHANCE (RANDOM VARIATION).

AS EXPECTED, THIS BLEAK VISION RAISED THEOLOGICAL HACKLES, BUT DARWIN'S FLAWLESS LOGIC AND AMPLE EVIDENCE PERSUADED SCIENTISTS. THEY EMBRACED THE THEORY OF EVOLUTION BY NATURAL SELECTION AND HAVE NEVER LET GO.

THUMP THUD
OOK OOK!
HAS SCIENCE EVER LOWERED A HACKLE?
WELL, MINE.

NATURAL SELECTION SHAPES ORGANISMS TO FIT THEIR ENVIRONMENT. WHAT HAPPENS, THEN, WHEN THE ENVIRONMENT **CHANGES?** CONSIDER THE CASE OF THE

peppered moth.

THIS PALE, SPECKLED MOTH WAS COMMON IN ENGLAND AT A TIME (1800) WHEN COAL-FIRED FACTORIES BEGAN RISING IN AND AROUND MANCHESTER.

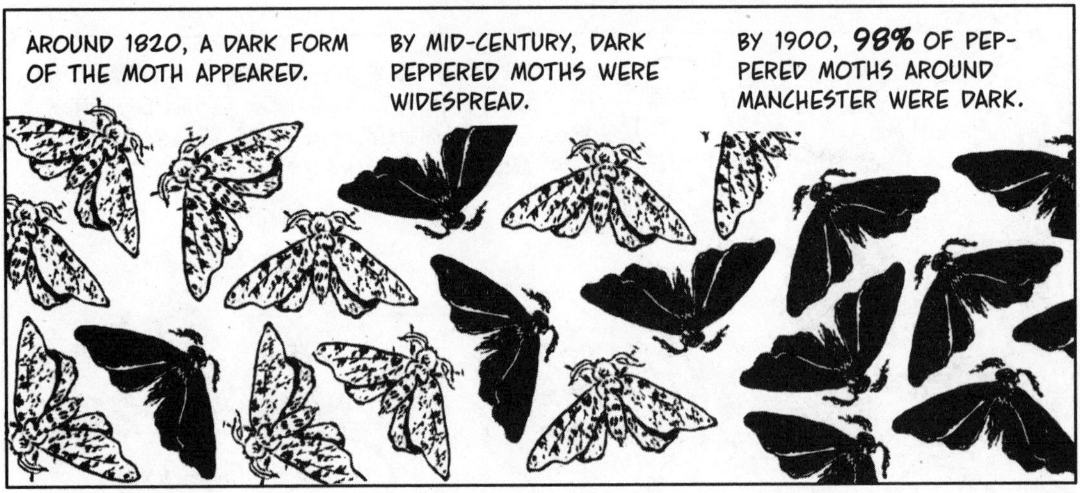
AROUND 1820, A DARK FORM OF THE MOTH APPEARED.
BY MID-CENTURY, DARK PEPPERED MOTHS WERE WIDESPREAD.
BY 1900, 98% OF PEPPERED MOTHS AROUND MANCHESTER WERE DARK.

WHAT HAPPENED? AS COAL SOOT GRADUALLY BLACKENED TREE TRUNKS AND WALLS, PALE MOTHS STOOD OUT; BIRDS ATE THEM; MORE DARK MOTHS SURVIVED.

UNDER SELECTIVE PRESSURE, THE PEPPERED MOTH EVOLVED.
FIT TODAY, DEAD TOMORROW!

WE CAN INTERPRET THIS CHANGE IN TERMS OF GENETICS, EVEN THOUGH DARWIN HIMSELF KNEW NOTHING ABOUT GENES.

HERE IS A SIMPLE MODEL: SUPPOSE PEPPERED MOTHS' WING COLOR IS GOVERNED BY A SINGLE GENE WITH TWO ALLELES: ***b*** MAKES PALE WINGS, AND ***B*** MAKES DARK ONES. WE MAY AS WELL ASSUME THAT ***B*** IS DOMINANT OVER ***b***.

WE WANT TO KNOW WHAT HAPPENS TO THE **RELATIVE FREQUENCY** OF THE TWO ALLELES IN THE POPULATION. WHAT FRACTION OF ALL ALLELES ARE ***B***, AND WHAT FRACTION ARE ***b*?** AND HOW DO THESE FRACTIONS CHANGE OVER TIME?

SAMPLING THE POPULATION AND SIMPLY COUNTING ALLELES, LET'S SAY WE DISCOVER THAT ***p***, THE FRACTION OF ***b***, IS 3/4, AND ***q***, THE FRACTION OF ***B***, IS 1/4.

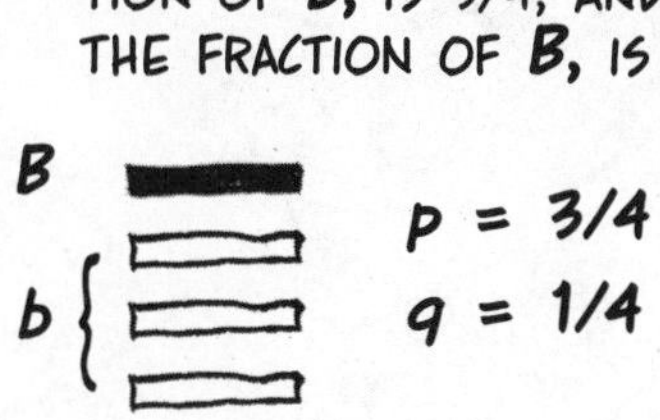

NOW LET'S GET ALGEBRAIC. IF p = FREQUENCY OF **b** (HERE 3/4), AND q = FREQUENCY OF **B** (HERE 1/4), A CROSSING SQUARE GIVES THE FREQUENCY OF EACH DIPLOID GENOTYPE (ASSUMING THE POPULATION IS LARGE AND INDIVIDUALS MATE RANDOMLY).

	p	q
p	bb: p^2	Bb: pq
q	Bb: pq	BB: q^2

IN OUR EXAMPLE,

bb HAS FREQUENCY $p^2 = \left(\frac{3}{4}\right)^2 = \frac{9}{16}$

Bb HAS FREQUENCY $2pq = 2\left(\frac{3}{4}\right)\left(\frac{1}{4}\right) = \frac{3}{8}$

BB HAS FREQUENCY $q^2 = \left(\frac{1}{4}\right)^2 = \frac{1}{16}$

IN A LARGE POPULATION IN WHICH EVERYONE MATES RANDOMLY, THESE PROPORTIONS TEND TO REMAIN **CONSTANT** OVER TIME, GENERATION AFTER GENERATION. THIS IS KNOWN AS THE **HARDY-WEINBERG PRINCIPLE.**

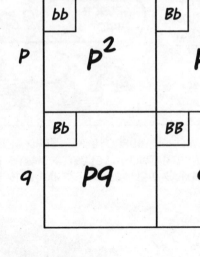

IN OTHER WORDS, A POPULOUS SPECIES UNDER NO SELECTIVE PRESSURE WILL TEND TO BE STABLE.

WHEN SELECTIVE PRESSURE WALLOPS THE POPULATION, EVERYTHING CHANGES. SUPPOSE SOME CATASTROPHE WIPES OUT ALL THE ***bb*** HOMOZYGOTES—9 OUT OF EVERY 16 IN THE POPULATION.

OF THOSE 16 REPRESENTATIVE INDIVIDUALS, ONLY 7 ARE LEFT, 6 ***Bb*** AND ONE ***BB.***

COUNTING ALLELES IN THOSE SEVEN PAIRS, WE SEE ***6*** OF 14 ARE ***b,*** AND ***8*** ARE ***B.***

THE FREQUENCY OF ***b*** HAS FALLEN FROM 3/4 TO ***3/7.***

THE FREQUENCY OF ***B*** HAS RISEN FROM 1/4 TO ***4/7.***

MATH LOVERS: SHOW THAT IN GENERAL, THESE FREQUENCIES WILL BE $p/(p+1)$ AND $1/(p+1)$. HINT: REMEMBER $p+q=1$.

AN EXAGGERATED SCENARIO, MAYBE, BUT IT MAKES THE POINT: GENE FREQUENCIES CHANGE WHEN SOME GENOTYPES ARE FAVORED.

SPECIATION

AS NATURAL SELECTION ALTERS POPULATIONS, DARWIN ARGUED, THE CHANGES COULD EVENTUALLY ACCUMULATE TO MAKE A NEW **SPECIES,** SO DIFFERENT THAT IT COULD NO LONGER INTERBREED WITH ITS DISTANT COUSINS.

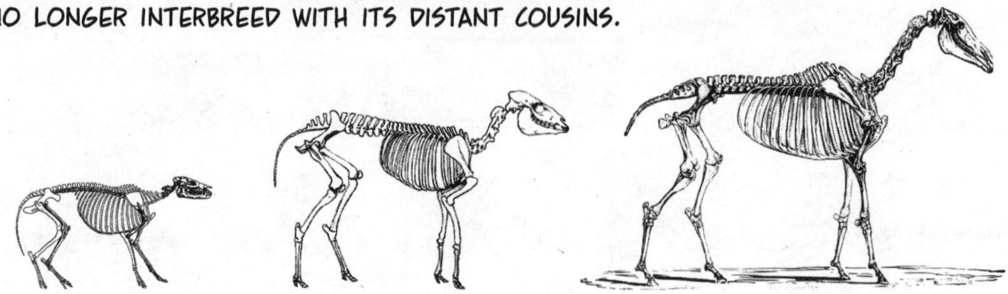

HYRACOTHERIUM MESOHIPPUS EQUUS (MODERN HORSE)

DARWIN SAW THIS EFFECT IN THE **FINCHES** HE COLLECTED IN THE GALÁPAGOS ISLANDS. EACH ISLAND HAS ITS OWN BREED, WHICH DIFFERS FROM THE OTHERS IN SIZE, HABITS, DIET, COLOR, BEAK LENGTH AND STRENGTH, ETC.
DARWIN'S EXPLANATION: LONG AGO, A FEW ANCESTRAL FINCHES RODE THE WIND OR FLOATED ON LOGS FROM THE MAINLAND. GROUPS THAT LANDED ON DIFFERENT ISLANDS EVOLVED TO ADAPT TO THE FOOD AND TERRAIN WHERE THEY LIVED.
YES! A NEW START!

WHEN A SINGLE GROUP GIVES RISE TO MANY NEW FORMS IN RESPONSE TO DIFFERENT ENVIRONMENTS, WE CALL THE PROCESS **ADAPTIVE RADIATION.**

NOTE THAT THE FINCHES WERE **GEOGRAPHICALLY ISOLATED** FROM EACH OTHER. SUCH ISOLATION CAN LEAD TO SPECIATION. GENERALLY, SMALL GROUPS HAVE DIFFERENT **ALLELE FREQUENCIES.*** BREEDING ONLY AMONG THEMSELVES, EACH GROUP'S GENETICS WILL SKEW AWAY FROM THE OTHERS.

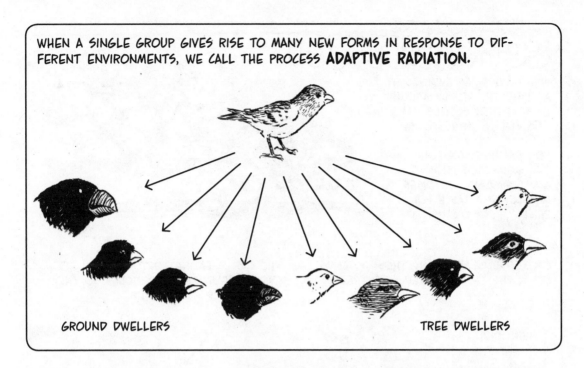

IF YOUR FORECOWS HAD COME TO MORE FAMILY REUNIONS, WE WOULDN'T BE IN THIS POSITION.

SINCE THEN, **EVERYTHING** WE HAVE LEARNED ABOUT GENETICS, THE CELL, AND LIFE'S CHEMISTRY SUPPORTS DARWIN'S IDEAS.

*SMALL SAMPLES TEND TO HAVE HIGHER VARIANCE. THIS IS A GENERAL STATISTICAL FACT, NOT ANYTHING PARTICULAR TO LIVING THINGS.

COEVOLUTION
WHEN A SPECIES EVOLVES, IT AFFECTS THE WORLD AROUND IT, SO OTHER SPECIES MAY COEVOLVE IN RESPONSE.
FOR EXAMPLE, ON FOUR OF THE GALÁPAGOS ISLANDS, GIANT GROUND TORTOISES DEVELOPED A TASTE FOR PRICKLY PEAR CACTUS.
GRRR...
ON THOSE ISLANDS, AND THOSE ISLANDS ONLY, THE CACTI EVOLVED INTO TREES.
HEH! DIDN'T I GET THE GOOD ALLELES...
THE CACTUS-EATING TORTOISE POPULATION EVOLVED ELEVATED NECK HOLES SO THEIR HEADS COULD RISE HIGHER.
#&*%!
BIRDS, BEES, AND FLOWERS OFFER MORE COOPERATIVE EXAMPLES. FLOWERS EVOLVED NECTAR TO LURE BEES TO THE PLANT'S SEX ORGANS. NECTAR NOURISHES THE BEES, WHICH INCIDENTALLY PICK UP POLLEN AND FLY IT TO OTHER FLOWERS.
SOME SPECIES HAVE COEVOLVED SO CLOSELY THAT THEY DEPEND ON EACH OTHER FOR SHEER SURVIVAL. THEY LITERALLY CAN'T LIVE WITHOUT EACH OTHER.

CONVERGENT EVOLUTION

WE USUALLY THINK OF CLOSELY RELATED SPECIES GROWING APART OVER TIME, BUT SOMETIMES DISTANTLY RELATED SPECIES MAY **CONVERGE** IN APPEARANCE.

DIVERGENCE
CONVERGENCE

THIS HAPPENS BECAUSE THE ENVIRONMENT STRONGLY MOLDS THE SHAPE OF ALL SPECIES. IF YOU SWIM IN THE WATER, YOU HAD BETTER BE STREAMLINED! AND SO IT WAS THAT THE EXTINCT **PLIOSAUR,** A REPTILE, RESEMBLED **WHALES** AND **DOLPHINS,** WHICH ARE MAMMALS, AND **SHARKS,** WHICH ARE FISH.

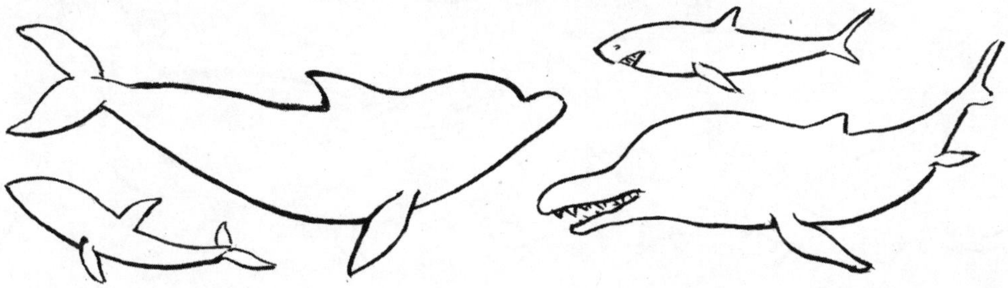

MIMICRY CAN ALSO AID SURVIVAL. THE VICEROY BUTTERFLY (LEFT) HAS EVOLVED TO RESEMBLE THE MONARCH (RIGHT), WHICH IS TOXIC TO BIRDS. BIRDS AVOID VICEROYS TOO!

MORAL: YOU CAN'T ALWAYS IDENTIFY CLOSE RELATIVES BY APPEARANCE ALONE, BUT BIRDS ARE TOO DUMB TO FIGURE THAT OUT.

BETTER SAFE THAN SORRY...

REPRODUCTION and EVOLUTION

TO CONTRIBUTE TO THE SPECIES GENE POOL, AN INDIVIDUAL OBVIOUSLY HAS TO SURVIVE. BUT SURVIVAL ALONE IS NOT ENOUGH.

WHAT, THEN?

TO "WIN" THE EVOLUTION GAME, AN ORGANISM ALSO HAS TO **REPRODUCE SUCCESSFULLY:** IT HAS TO PASS ITS GENES TO OFFSPRING, WHICH THEMSELVES MUST SURVIVE TO REPRODUCTIVE AGE.

YES, GENES, WHATEVER YOU SAY!

THIS ACCOUNTS FOR **PARENTAL CARE** OF THE YOUNG. IN MANY SPECIES, PARENTS CARRY OR INCUBATE THE EGGS AND THEN FEED AND GUARD THE NEWBORNS.

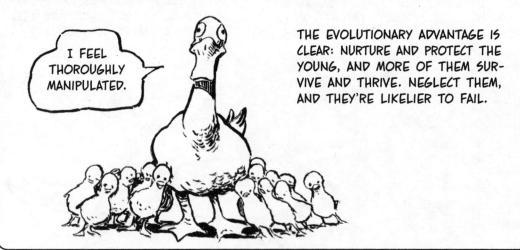

I FEEL THOROUGHLY MANIPULATED.

THE EVOLUTIONARY ADVANTAGE IS CLEAR: NURTURE AND PROTECT THE YOUNG, AND MORE OF THEM SURVIVE AND THRIVE. NEGLECT THEM, AND THEY'RE LIKELIER TO FAIL.

THIS BEHAVIOR IS **INSTINCTIVE,** GENETICALLY HARD-WIRED INTO ANIMALS' BRAINS AND CHEMISTRY. BEING GENETIC, INSTINCT IS **HERITABLE,** SO IT EVOLVES TO FAVOR SUCCESSFUL REPRODUCTION.
I'M EVEN PROGRAMMED TO **LIKE** IT...
PARENTS RISK THEIR OWN SURVIVAL TO PROTECT THEIR YOUNG; THE MOTHER SAGE GROUSE WILL FAKE A WOUNDED WING AND SQUAWK WILDLY TO DECOY A PREDATOR AWAY FROM HER CHICKS.
DEVOTING ENERGY TO PROTECTING THE YOUNG IS CALLED A *K* STRATEGY. OTHER SPECIES DISPLAY AN ALTERNATIVE ("*r*") STRATEGY: THEY INVEST THEIR LIMITED ENERGY IN SPEWING OUT HUGE NUMBERS OF EGGS THAT HAVE TO FEND FOR THEMSELVES.
I'M NOT CALLOUS; I'M **TIRED...**

BESIDES COMPETING FOR SURVIVAL, INDIVIDUALS ALSO COMPETE FOR **OPPORTUNITIES TO MATE.** CHOOSING THE WINNERS IS CALLED

SEXUAL SELECTION.

(ALFRED WALLACE STRONGLY AND WRONGLY OPPOSED THIS IDEA OF DARWIN'S, BY THE WAY.)

ALL THIS STRUGGLE GETS MY GOAT...

THE DYNAMIC STARTS HERE: SPERM ARE **CHEAP,** WHILE EGGS ARE **EXPENSIVE.** A MALE BLASTS OUT BILLIONS OF SPERM, WHILE A FEMALE, EVEN ONE FROM A VERY "EGGY" SPECIES, MAKES FAR FEWER. AT THE CELLULAR LEVEL, HORDES OF SPERM CHASE A SPARSE SET OF TARGETS.

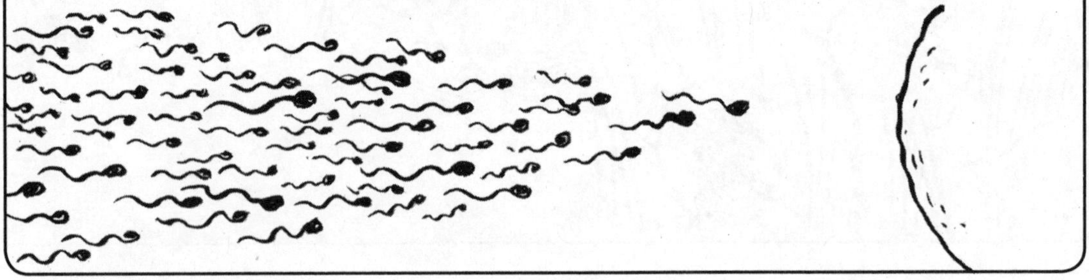

FEMALES THEN FACE ADDED BURDENS: PREGNANCY, EGG-LAYING AND BROODING, OR GIVING LIVE BIRTH AND NURSING. AS A GENERAL RULE, THEN, FEMALES ARE MORE RELUCTANT TO MATE. THEY WANT TO MAKE SURE THEY GET A "GOOD" MALE, AND MALES MUST COMPETE TO WIN FEMALES.

HMM... HMMM...

SOMETIMES COMPETITION IS PEACEFUL, WITH **FEMALE CHOICE.** IN MANY BIRD SPECIES, FEMALES CHOOSE A MATE BASED ON HIS BRIGHT PLUMAGE, FANCY FEATHERS, OR COMPETITIVE SINGING.

LOVE THE AMPLIFIER.

FEMALE MOUNTAIN SHEEP HAVE LESS AGENCY. RAMS BUTT HEADS UNTIL ONE VANQUISHES ALL OTHERS, AND HE GETS A HAREM OF EWES.

Result:

HIGH-COMPETITION SPECIES HAVE FEATURES THAT WIN COMPETITIONS. MALES TEND TO BE BIGGER AND STRONGER; AND THEY MAY SPORT SPECIAL HORNS OR ANTLERS, BRIGHT COLORS, AND OTHER DISPLAYS. THIS ACCOUNTS FOR THE **SEXUAL DIMORPHISM** ("TWO SHAPES"), MENTIONED ON P. 198.

$#%! NOT AGAIN!

SOME OF THESE MALE FEATURES MAY ACTUALLY SHORTEN LIFE! BUT IN THE END, NATURAL SELECTION FAVORS TRAITS THAT **REPRODUCE.**

THIS BRINGS US BACK TO THE QUESTION RAISED AT THE END OF THE LAST CHAPTER: WHY DO SO MANY SPECIES BURN SO MUCH ENERGY IN THIS TROUBLESOME, INEFFICIENT BUSINESS OF SPERM-CHASES-EGG? WHY IS THERE SEX AT ALL?

COULDN'T WE JUST ELOPE AND BE CLONED?

EVOLUTIONARY THINKING OFFERS A FRESH PERSPECTIVE HERE. THE GENES FOR SEX MUST BE SUPER AT REPRODUCING THEMSELVES! SEXUAL BEINGS—THOSE WITH GENES FOR MEIOSIS AND THE REST OF IT—MUST HAVE A **REPRODUCTIVE ADVANTAGE** OVER ASEXUAL COMPETITORS. YOU MIGHT SAY THAT REPRODUCTION CAUSES SEX.

STRANGE, I ALWAYS THOUGHT IT WAS THE OTHER WAY AROUND...

TO PUT IT ANOTHER WAY: THE EVOLUTIONARY BENEFITS OF SEX OUTWEIGH ITS COSTS.

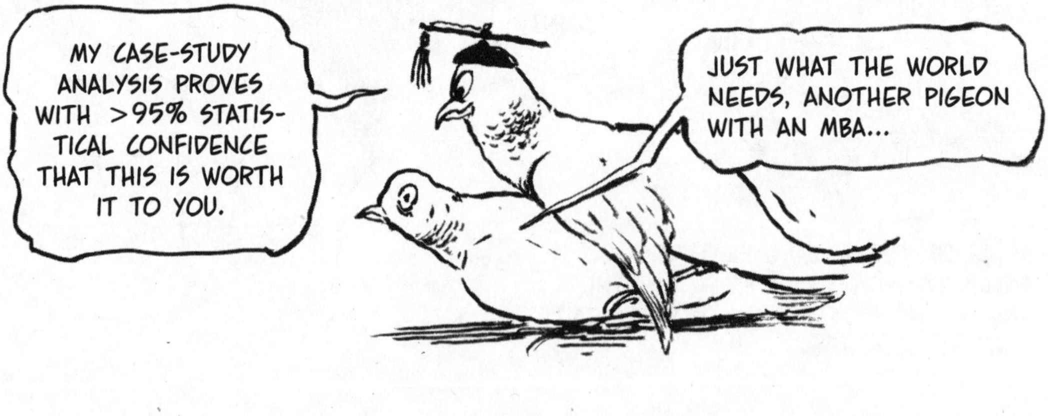
MY CASE-STUDY ANALYSIS PROVES WITH >95% STATISTICAL CONFIDENCE THAT THIS IS WORTH IT TO YOU.
JUST WHAT THE WORLD NEEDS, ANOTHER PIGEON WITH AN MBA...

WHAT ARE THE BENEFITS? SURPRISINGLY, DESPITE MANY SYMPOSIA, PAPERS, AND POSTERS, SCIENCE OFFERS NO SINGLE DEFINITIVE ANSWER. STILL, MOST PEOPLE WOULD AGREE ON THESE TWO:

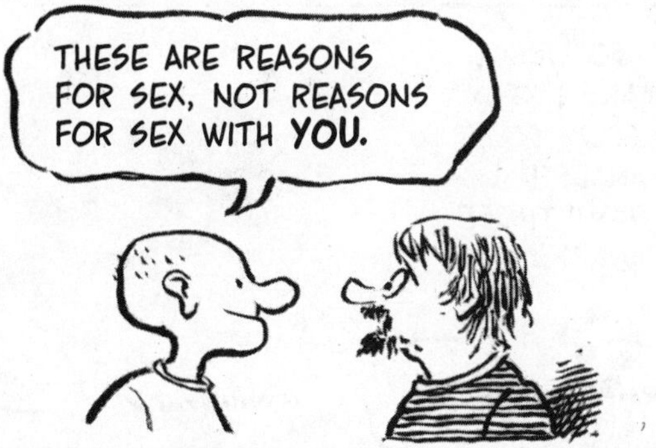

THESE ARE REASONS FOR SEX, NOT REASONS FOR SEX WITH **YOU.**

SEX SPEEDS THE **SPREAD OF ADVANTAGEOUS MUTATIONS.** WITH SEX, MANY INDIVIDUALS CAN MATE WITH A LONE "IMPROVED" ONE.

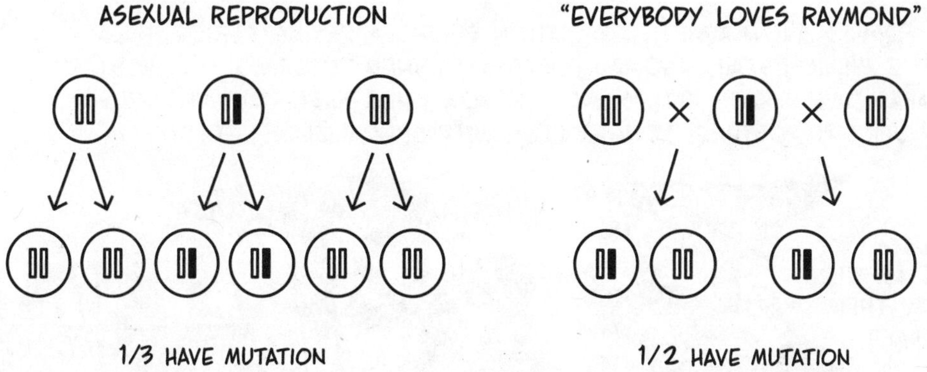

SEX INCREASES VARIATION. BY SHUFFLING THE GENOME, SEX "TRIES OUT" NEW VARIANTS AND SO INCREASES ADAPTABILITY TO A DYNAMIC WORLD.

MAN, SINCE THOSE FLYING CHAINSAWS SHOWED UP, **EVERYBODY** LOVES **ME**...

SO GREAT ARE THE BENEFITS THAT NATURE SELECTS SPECIAL TRAITS TO **ENCOURAGE** SEX. WHICH IS WHY, IN A CARTOONIST'S IMMORTAL WORDS, SEX "FEELS GOOD **NOW** AND CAN ONLY FEEL BETTER **TOMORROW.**"

FUTURE GENERATIONS ARE THE LUCKY ONES!

DARWIN ALSO WROTE THAT HUMANS LIKELY EVOLVED FROM SOME KIND OF ANCESTRAL APE. UP WENT THOSE HACKLES AGAIN...
HAHA HAHA

IN FACT, FOSSILS SHOW ADAPTIVE RADIATION DURING A DRYING PERIOD IN EAST AFRICA FIVE MILLION YEARS AGO, AS FORESTS THINNED TO GRASSLAND. WESTERN FOREST APES GAVE RISE TO CHIMPS AND BONOBOS, WHILE EASTERN PLAINS APES EVOLVED UPRIGHT POSTURE, LONGER LEGS WITH SPECIALIZED FEET, ETC.

IT'S OK, I GUESS, AS LONG AS THEY KNOW THEIR PLACE.

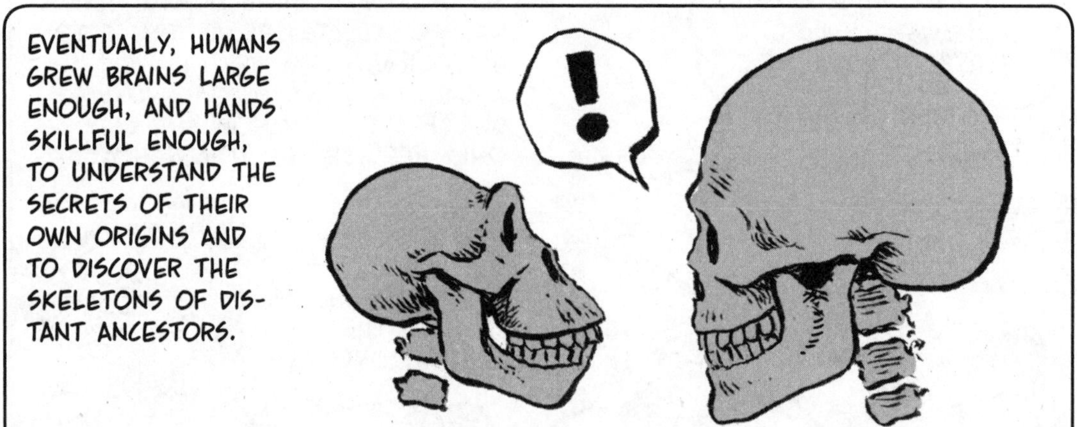
EVENTUALLY, HUMANS GREW BRAINS LARGE ENOUGH, AND HANDS SKILLFUL ENOUGH, TO UNDERSTAND THE SECRETS OF THEIR OWN ORIGINS AND TO DISCOVER THE SKELETONS OF DISTANT ANCESTORS.
!

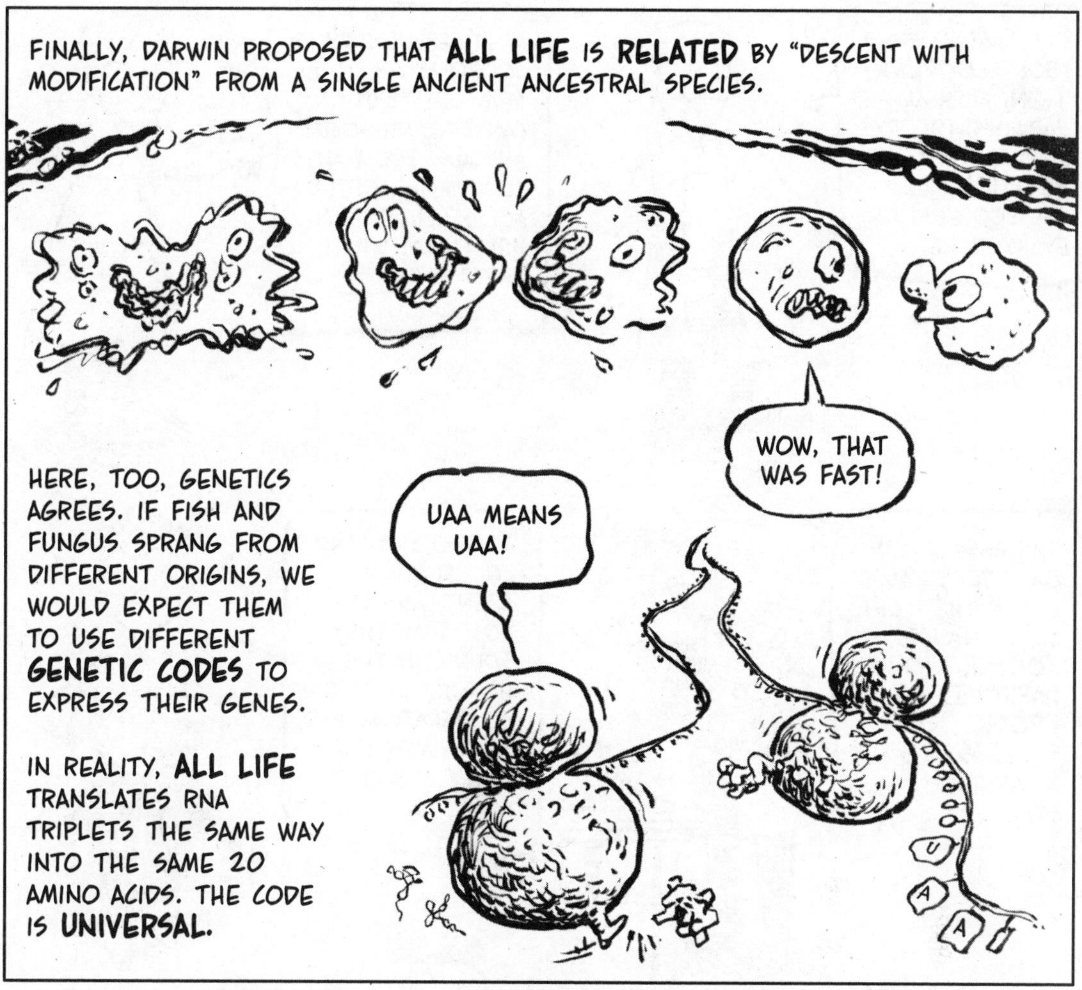
FINALLY, DARWIN PROPOSED THAT ALL LIFE IS RELATED BY "DESCENT WITH MODIFICATION" FROM A SINGLE ANCIENT ANCESTRAL SPECIES.
WOW, THAT WAS FAST!
HERE, TOO, GENETICS AGREES. IF FISH AND FUNGUS SPRANG FROM DIFFERENT ORIGINS, WE WOULD EXPECT THEM TO USE DIFFERENT GENETIC CODES TO EXPRESS THEIR GENES.
UAA MEANS UAA!
IN REALITY, ALL LIFE TRANSLATES RNA TRIPLETS THE SAME WAY INTO THE SAME 20 AMINO ACIDS. THE CODE IS UNIVERSAL.

IN THE NEXT CHAPTER, WE'LL SEE HOW MODERN BIOLOGY FILLS IN THE MANY BRANCHES OF LIFE'S ENORMOUS EXTENDED-FAMILY TREE. BUT FIRST, A FINAL THOUGHT:

OUR LAST WORD IS ABOUT **COOPERATION.** ALTHOUGH DARWIN'S THEORY EMPHASIZES **COMPETITION,** WE SEE COOPERATION EVERYWHERE.
MULTICELLULARITY IS COOPERATIVE; SO IS SEX; COEVOLUTION CAN BE; AND SOCIAL ANIMALS, FROM ANTS TO HUMANS, OFTEN ACT SELFLESSLY AND HELPFULLY. WHY IS THAT?
THE ANSWER CAN ONLY BE: BECAUSE UNDER THE RIGHT CIRCUMSTANCES, **COOPERATION IMPROVES REPRODUCTIVE SUCCESS.** IT'S A WINNING STRATEGY!
RYE KNOTS
COMPETITION AND COOPERATION AREN'T AS OPPOSITE AS THEY SOUND. IN THE HISTORY OF LIFE, COOPERATION IS SOMETIMES THE **ONLY** WAY TO COMPETE.
JUST LOOK AT SPORTS! IT TAKES **TEAMWORK** TO WIN!
AND THE **REPRODUCTIVE ADVANTAGE** OF **ATHLETES**— WELL, LET'S NOT EVEN **GO** THERE!!!

Chapter 15
CLASSIFICATION

BIOLOGISTS LOVE TO CLASSIFY, NEED TO CLASSIFY, HAVE AN OVERWHELMING URGE TO CLASSIFY. LIVING THINGS, LIKE SO MUCH ELSE, ARE BEST UNDERSTOOD BY PUTTING THEM INTO LABELED BOXES.

IT'S A START.
ALLIGATORS

THE TECHNICAL WORD FOR A BOX IS A **TAXON** (PL. *TAXA*), AND THE BUSINESS OF BOXING AND NAMING THINGS IS CALLED **TAXONOMY.** A TAXON CAN HAVE SUBTAXA, AS IN THIS VERY SIMPLE TAXONOMY OF CARS.

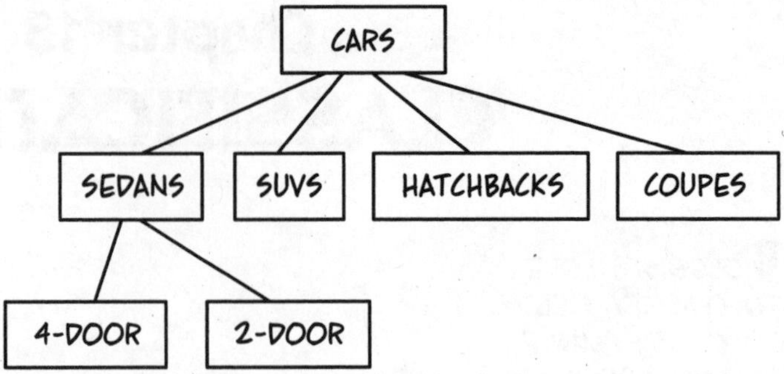

IN BIOLOGY, THE GROUND-FLOOR TAXON IS THE **SPECIES,** A NATURAL GROUPING OF INDIVIDUALS BASED ON INTERBREEDING. THE NEXT-HIGHER TAXON, A GROUP OF SIMILAR SPECIES, IS CALLED A **GENUS** (PLURAL *GENERA*).

HERE, FOR INSTANCE, ARE A FEW MEMBERS OF THE GENUS ***MUS,*** THE **MICE.** *MUS* INCLUDES DOZENS OF SPECIES OF SMALL, FURRY, NAKED-TAILED, GNAWING CRITTERS.

ALGERIAN MOUSE
NEAVE'S MOUSE
DESERT PYGMY MOUSE
LITTLE INDIAN FIELD MOUSE
STEPPE MOUSE
MAYOR'S MOUSE
CYPRIOT MOUSE
AFRICAN PYGMY MOUSE
MATTHEY'S MOUSE
HOUSE MOUSE
DELICATE MOUSE
RYUKYU MOUSE

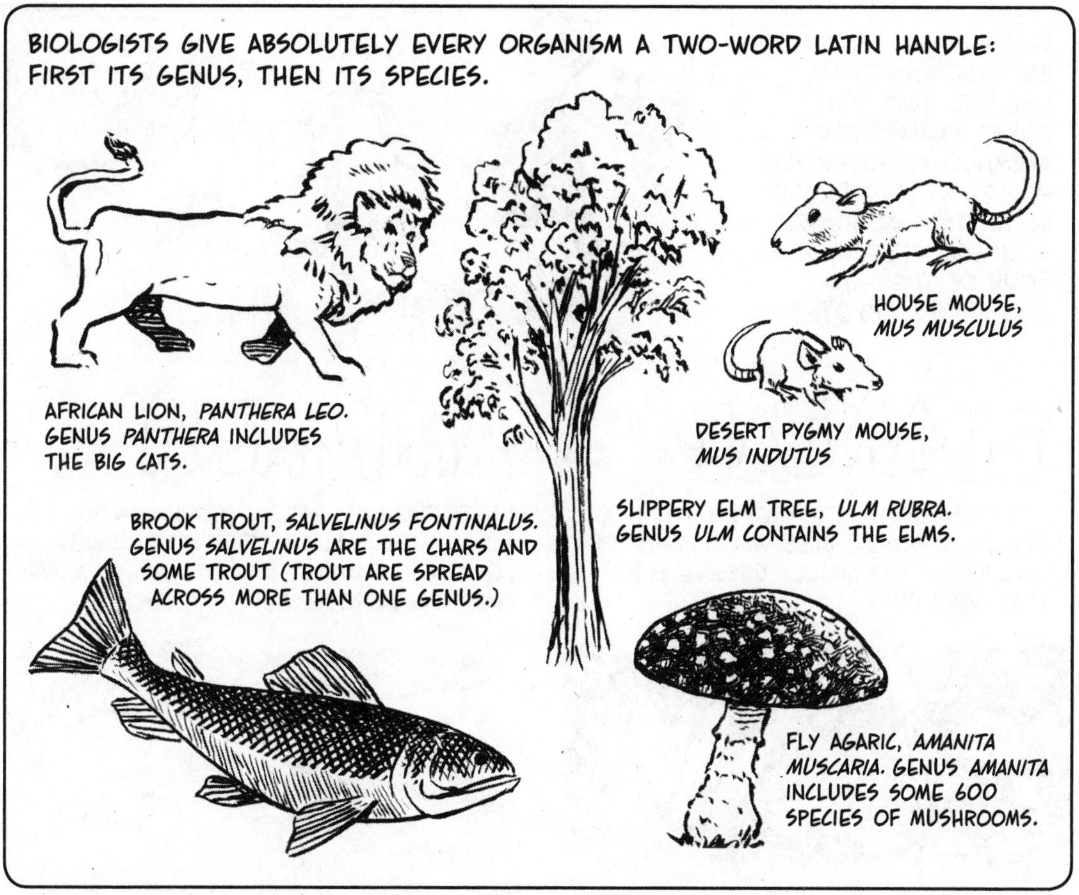
BIOLOGISTS GIVE ABSOLUTELY EVERY ORGANISM A TWO-WORD LATIN HANDLE: FIRST ITS GENUS, THEN ITS SPECIES.
HOUSE MOUSE, *MUS MUSCULUS*
AFRICAN LION, *PANTHERA LEO.* GENUS *PANTHERA* INCLUDES THE BIG CATS.
DESERT PYGMY MOUSE, *MUS INDUTUS*
BROOK TROUT, *SALVELINUS FONTINALUS.* GENUS *SALVELINUS* ARE THE CHARS AND SOME TROUT (TROUT ARE SPREAD ACROSS MORE THAN ONE GENUS.)
SLIPPERY ELM TREE, *ULM RUBRA.* GENUS *ULM* CONTAINS THE ELMS.
FLY AGARIC, *AMANITA MUSCARIA.* GENUS *AMANITA* INCLUDES SOME 600 SPECIES OF MUSHROOMS.

THIS "BINOMIAL NOMENCLATURE" SPRANG FROM ABOVE THE EARS OF CARL **LINNAEUS** (1707–1778), THE GRANDDADDY OF BIOLOGICAL TAXONOMY.

HUMAN BEING, *HOMO SAPIENS,* WITH SOMEONE ELSE'S HAIR.

THE LINNAEAN PROGRAM: CAREFULLY COMPARE THE SHAPES AND STRUCTURES OF ALL ORGANISMS, GROUP THEM INTO SPECIES, GROUP SIMILAR SPECIES INTO GENERA, GROUP SIMILAR GENERA, AND SO ON, UNTIL EVERY LIVING CREATURE IS BOXED.

THIS COULD TAKE A WHILE.

LINNAEUS, WHO FORGOT BACTERIA, DIVIDED ALL LIFE INTO TWO "KINGDOMS," **PLANTS** AND **ANIMALS.** ALTHOUGH FEW MODERN BIOLOGISTS ARE ROYALISTS, TAXONOMISTS STILL USE KINGDOMS—**FOUR** OF THEM—TO CLASSIFY **EUKARYOTES.**

AND WE KEEP THE LATIN, OUTA RESPECT...

PLANTAE

MULTICELLULAR; AUTOTROPHIC (WITH A FEW EXCEPTIONS); CELL WALLS MADE OF CELLULOSE; GENERALLY SESSILE (I.E., THEY STAY PUT)

ANIMALIA

MULTICELLULAR; HETEROTROPHIC; NO CELL WALL; GENERALLY MOTILE (THEY SWIM, LEAP, FLY, WADDLE, AND OTHERWISE MOVE AROUND); AEROBIC RESPIRATORS

FUNGI

MAY HAVE ONE CELL OR MANY; HETEROTROPHIC; SESSILE; CELL WALL MADE OF **CHITIN,** A NITROGENOUS POLYMER

PROTISTA

THE GRAB-BAG KINGDOM OF ALL ONE-CELLED EUKARYOTES EXCEPT FUNGI

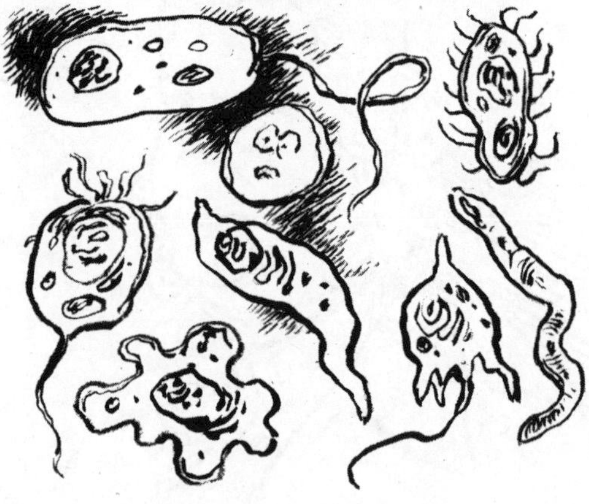

TRADITIONALLY, FIVE LAYERS OF TAXA STEP DOWN BETWEEN KINGDOM AND SPECIES. IN DESCENDING ORDER, THEY ARE **PHYLUM, CLASS, ORDER, FAMILY,** AND **GENUS.**

TO ILLUSTRATE, HERE IS HOW TAXONOMISTS DISTINGUISH BETWEEN HUMANS AND HOUSE MICE:

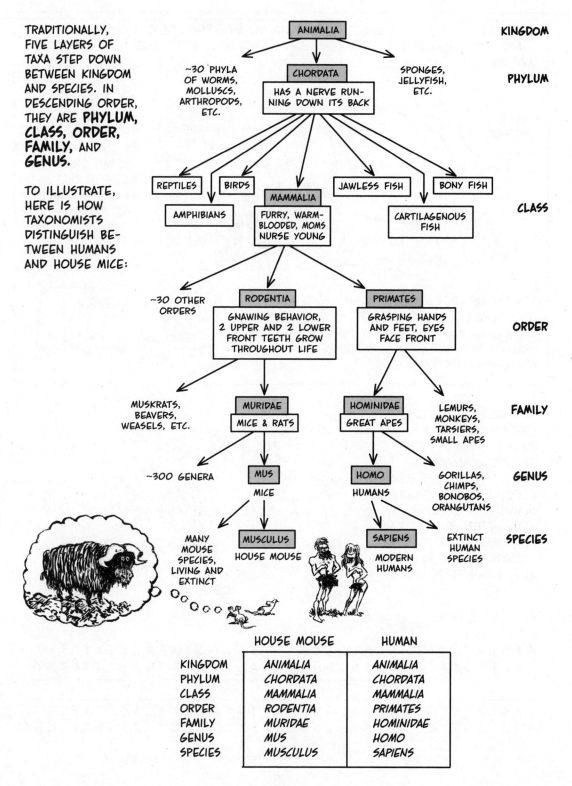

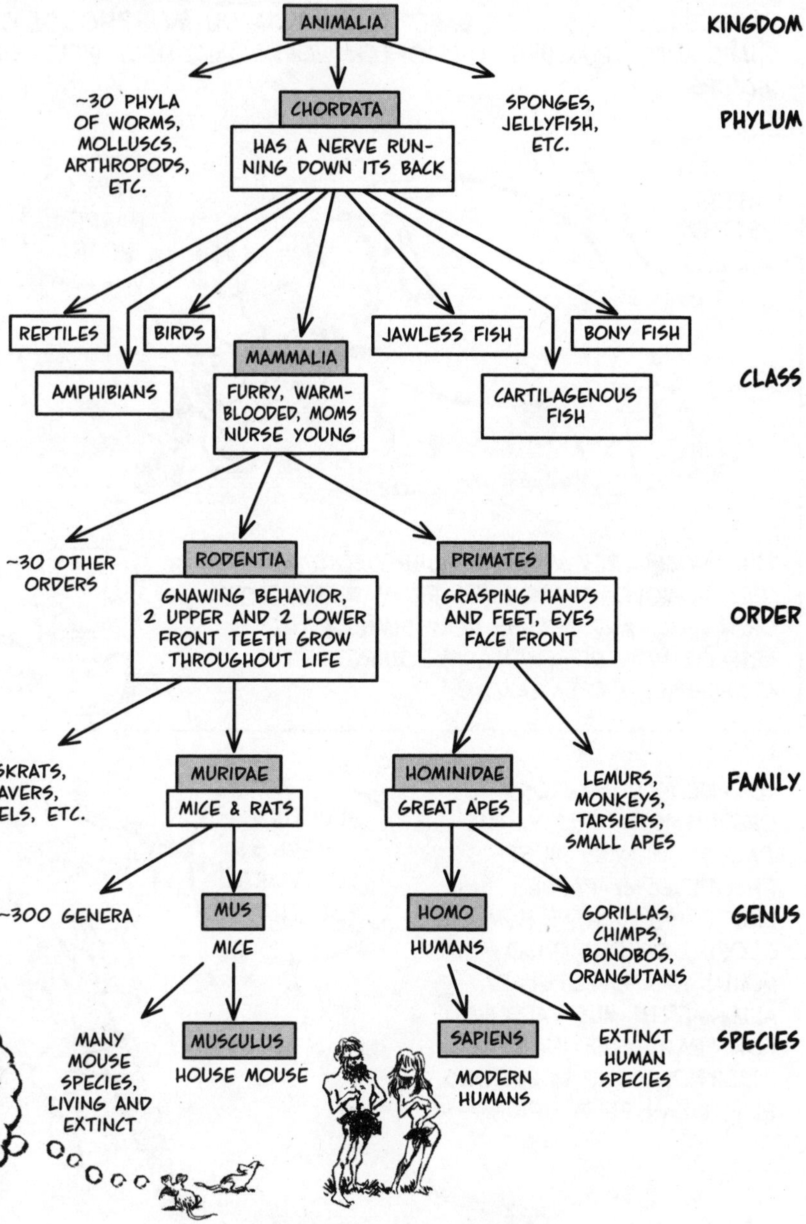
ANIMALIA
KINGDOM
~30 PHYLA OF WORMS, MOLLUSCS, ARTHROPODS, ETC.
CHORDATA
HAS A NERVE RUNNING DOWN ITS BACK
SPONGES, JELLYFISH, ETC.
PHYLUM
REPTILES
BIRDS
AMPHIBIANS
MAMMALIA
FURRY, WARM-BLOODED, MOMS NURSE YOUNG
JAWLESS FISH
BONY FISH
CARTILAGENOUS FISH
CLASS
~30 OTHER ORDERS
RODENTIA
GNAWING BEHAVIOR, 2 UPPER AND 2 LOWER FRONT TEETH GROW THROUGHOUT LIFE
PRIMATES
GRASPING HANDS AND FEET, EYES FACE FRONT
ORDER
MUSKRATS, BEAVERS, WEASELS, ETC.
MURIDAE
MICE & RATS
HOMINIDAE
GREAT APES
LEMURS, MONKEYS, TARSIERS, SMALL APES
FAMILY
~300 GENERA
MUS
MICE
HOMO
HUMANS
GORILLAS, CHIMPS, BONOBOS, ORANGUTANS
GENUS
MANY MOUSE SPECIES, LIVING AND EXTINCT
MUSCULUS
HOUSE MOUSE
SAPIENS
MODERN HUMANS
EXTINCT HUMAN SPECIES
SPECIES

	HOUSE MOUSE	HUMAN
KINGDOM	*ANIMALIA*	*ANIMALIA*
PHYLUM	*CHORDATA*	*CHORDATA*
CLASS	*MAMMALIA*	*MAMMALIA*
ORDER	*RODENTIA*	*PRIMATES*
FAMILY	*MURIDAE*	*HOMINIDAE*
GENUS	*MUS*	*HOMO*
SPECIES	*MUSCULUS*	*SAPIENS*

LINNAEUS AND HIS HEIRS BASED THEIR TAXA ON **MORPHOLOGY,** A CREATURE'S SHAPE AND STRUCTURE. THIS OFTEN WORKS AMAZINGLY WELL, DESPITE OCCASIONAL HICCUPS.
IT'S AN UMBRELLA STAND.
IT'S A GARDEN HOSE.
BUT TODAY WE HAVE DNA SEQUENCING KITS IN OUR LABS AND EVOLUTIONARY IDEAS IN OUR MINDS. WHY NOT USE THEM?
THE PROCEDURE: COMPARE CORRESPONDING (I.E., HOMOLOGOUS) GENES FROM DIFFERENT ORGANISMS AND ASSESS HOW SIMILAR THEY ARE—OR HOW DIFFERENT, IF YOU'RE IN THE GLASS-HALF-EMPTY TAXON.
THERE ARE TWO TAXA OF PEOPLE, APPARENTLY.

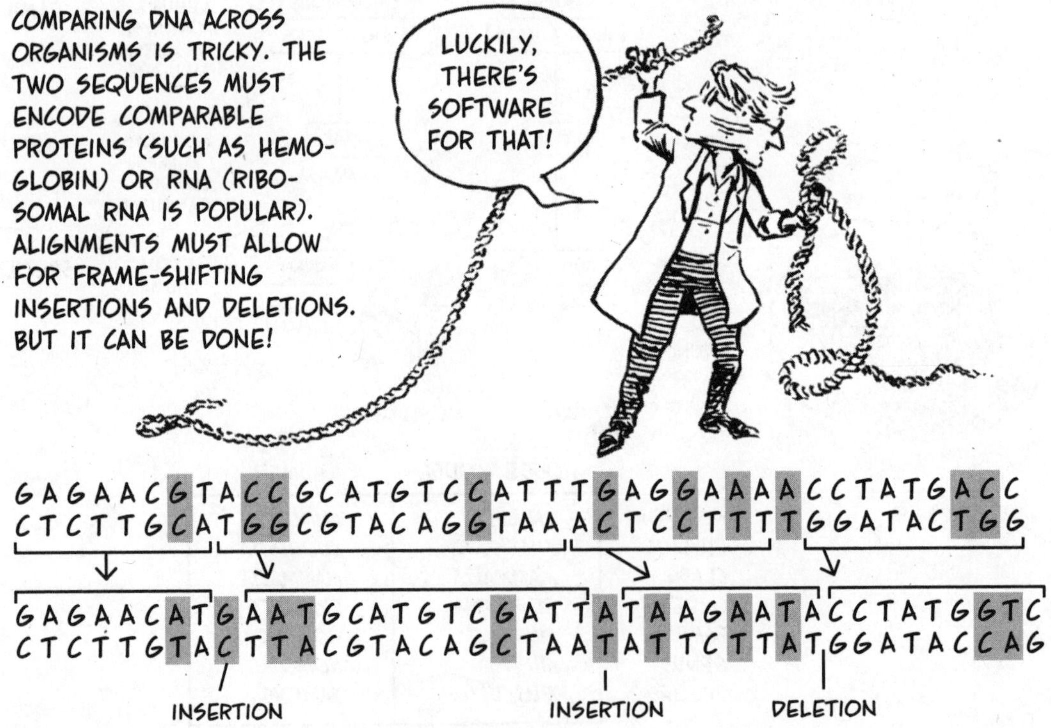
COMPARING DNA ACROSS ORGANISMS IS TRICKY. THE TWO SEQUENCES MUST ENCODE COMPARABLE PROTEINS (SUCH AS HEMOGLOBIN) OR RNA (RIBOSOMAL RNA IS POPULAR). ALIGNMENTS MUST ALLOW FOR FRAME-SHIFTING INSERTIONS AND DELETIONS. BUT IT CAN BE DONE!
LUCKILY, THERE'S SOFTWARE FOR THAT!
GAGAACGTACCGCATGTCCATTTGAGGAAAACCTATGACC
CTCTTGCATGGCGTACAGGTAAACTCCTTTTGGATACTGG
GAGAACATGAATGCATGTCGATTATAAGAATACCTATGGTC
CTCTTGTACTTACGTACAGCTAATATTCTTATGGATACCAG
INSERTION
INSERTION
DELETION

SINCE MUTATIONS PILE UP OVER TIME, MORE-SIMILAR SEQUENCES DIVERGED MORE RECENTLY FROM A COMMON ANCESTOR. BY COMPARING GENES FROM MANY ORGANISMS, WE CAN BUILD A **PHYLOGENETIC TREE,** OR GENEALOGY. EACH BRANCHING POINT SHOWS THE LAST COMMON ANCESTOR OF ALL ITS BRANCHES.

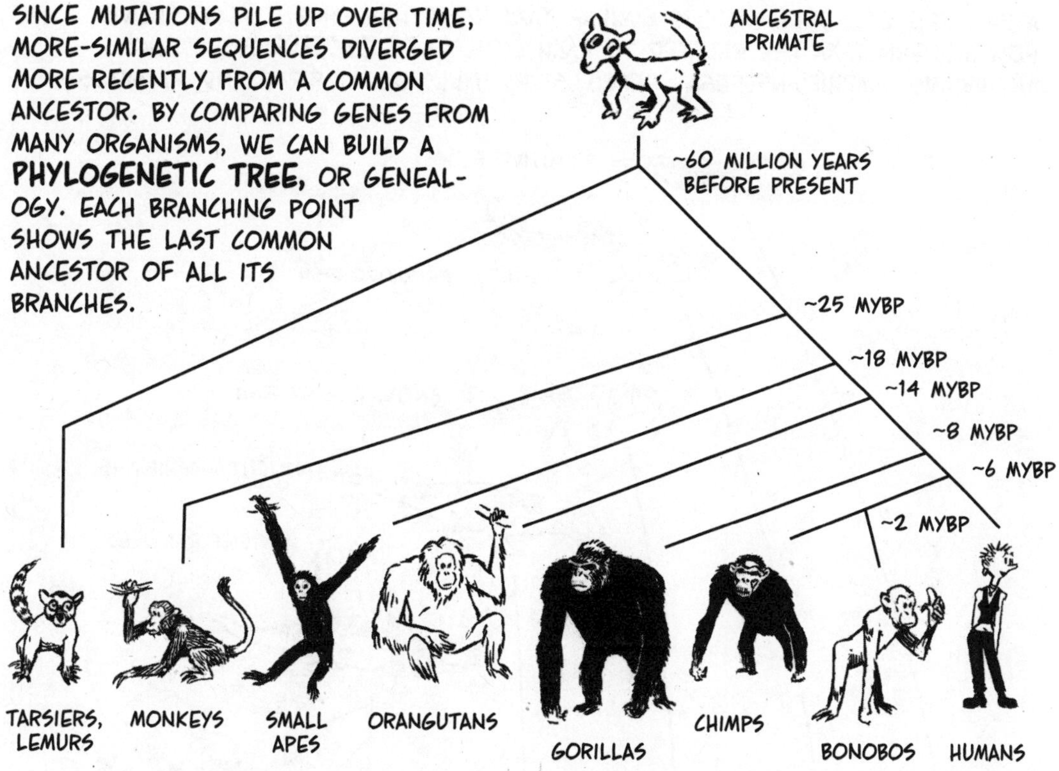
ANCESTRAL PRIMATE
~60 MILLION YEARS BEFORE PRESENT
~25 MYBP
~18 MYBP
~14 MYBP
~8 MYBP
~6 MYBP
~2 MYBP
TARSIERS, LEMURS
MONKEYS
SMALL APES
ORANGUTANS
GORILLAS
CHIMPS
BONOBOS
HUMANS

AS THE NUMBERS SUGGEST, WE CAN EVEN ESTIMATE **HOW LONG AGO** TWO LINES DIVERGED. MUTATIONS HAPPEN AT A MORE OR LESS STEADY RATE, SO THE DEGREE OF DIFFERENCE BETWEEN GENES BECOMES A **GENETIC CLOCK.**

TICK
TICK
TICK

A PHYLOGENETIC TREE GROUPS SIMILAR TAXA TOGETHER—AND IT ALSO SHOWS HOW MODERN TAXA ARE RELATED TO EACH OTHER BY **DESCENT.** FISH BEGAT AMPHIBIANS; AMPHIBIANS BEGAT REPTILES; REPTILES BEGAT BIRDS AND MAMMALS.

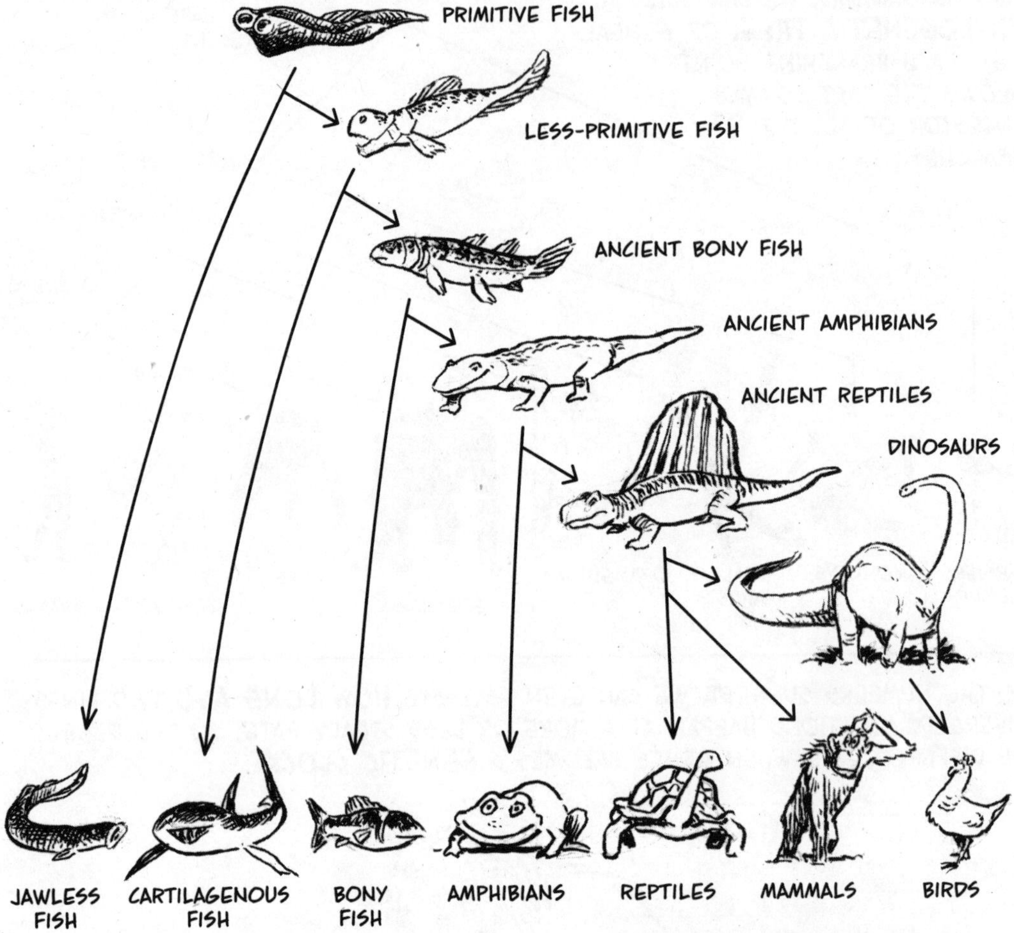

PRIMITIVE FISH
LESS-PRIMITIVE FISH
ANCIENT BONY FISH
ANCIENT AMPHIBIANS
ANCIENT REPTILES
DINOSAURS
JAWLESS FISH
CARTILAGENOUS FISH
BONY FISH
AMPHIBIANS
REPTILES
MAMMALS
BIRDS

TAXONOMISTS NOW ALWAYS PREFER PHYLOGENY TO MORPHOLOGY; GENETICS IS MORE PRECISE, MORE RELIABLE, AND MORE INFORMATIVE THAN BODILY STRUCTURES.
IF IT ISN'T A GARDEN HOSE, WHAT IS IT?
A PIG'S COUSIN.

PHYLOGENY REVEALS **HISTORY.** BY COMPARING DNA FROM LIVING PEOPLE, GENETICISTS CAN SKETCH THE MIGRATION ROUTES OF *HOMO SAPIENS'* ANCESTORS AS THEY COLONIZED THE WORLD.

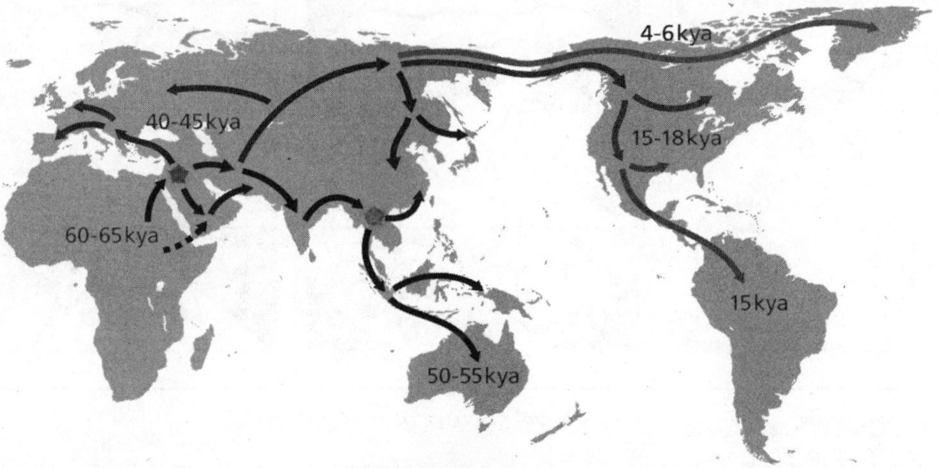

GENETICS FOUND THE LIKELY SOURCE OF **HIV,** THE VIRUS THAT CAUSES **AIDS.** THE MOST COMMON FORM MUTATED FROM A CHIMPANZEE VIRUS SOMETIME AFTER 1910 AND PROBABLY MIGRATED TO HUMANS MORE THAN ONCE (BY THE HUNTING AND/OR EATING OF CHIMPS).

WE'D ALL BE BETTER OFF IF THEY'D JUST LEAVE US ALONE.

DNA CAN ALSO HELP SETTLE TAXONOMIC PUZZLES. BIOLOGISTS USED TO ARGUE ABOUT WHETHER THE **GIANT PANDA** WAS A BEAR OR A RACCOON, UNTIL GENETIC STUDIES AT THE U.S. NATIONAL ZOO ANSWERED THE QUESTION. IT'S A BEAR.

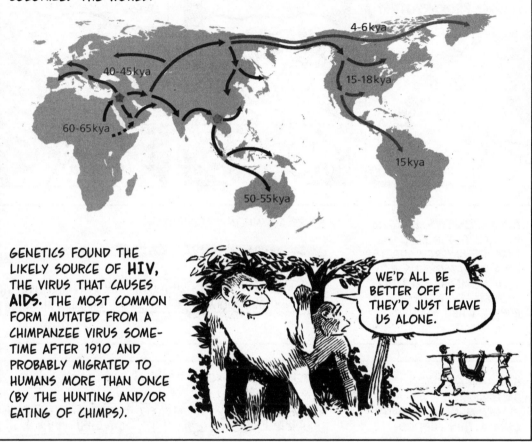

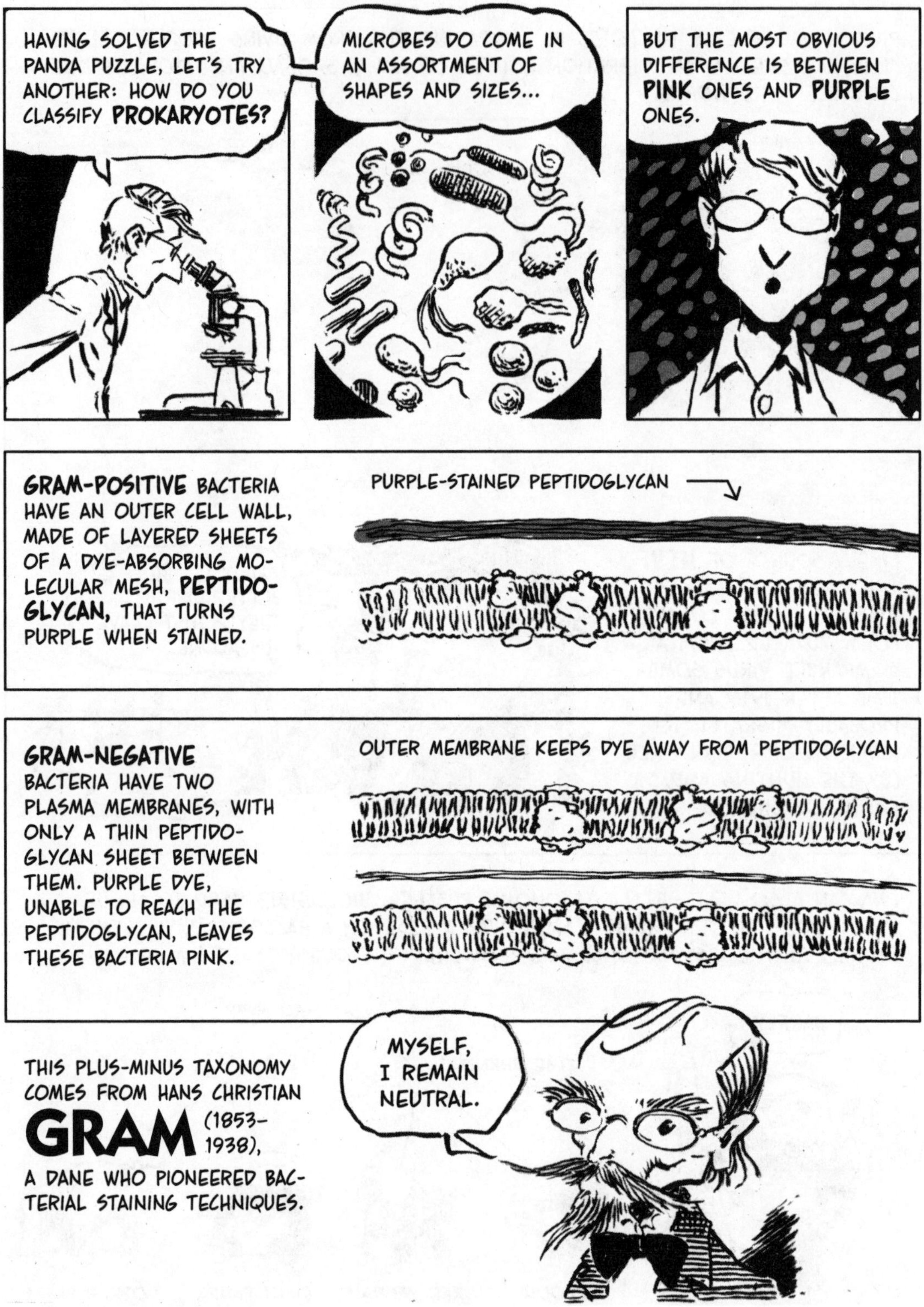
HAVING SOLVED THE PANDA PUZZLE, LET'S TRY ANOTHER: HOW DO YOU CLASSIFY PROKARYOTES?
MICROBES DO COME IN AN ASSORTMENT OF SHAPES AND SIZES...
BUT THE MOST OBVIOUS DIFFERENCE IS BETWEEN PINK ONES AND PURPLE ONES.
GRAM-POSITIVE BACTERIA HAVE AN OUTER CELL WALL, MADE OF LAYERED SHEETS OF A DYE-ABSORBING MOLECULAR MESH, PEPTIDOGLYCAN, THAT TURNS PURPLE WHEN STAINED.
PURPLE-STAINED PEPTIDOGLYCAN
GRAM-NEGATIVE BACTERIA HAVE TWO PLASMA MEMBRANES, WITH ONLY A THIN PEPTIDOGLYCAN SHEET BETWEEN THEM. PURPLE DYE, UNABLE TO REACH THE PEPTIDOGLYCAN, LEAVES THESE BACTERIA PINK.
OUTER MEMBRANE KEEPS DYE AWAY FROM PEPTIDOGLYCAN
THIS PLUS-MINUS TAXONOMY COMES FROM HANS CHRISTIAN GRAM (1853–1938), A DANE WHO PIONEERED BACTERIAL STAINING TECHNIQUES.
MYSELF, I REMAIN NEUTRAL.

WHAT ABOUT PROKARYOTIC PHYLOGENY? IN THE LATE 1970S, U. OF ILLINOIS BIOLOGIST CARL

WOESE

(1928–2012) EXPLORED THE SUBJECT. FOR COMPARISON PURPOSES, HE CHOSE A GENE SHARED, IN SOME FORM, BY ALL PROKARYOTES.

IT ENCODES A LONG (~1,540 BASES) RNA MOLECULE FOUND IN THE RIBOSOME'S SMALL SUBUNIT (SSU). THE MOLECULE, KNOWN AS **16S rRNA,** PLAYS A CRITICAL ROLE IN MAKING PROTEINS.

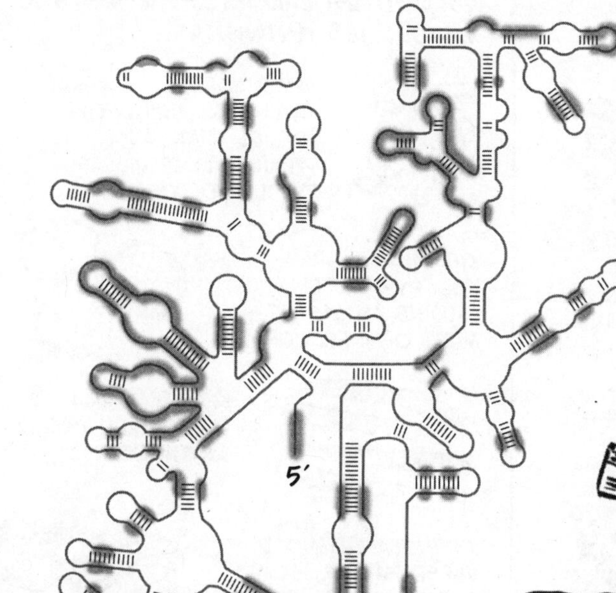

THE CELL DIES WITHOUT "GOOD" SSU 16S rRNA; SOME OF THE MOLECULE'S SEQUENCES MUST **NEVER CHANGE.** MUTATIONS AT THOSE SITES WOULD BE FATAL.

ON THE OTHER HAND, SOME PARTS **CAN** VARY, AND THESE VARIABLE REGIONS ARE INTERESTING.

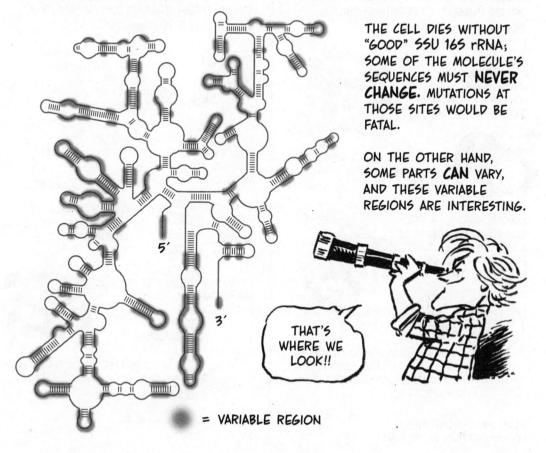

TO WOESE'S SURPRISE, THESE SEQUENCES FELL INTO TWO GROUPS, EACH CONSISTENT WITHIN ITSELF, BUT DISTINCT FROM THE OTHER. THE BIOLOGIST ANNOUNCED THAT PROKARYOTES SHOULD BE **SPLIT** INTO TWO SEPARATE "DOMAINS."

NOTHING NEGATIVE ABOUT YOU, HANS CHRISTIAN! I'M POSI-TIVE! NO STAIN ON YOUR REPUTATION!
GRAM

ARCHAEA

NO PEPTIDOGLYCAN; FREAKISH MEMBRANE CHEMISTRY; SOME PRODUCE METHANE; OFTEN INHABIT EXTREME ENVIRONMENTS.

EURYARCHAEOTA: A DIVERSE PHYLUM, IN WHICH SOME SPECIES LOVE HIGH HEAT AND OTHERS NEED SALT.

"A.R.M.A.N.": FOUND IN EXTREMELY ACIDIC MINING WASTEWATER.

LOKIARCHAEOTA: A DEEP-SEA DWELLER FOUND AT A HOT OCEAN-FLOOR FORMATION CALLED LOKI'S CASTLE.

THORARCHAEOTA: ANOTHER OF THE SUPERPHYLUM *ASGARD*, THE ARCHAEAN GROUP CLOSEST TO EUKARYOTES.

PLUS MANY MORE PHYLA (CLASSIFICATION IS IN FLUX).

BACTERIA

PEPTIDOGLYCAN CELL WALLS, USUALLY; CELLULAR MEMBRANES LIKE THOSE OF EUKARYOTES; CHARACTERISTIC BASES AT VARIOUS 16S rRNA SITES.

SPIROCHAETAE: A DIVERSE PHYLUM INCLUDING THE SPECIES THAT CAUSE SYPHILIS, LYME DISEASE, AND LEPTOSPOROSIS.

CHLAMYDIAE: GRAM-NEGATIVE BACTERIA CAUSING AN ASSORTMENT OF INFECTIONS.

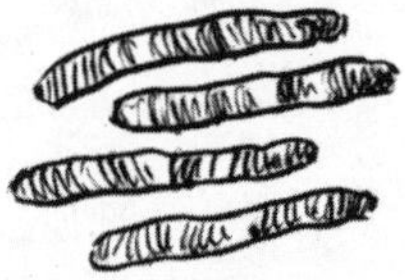

CYANOBACTERIA: BLUE-GREEN PHOTOSYNTHESIZERS.

ACTINOBACTERIA: A DIVERSE PHYLUM; SOME SPECIES MAINTAIN SOIL HEALTH; ANOTHER CAUSES TUBERCULOSIS.

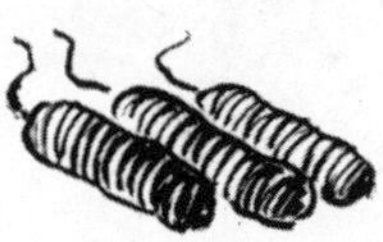

PROTEOBACTERIA: INCLUDES *E. COLI*, *VIBRIO*, *SALMONELLA*, AND OTHERS.

PLUS 24 MORE PHYLA (BY ONE COUNT).

WEIRDLY ENOUGH, ARCHAEA ARE GENETICALLY CLOSER TO **EUKARYOTES** THAN TO **BACTERIA.** WOESE'S THREE DOMAINS HAVE THIS PHYLOGENETIC TREE.

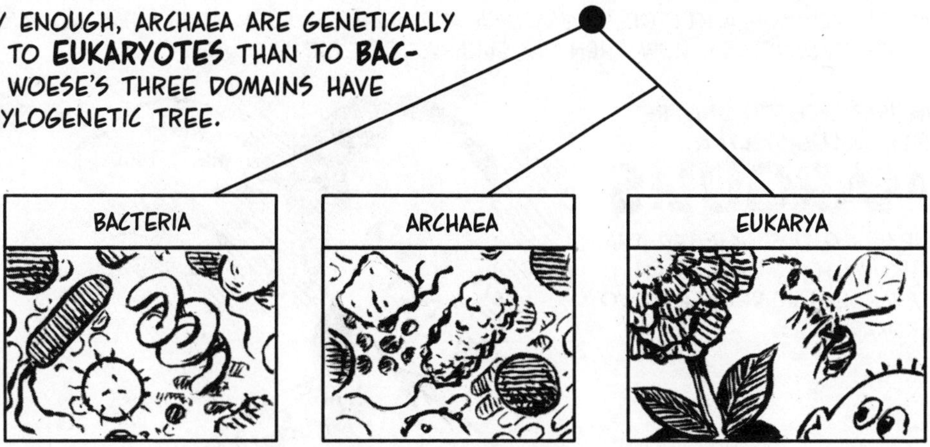

WHAT IS THAT TOP DOT? WHAT SPECIES DIVERGED INTO THE TWO DIFFERENT LINES? BIOLOGISTS CALL THE THING **LUCA,** THE **L**AST **U**NIVERSAL **C**OMMON **A**NCESTOR, AND DATE IT TO MORE THAN **3.5 BILLION** YEARS AGO. EVERY ORGANISM ON EARTH COMES FROM LUCA.

LUCA WAS AN EARLY LIFE FORM, BUT PRESUMABLY NOT THE FIRST. IT HAD AUNTS AND COUSINS, BUT ONLY LUCA'S DIRECT LINE SURVIVED. THE ORIGIN OF LIFE LIES EVEN FARTHER BACK IN THE MYSTERIOUS PAST.

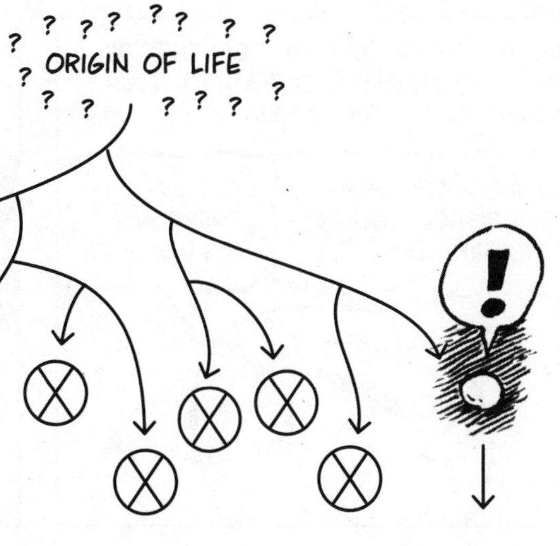

AS THE COMMON ANCESTOR OF BACTERIA AND ARCHAEA, LUCA SURELY HAD NO NUCLEUS OR OTHER ORGANELLES. HOW THEN DID EUKARYOTES EVER EVOLVE?

IN 1967, BOSTON UNIVERSITY BIOLOGIST LYNN **MARGULIS** (1938–2011) SUGGESTED THAT EUKARYOTES BEGAN AS AN ARRANGEMENT BETWEEN TWO PROKARYOTES.

KISS ME, STUPID!

ONE PROKARYOTE ENGULFED, INVADED, OR MERGED WITH ANOTHER ONE.

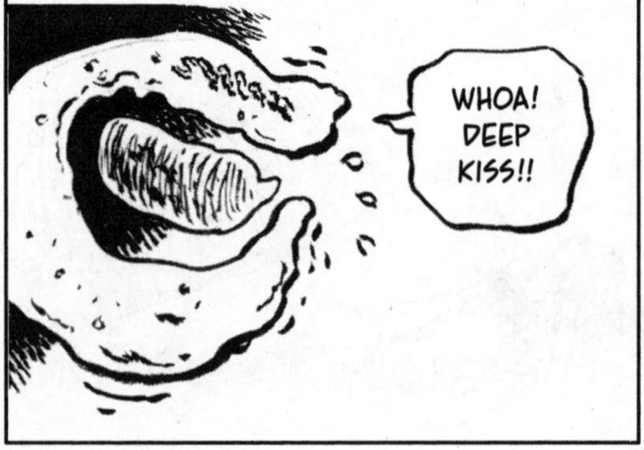

WHOA! DEEP KISS!!

THE TWO CELLS EVIDENTLY FOUND IT EASIER TO PARTNER THAN TO FIGHT.

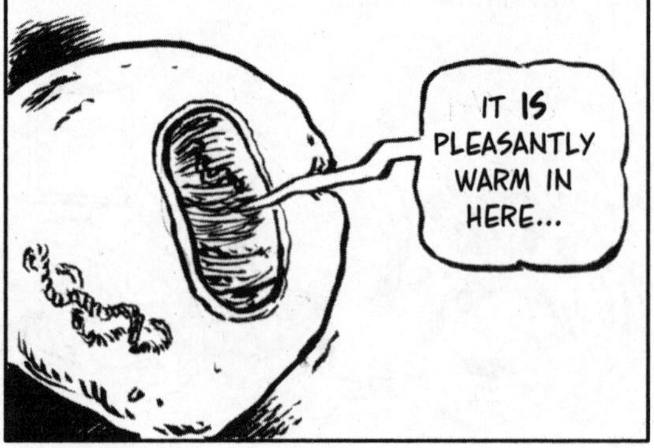

IT **IS** PLEASANTLY WARM IN HERE...

DIVIDING UP RESPONSIBILITIES, THEY BECAME A **COMPOSITE ORGANISM,** EACH DEPENDING ON THE OTHER.

I'LL MAKE ATP AND WASH DISHES; YOU DO EVERYTHING ELSE.
OK! BUT... THERE ARE NO DISHES...

MARGULIS CALLED THEIR RELATIONSHIP **ENDOSYMBIOSIS** ("INSIDE TOGETHER LIVING").

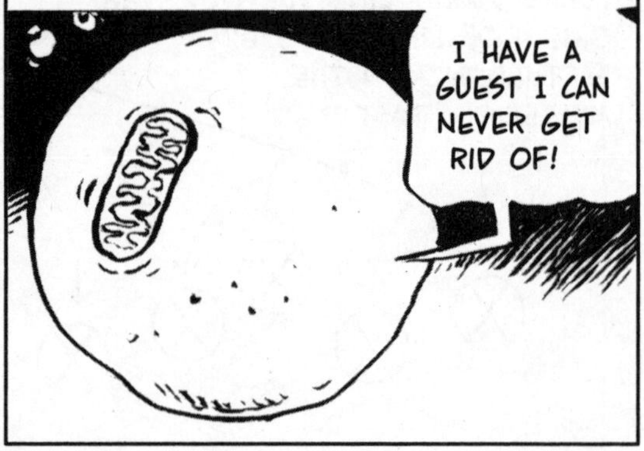

I HAVE A GUEST I CAN NEVER GET RID OF!

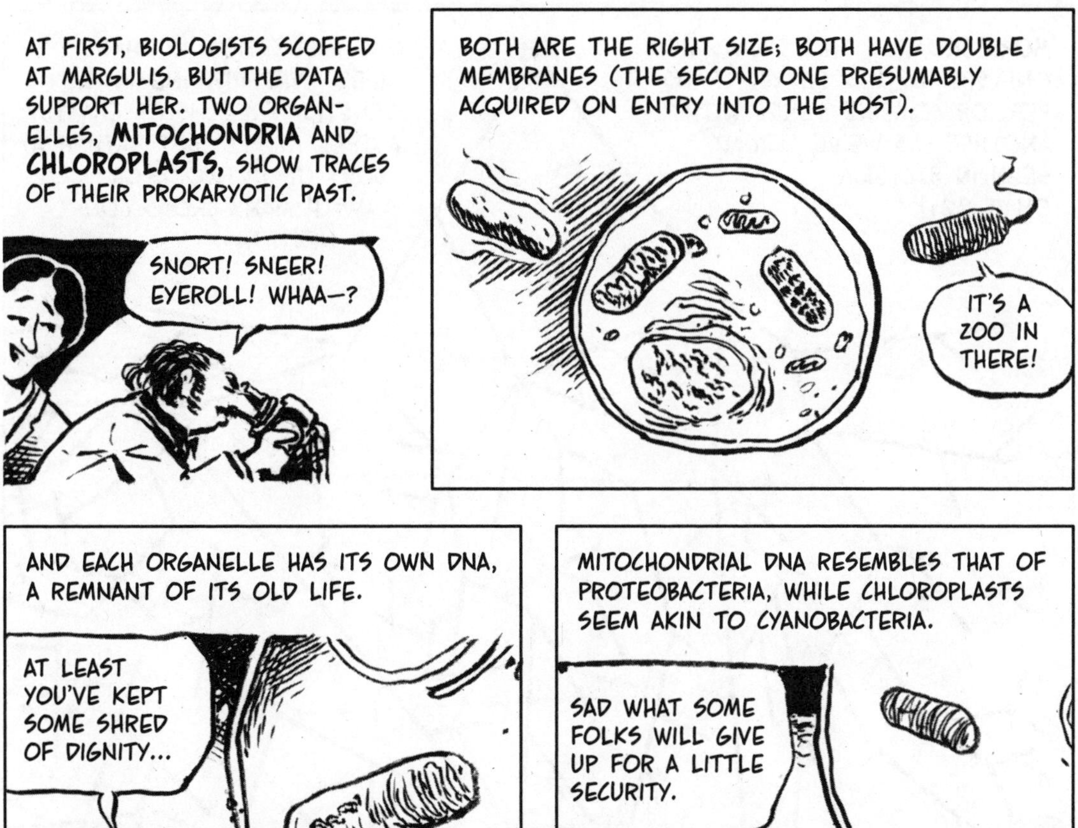
AT FIRST, BIOLOGISTS SCOFFED AT MARGULIS, BUT THE DATA SUPPORT HER. TWO ORGANELLES, **MITOCHONDRIA** AND **CHLOROPLASTS,** SHOW TRACES OF THEIR PROKARYOTIC PAST.
SNORT! SNEER! EYEROLL! WHAA—?
BOTH ARE THE RIGHT SIZE; BOTH HAVE DOUBLE MEMBRANES (THE SECOND ONE PRESUMABLY ACQUIRED ON ENTRY INTO THE HOST).
IT'S A ZOO IN THERE!
AND EACH ORGANELLE HAS ITS OWN DNA, A REMNANT OF ITS OLD LIFE.
AT LEAST YOU'VE KEPT SOME SHRED OF DIGNITY...
MITOCHONDRIAL DNA RESEMBLES THAT OF PROTEOBACTERIA, WHILE CHLOROPLASTS SEEM AKIN TO CYANOBACTERIA.
SAD WHAT SOME FOLKS WILL GIVE UP FOR A LITTLE SECURITY.
BEST GUESS: OUR CELLS GO BACK TO A 2-BILLION-YEAR-OLD BACTERIAL MOVE-IN THAT MADE A MITOCHONDRION.
YEAH, IT **MITO** HAPPENED THAT WAY...
OY.
HALF A BILLION YEARS LATER, THE ANCESTRAL CHLOROPLAST BARGED IN AND CREATED A PHOTOSYNTHETIC EUKARYOTE.
BY THE WAY, WHERE'S THE NUCLEUS?
WE'RE STILL WORKING THAT ONE OUT...

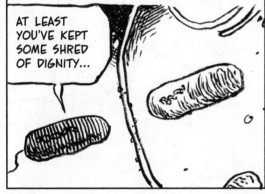

MORE BROADLY, THIS SHOWS HOW ONE SPECIES MAY SHARE, TRANSFER, OR COMBINE GENES WITH ANOTHER (AS WE'VE ALREADY SEEN IN BACTERIA ON P. 221).
THE IMPLICATION IS THAT LIFE'S EVER-DIVIDING PHYLOGENETIC "TREE" ISN'T EXACTLY A TREE AT ALL, BUT RATHER A SORT OF NETWORK OR WEB WITH MANY CROSS CONNECTIONS.
WE NEXT EXPLORE SOME OTHER WEBS OF INTERACTION.

Chapter 16
THE WORLDWIDE WEB

OF LIFE, THAT IS

EVERY SPECIES ADAPTS TO ITS ENVIRONMENT, AND ITS ENVIRONMENT INCLUDES OTHER SPECIES. BESIDES LIVING WITH RAINFALL, ALTITUDE OR DEPTH, TEMPERATURE, AND OTHER INORGANIC FACTORS, LIVING THINGS LIVE WITH **EACH OTHER**.
BEWARE THE GIANT FOOT OF HEAVEN!

AN INTIMATE RELATIONSHIP BETWEEN TWO SPECIES IS CALLED

SYMBIOSIS.

(ENDOSYMBIOSIS IS THE SPECIAL CASE OF ONE CREATURE LIVING INSIDE ANOTHER.)

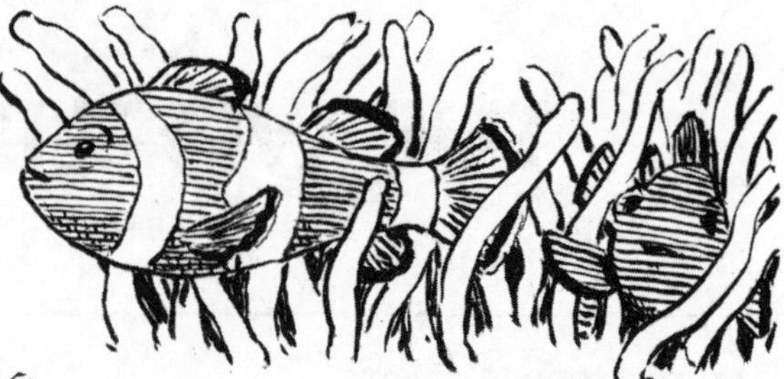

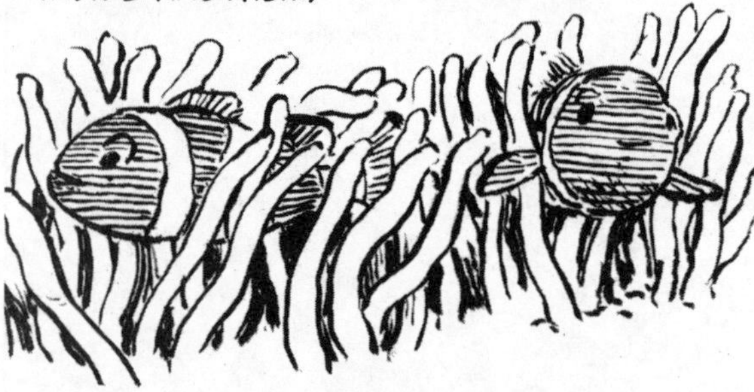

AS SEEN IN THE MOVIE *FINDING NEMO*, FOR EXAMPLE, **CLOWNFISH** MAKE A HOME AMONG THE STINGING TENTACLES OF **SEA ANEMONES.**

PROTECTED FROM PREDATORS BY ANEMONE STINGERS (CLOWNFISH HAVE AN ANTI-STING MUCUS COATING), THE FISH ALSO FIND FOOD IN THE ANEMONES' LEFTOVERS.

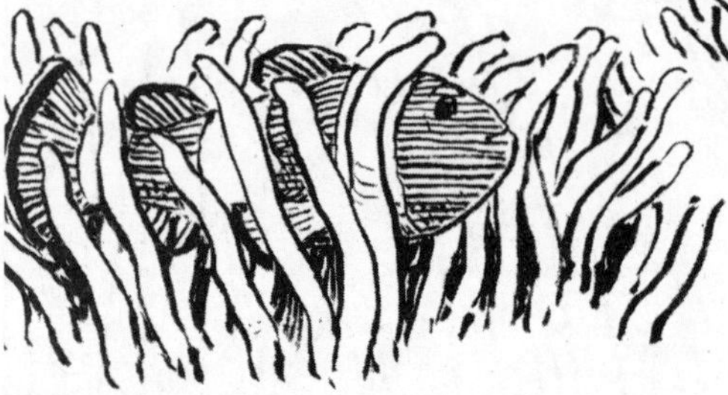

ANEMONES, MEANWHILE, FIND NOURISHMENT IN CLOWNFISH EXCREMENT AND ENJOY BETTER HEALTH BECAUSE THE FISHES' BEATING FINS CIRCULATE WATER.

A SYMBIOTIC RELATIONSHIP THAT BENEFITS BOTH PARTIES IS CALLED **MUTUALISTIC.**

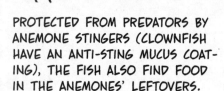

YOU WANT ME TO **WHAT** ON YOU?
OW!
OW!
OW!
PLEASE?

SYMBIOSIS IS SAID TO BE **COMMENSAL** WHEN IT BENEFITS ONE PARTY BUT IS MORE OR LESS NEUTRAL FOR THE OTHER.

A **CATTLE EGRET** FOLLOWS LIVESTOCK AROUND THE FIELD, BECAUSE THE BIG MAMMALS STIR UP INSECTS AND RODENTS THAT THE BIRDS LIKE TO EAT. THE COW BARELY NOTICES.

AND THEY CALL US BIRD-BRAINS!

IN A **PARASITIC** RELATIONSHIP, ONE SPECIES GAINS WHILE THE OTHER LOSES. A FLEA LIVES OFF A DOG'S BLOOD, WHILE THE DOG GETS NOTHING BUT AN ITCH OR A FLEA-BORNE DISEASE. DRAINING OFF BLOOD NEVER HELPED ANYONE, WHATEVER DOCTORS USED TO SAY.

DOCTOR, YOU'RE A REGULAR LEECH!
YOU SHOULD SEE THE INSURANCE COMPANY!

SEVERAL SPECIES CAN JOIN FORCES TO BUILD A COMMUNITY. THE SCUM IN YOUR SHOWER DRAIN IS A MICROBIAL EXAMPLE.

YUCK!

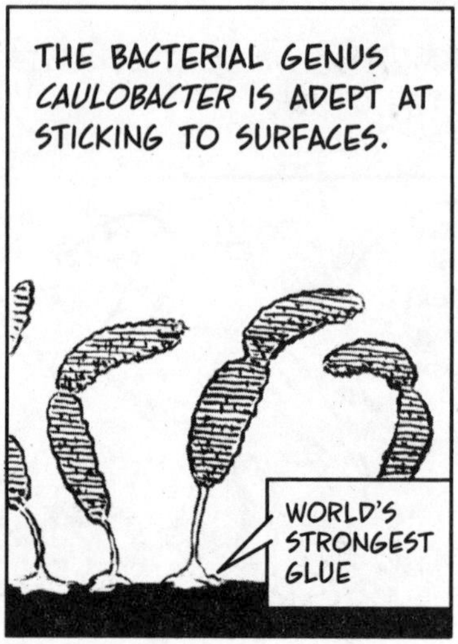
THE BACTERIAL GENUS *CAULOBACTER* IS ADEPT AT STICKING TO SURFACES.
WORLD'S STRONGEST GLUE

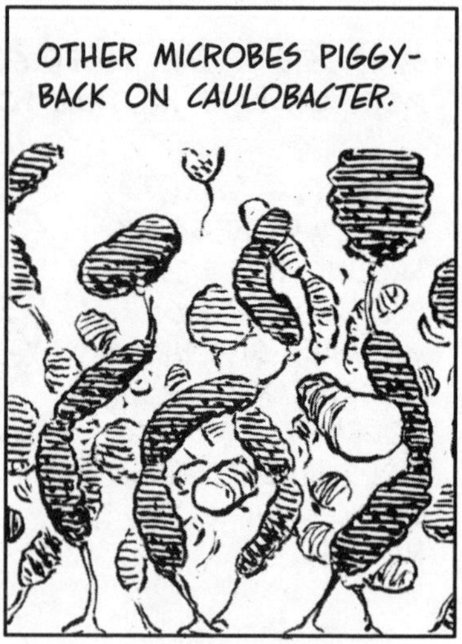
OTHER MICROBES PIGGY-BACK ON *CAULOBACTER*.

TOGETHER THEY SECRETE A GEL-LIKE, PROTECTIVE POLYSACCHARIDE MESH PIERCED BY CHANNELS THAT PIPE NUTRIENTS IN AND WASTE OUT. THE RESULTING **BIOFILM,** LIKE A CITY IN MINIATURE, HOSTS FAR MORE LIFE THAN COULD POSSIBLY SUBSIST THERE WITHOUT COMMUNITY SUPPORT.

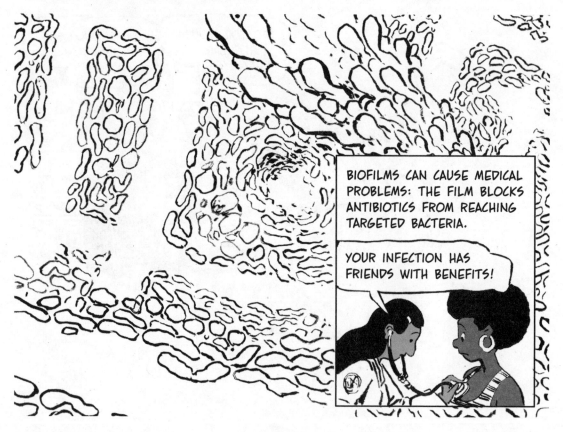

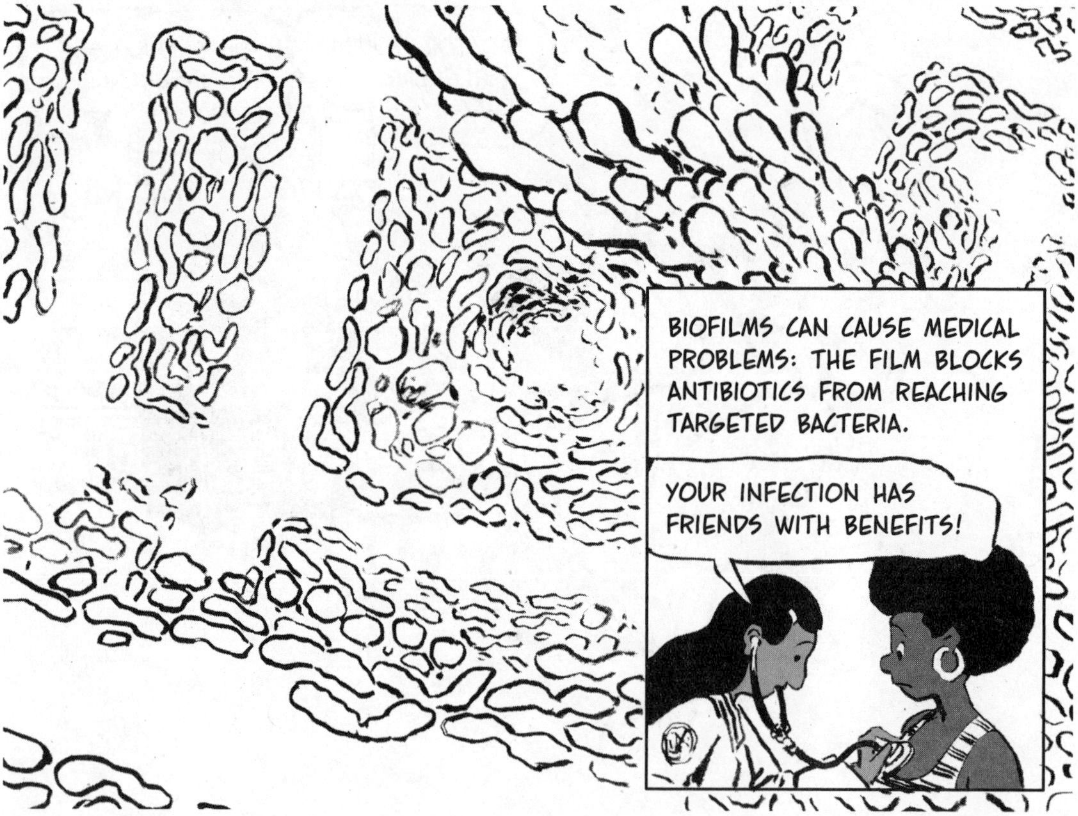
BIOFILMS CAN CAUSE MEDICAL PROBLEMS: THE FILM BLOCKS ANTIBIOTICS FROM REACHING TARGETED BACTERIA.
YOUR INFECTION HAS FRIENDS WITH BENEFITS!

COMMUNITIES OF BACTERIA ALSO LIVE IN ANIMAL INTESTINES. **TERMITES,** FOR EXAMPLE, WOULD STARVE WITHOUT THEIR GUT MICROBES, WHICH DO ALL THE WORK OF DIGESTING **WOOD,** THE TERMITE'S ONLY FOOD(!).

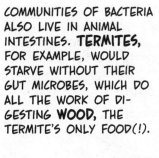

WE'RE PASSENGERS IN A WOOD DELIVERY SYSTEM...

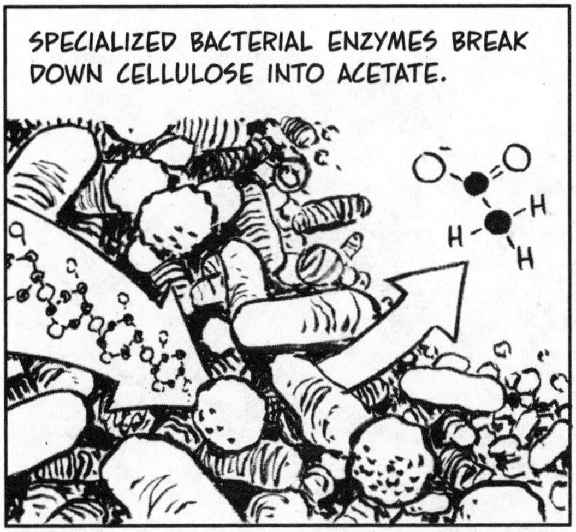
SPECIALIZED BACTERIAL ENZYMES BREAK DOWN CELLULOSE INTO ACETATE.

THE TERMITE STEALS THE ACETATE FOR ITS OWN KREBS CYCLE, SO THE BACTERIA ARE FORCED TO WORK OVERTIME.
DUDE, WHERE'S MY ACETATE?

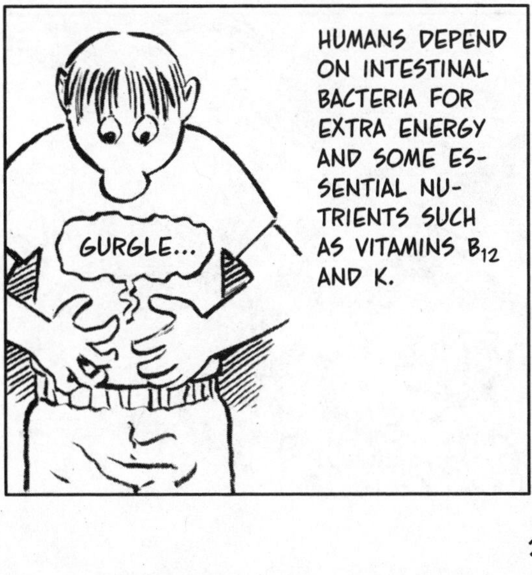
HUMANS DEPEND ON INTESTINAL BACTERIA FOR EXTRA ENERGY AND SOME ESSENTIAL NUTRIENTS SUCH AS VITAMINS B_{12} AND K.
GURGLE...

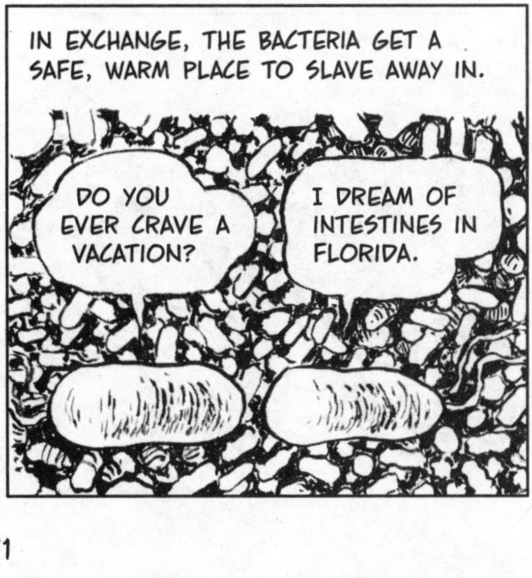
IN EXCHANGE, THE BACTERIA GET A SAFE, WARM PLACE TO SLAVE AWAY IN.
DO YOU EVER CRAVE A VACATION?
I DREAM OF INTESTINES IN FLORIDA.

THE GROUP OF ALL ORGANISMS THAT SHARE AN ENVIRONMENT IS CALLED A

BIOME.

IF THE ORGANISMS ARE MICROBES, THE BIOME IS A **MICROBIOME**.

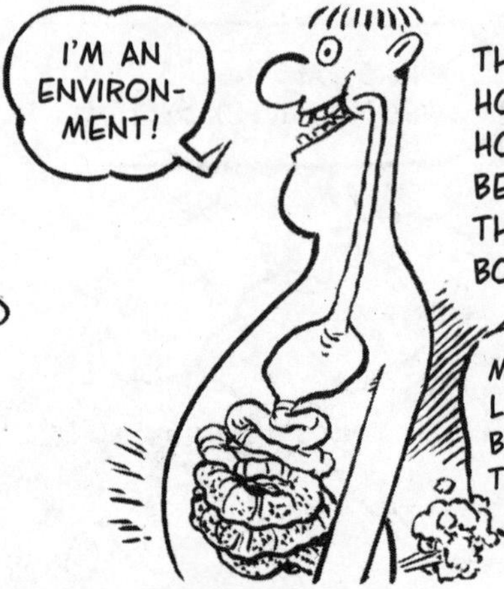

I'M AN ENVIRONMENT!

THE HUMAN MICROBIOME IS A HOT TOPIC TODAY, AS WE LEARN HOW HEALTH, ATTITUDES, AND BEHAVIOR (!) CAN DEPEND ON THE MICROBIAL MIX IN SOMEBODY'S GUTS.

MY MICROBIOME LEANS LEFT, BUT I HAVE BIG-TENT GENETICS.

SOME BIGGER BIOMES:

TROPICAL FOREST

DEEP-OCEAN VOLCANIC VENT

FRESHWATER MARSH

ALSO: TEMPERATE FORESTS, BOREAL (COLD-WEATHER) FORESTS, PLAINS, MOUNTAINS, ARCTIC OCEANS, AND MORE, NOT TO MENTION CITIES, FARMS, AND OTHER ARTIFICIAL ENVIRONMENTS.

ECOLOGY
IS THE STUDY OF ENVIRONMENTS, THEIR BIOMES, AND ALL THE INTERACTIONS WITHIN THEM, IN SHORT, THE STUDY OF ECOSYSTEMS.
I LOVE ECOLOGISTS! THEY NEVER SWAT!

SOME INTERACTIONS ARE NOT SO COOPERATIVE. TWO SPECIES MAY **COMPETE** FOR RESOURCES. THIS CAN HAPPEN IN AT LEAST TWO WAYS:

INTERFERENCE IS THE DIRECT OBSTRUCTION OF A COMPETITOR. EUCALYPTUS LEAVES, FOR INSTANCE, POISON THE GROUND AND STOP COMPETING PLANTS.

IN **EXPLOITATION COMPETITION,** TWO SPECIES BASICALLY TRY TO OUT-EAT EACH OTHER. VULTURES COMPETE WITH BACTERIAL DECOMPOSERS OVER ANIMAL CARCASSES.
DO VULTURES EAT DEAD VULTURES?
ASKING FOR A FRIEND.

AND OF COURSE, SOMETIMES ONE SPECIES SIMPLY **EATS** ANOTHER ONE. THAT'S AN INTERACTION, TOO!
HERBIVORES ARE THE PLANT-EATING ANIMALS, SUCH AS COWS, MICE, ANTELOPES, AND THIS VERY HUNGRY CATERPILLAR.
PREDATORS EAT OTHER ANIMALS. (YES, ANTS ARE ANIMALS, TOO.)
SCATTER!

PREDATION STIMULATES A SORT OF ARMS RACE: PREY SPECIES EVOLVE BETTER DEFENSES, WHILE PREDATORY SPECIES HONE THEIR ATTACK CAPABILITIES.

HERDING AND FLOCKING BEHAVIOR
SHARP ANTLERS OR HORNS
TALONS
POWERFUL JAWS, LIMBS, FANGS, CLAWS
ACUTE SENSES ON HIGH ALERT
ARMOR
VENOM
CAMOUFLAGE
SPECIALIZED SNOUT
HIDING

PLANTS AND FUNGI MAY ALSO FIGHT OFF HERBIVORES WITH THORNS AND TOXINS. THE DEATH-CAP MUSHROOM, *AMANITA PHALLOIDES*, MAKES A LETHAL POISON, α-AMANITIN, THAT **DISABLES RNA POLYMERASE.** (THE MUSHROOM ITSELF SURVIVES BY DEPLOYING A VARIANT RNA POLYMERASE MOLECULE.)

AMANITA PHALLOIDES, BY THE WAY, HAS A COMMENSAL SYMBIONT, A FRUIT FLY WITH ITS OWN α-AMANITIN-RESISTANT RNA POLYMERASE. THE FLY HATCHES ITS EGGS UNDER *AMANITA'S* DEADLY, PROTECTIVE CAP.

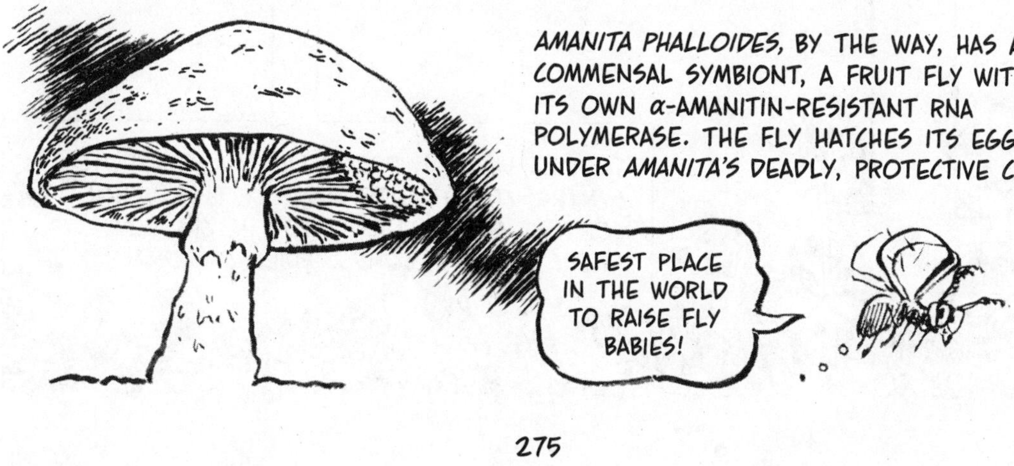
SAFEST PLACE IN THE WORLD TO RAISE FLY BABIES!

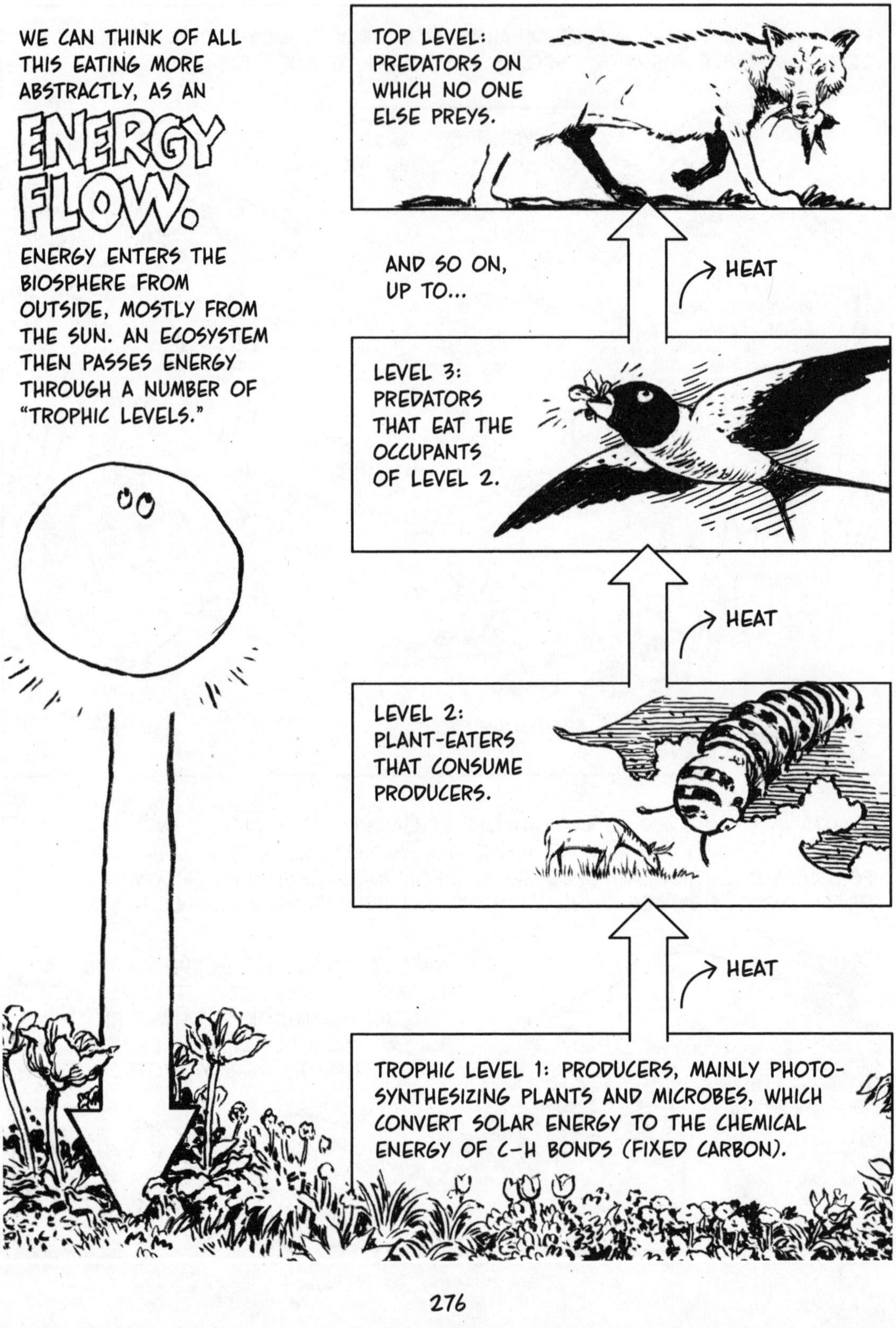

WE CAN THINK OF ALL THIS EATING MORE ABSTRACTLY, AS AN

ENERGY FLOW.

ENERGY ENTERS THE BIOSPHERE FROM OUTSIDE, MOSTLY FROM THE SUN. AN ECOSYSTEM THEN PASSES ENERGY THROUGH A NUMBER OF "TROPHIC LEVELS."

TOP LEVEL: PREDATORS ON WHICH NO ONE ELSE PREYS.

AND SO ON, UP TO...

→ HEAT

LEVEL 3: PREDATORS THAT EAT THE OCCUPANTS OF LEVEL 2.

→ HEAT

LEVEL 2: PLANT-EATERS THAT CONSUME PRODUCERS.

→ HEAT

TROPHIC LEVEL 1: PRODUCERS, MAINLY PHOTO-SYNTHESIZING PLANTS AND MICROBES, WHICH CONVERT SOLAR ENERGY TO THE CHEMICAL ENERGY OF C-H BONDS (FIXED CARBON).

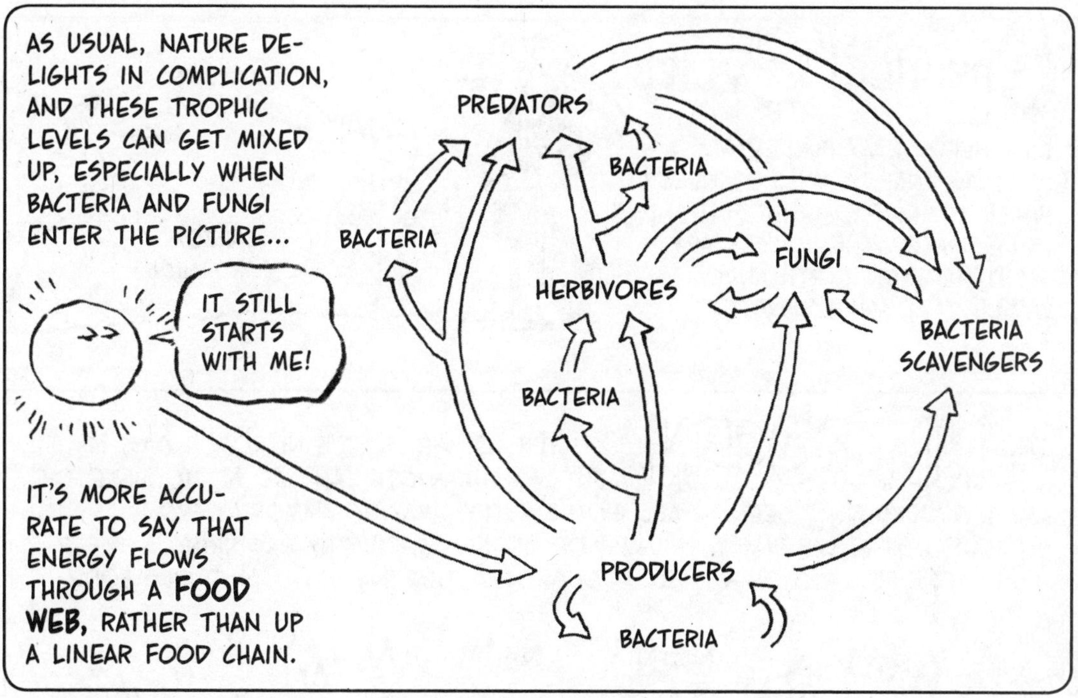
AS USUAL, NATURE DELIGHTS IN COMPLICATION, AND THESE TROPHIC LEVELS CAN GET MIXED UP, ESPECIALLY WHEN BACTERIA AND FUNGI ENTER THE PICTURE...
IT STILL STARTS WITH ME!
PREDATORS
BACTERIA
BACTERIA
FUNGI
HERBIVORES
BACTERIA
SCAVENGERS
BACTERIA
PRODUCERS
BACTERIA
IT'S MORE ACCURATE TO SAY THAT ENERGY FLOWS THROUGH A **FOOD WEB,** RATHER THAN UP A LINEAR FOOD CHAIN.

ALSO NOTE: EATERS ALWAYS HAVE LESS ENERGY THAN WHAT THEY EAT. ENERGY ESCAPES AT EVERY MEAL. NO ENERGY TRANSFER IS EVER 100% EFFICIENT.

FOR EXAMPLE, ONE SQUARE KILOMETER CAN SUPPORT UPWARD OF 200 MILLION WHEAT PLANTS, WITH AN ENORMOUS CALORIE CONTENT.

THAT SAME AREA WILL SUPPORT NO MORE THAN TWO COYOTES, A TOP PREDATOR THAT DEPENDS ON RODENTS AND BIRDS THAT LIVE OFF THE VEGETATION. NOT MUCH ENERGY IN TWO COYOTES!

THREE'S A CROWD.

Chemical Cycles

LIFE NEEDS **RAW MATERIALS** AS WELL AS ENERGY. UNLIKE ENERGY, WHICH COMES FROM THE SUN AND IS DISSIPATED, CHEMICALS START RIGHT HERE ON EARTH AND **CYCLE** THROUGH ECOSYSTEMS.

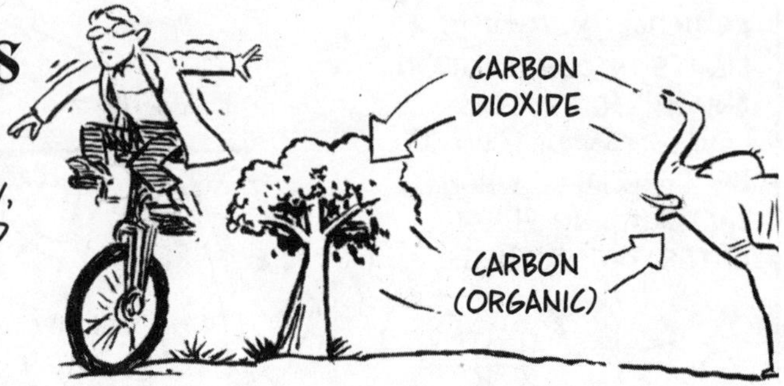

TAKE **NITROGEN**, OR TRY TO! AIR IS 80% NITROGEN GAS, N_2, BUT NO EUKARYOTE CAN USE N_2 TO MAKE THE AMINE GROUPS NH_2^- SEEN IN ALL BIOCHEMISTRY. (MAKING AMMONIA "FROM SCRATCH" IN THE LAB WAS A TRIUMPH OF MODERN CHEMISTRY REQUIRING A SUPER-HIGH-PRESSURE REACTION VESSEL.) TOUGH NUT, THAT N_2 WITH ITS TRIPLE BOND.

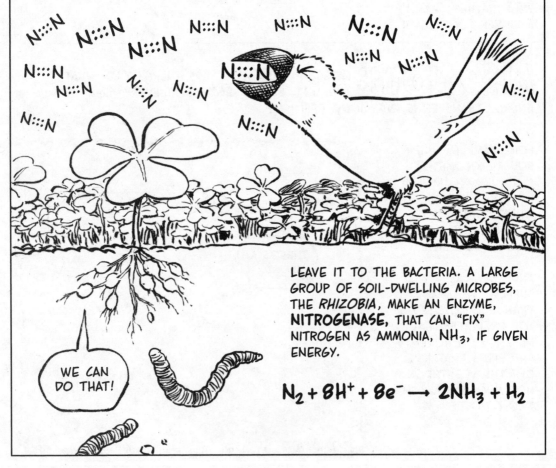

LEAVE IT TO THE BACTERIA. A LARGE GROUP OF SOIL-DWELLING MICROBES, THE *RHIZOBIA*, MAKE AN ENZYME, **NITROGENASE,** THAT CAN "FIX" NITROGEN AS AMMONIA, NH_3, IF GIVEN ENERGY.

$$N_2 + 8H^+ + 8e^- \longrightarrow 2NH_3 + H_2$$

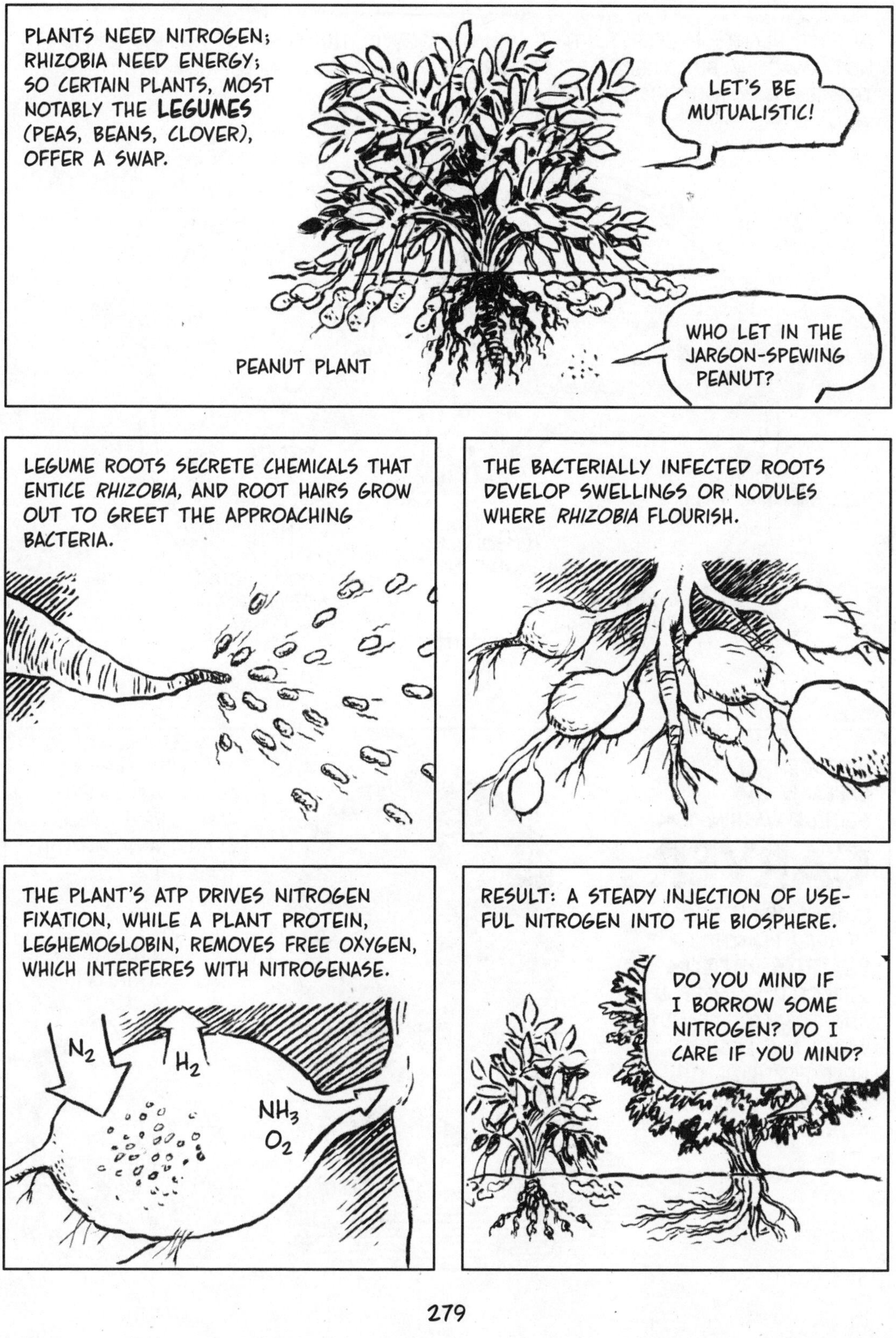
PLANTS NEED NITROGEN; RHIZOBIA NEED ENERGY; SO CERTAIN PLANTS, MOST NOTABLY THE **LEGUMES** (PEAS, BEANS, CLOVER), OFFER A SWAP.
LET'S BE MUTUALISTIC!
PEANUT PLANT
WHO LET IN THE JARGON-SPEWING PEANUT?
LEGUME ROOTS SECRETE CHEMICALS THAT ENTICE *RHIZOBIA*, AND ROOT HAIRS GROW OUT TO GREET THE APPROACHING BACTERIA.
THE BACTERIALLY INFECTED ROOTS DEVELOP SWELLINGS OR NODULES WHERE *RHIZOBIA* FLOURISH.
THE PLANT'S ATP DRIVES NITROGEN FIXATION, WHILE A PLANT PROTEIN, LEGHEMOGLOBIN, REMOVES FREE OXYGEN, WHICH INTERFERES WITH NITROGENASE.
N_2
H_2
NH_3
O_2
RESULT: A STEADY INJECTION OF USEFUL NITROGEN INTO THE BIOSPHERE.
DO YOU MIND IF I BORROW SOME NITROGEN? DO I CARE IF YOU MIND?

IN FACT, PLANTS PREFER TO TAKE UP NITROGEN IN THE FORM OF **NITRATE** (NO_3^-), NOT AMMONIA, BUT THERE ARE BACTERIA FOR THAT TOO. ONCE FIXED, NITROGEN TRAVELS THROUGH THIS CYCLE:

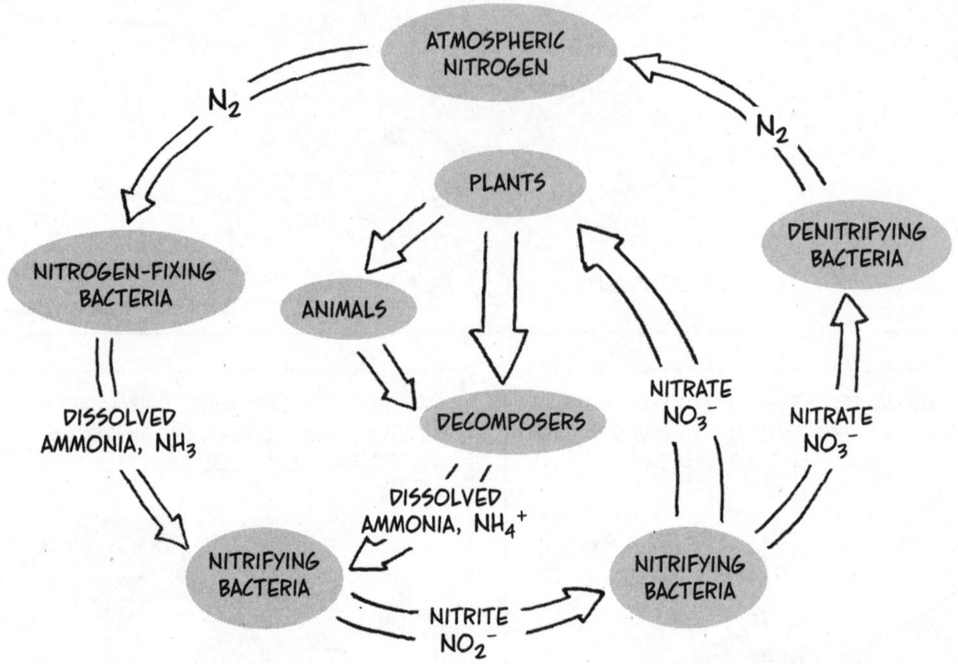

ATMOSPHERIC NITROGEN
N_2
N_2
PLANTS
DENITRIFYING BACTERIA
NITROGEN-FIXING BACTERIA
ANIMALS
DECOMPOSERS
NITRATE NO_3^-
NITRATE NO_3^-
DISSOLVED AMMONIA, NH_3
DISSOLVED AMMONIA, NH_4^+
NITRIFYING BACTERIA
NITRIFYING BACTERIA
NITRITE NO_2^-

NITROGEN FIXATION EXPLAINS WHY GEORGE WASHINGTON **CARVER** (C.1864–1943) RECOMMENDED PLANTING **PEANUTS** IN FIELDS DEPLETED BY COTTON AND TOBACCO. PEANUTS, WHICH ARE LEGUMES, HELP FERTILIZE THE SOIL WITH USABLE NITROGEN.
AND WHY PRE-COLUMBIAN AMERICANS GREW CORN, BEANS, AND SQUASH TOGETHER.
OUR HOLY TRINITY!

CARBON,

AS WE'VE SEEN, ENTERS THE BIOSPHERE WHEN PHOTOSYNTHESIZERS REDUCE CO_2. EATING, RESPIRATION, AND OTHER PROCESSES MOVE CARBON AROUND. HERE ARE SOME PATHWAYS IN THE CARBON CYCLE.

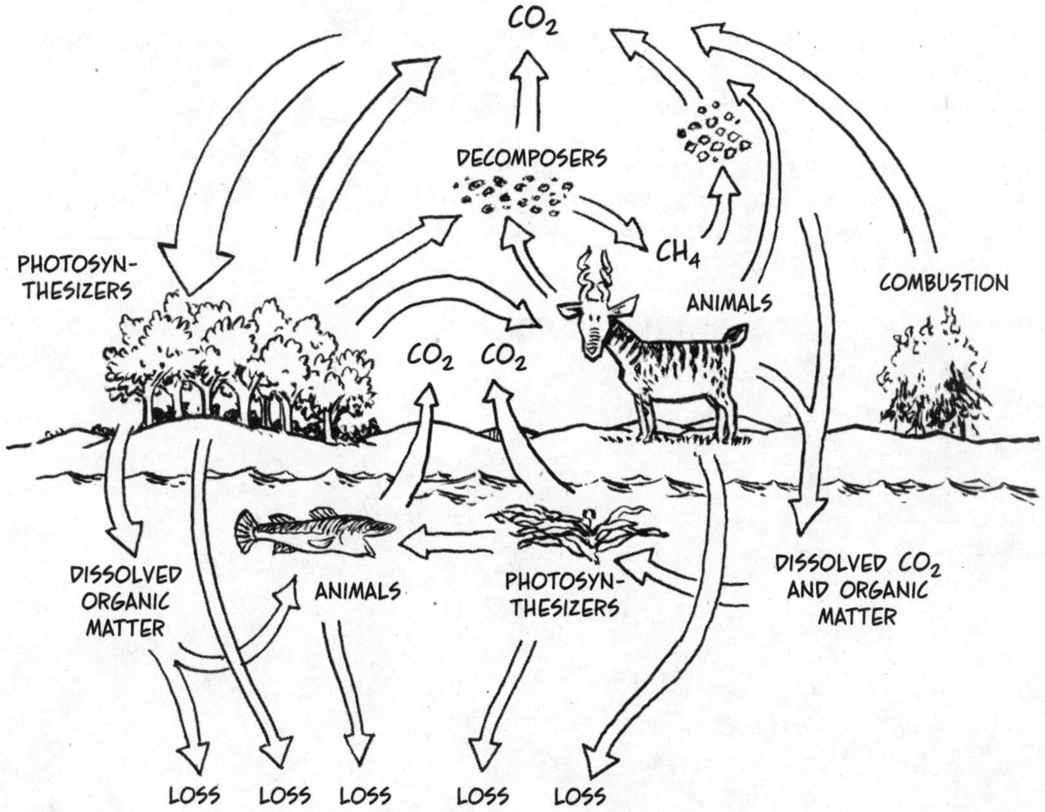

CO_2
DECOMPOSERS
CH_4
PHOTOSYNTHESIZERS
ANIMALS
COMBUSTION
CO_2 CO_2
DISSOLVED ORGANIC MATTER
ANIMALS
PHOTOSYNTHESIZERS
DISSOLVED CO_2 AND ORGANIC MATTER
LOSS LOSS LOSS LOSS LOSS

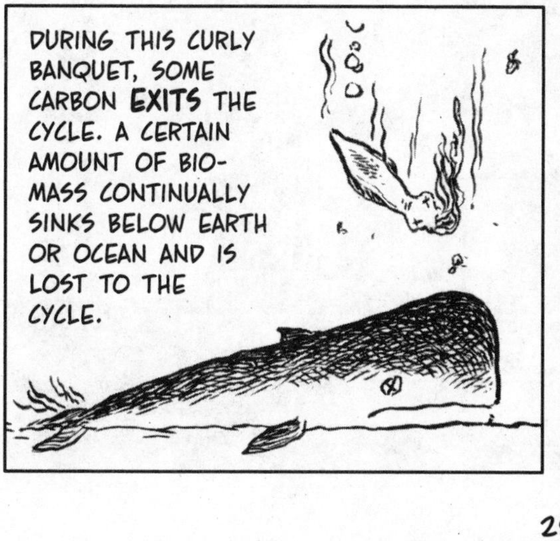

DURING THIS CURLY BANQUET, SOME CARBON **EXITS** THE CYCLE. A CERTAIN AMOUNT OF BIOMASS CONTINUALLY SINKS BELOW EARTH OR OCEAN AND IS LOST TO THE CYCLE.

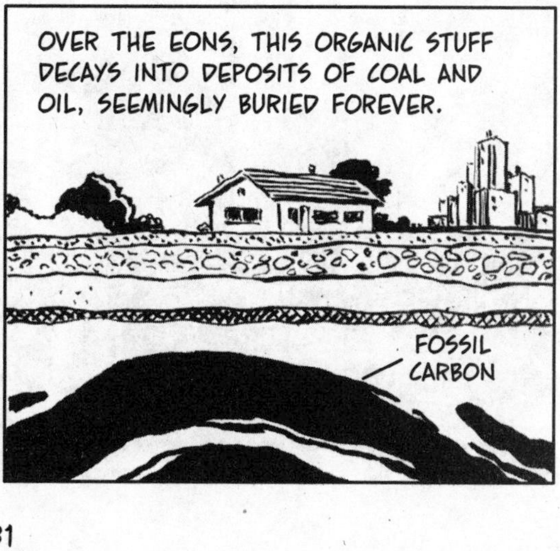

OVER THE EONS, THIS ORGANIC STUFF DECAYS INTO DEPOSITS OF COAL AND OIL, SEEMINGLY BURIED FOREVER.
FOSSIL CARBON

UNTIL, THAT IS, ONE SPECIES DEVOTED PART OF ITS CONSIDERABLE INGENUITY TO FINDING FOSSIL CARBON, DIGGING IT UP, AND BURNING IT, ALL WITHIN A SPAN OF TWO CENTURIES.

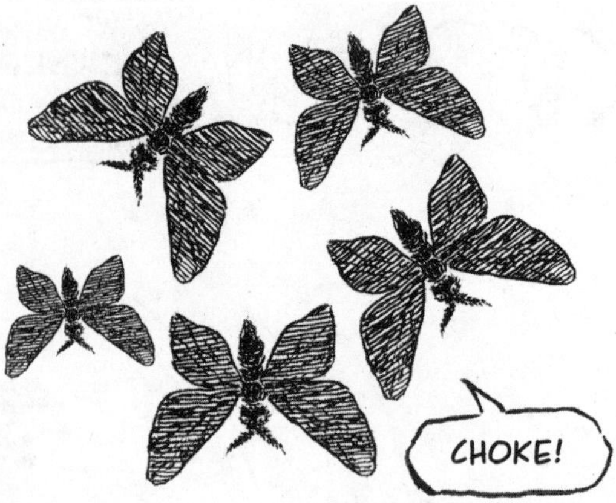

HOW DOES LIFE RESPOND TO A SUDDEN INFLUX OF ENERGY AND ITS DIVERSION TO SUPPORT A SINGLE SPECIES?

WILL NATURAL CYCLES ACCOMMODATE THE SHOCK AND FIND A NEW EQUILIBRIUM? IN OUR FINAL CHAPTER, WE EXPLORE DISRUPTIONS TO HOMEOSTASIS.

Chapter 17
DISRUPTION

WE'VE OFTEN SAID THAT CELLS AND ORGANISMS HAVE WAYS TO KEEP THEIR SYSTEMS IN TOP WORKING ORDER. LIVING THINGS CAN ADJUST TO CHANGES IN THEMSELVES AND THEIR SURROUNDINGS—UP TO A POINT, THAT IS.

TAKE NO NOTICE... CARRY ON... EVERY-THING'S FINE... NO NEED TO PANIC...

THIS CHAPTER IS ABOUT LIFE'S RESPONSES TO EXTREME STRESS.

BACTERIA, WHEN HEATED, EXPRESS GENES FOR SPECIAL PROTECTIVE PROTEINS (SEE P. 151), BUT NOT MANY BACTERIA CAN SURVIVE BOILING. THAT'S ONE REASON WE COOK SOUP.

WARM TEMPERATURES PUT STRESS ON HUMANS, TOO. WE DEAL WITH HEAT BY **PERSPIRING,** SECRETING SWEAT WHOSE EVAPORATION COOLS US DOWN.

BUT WITHOUT REPLACING THE LOST SALT AND WATER, WE MAY GET **HEAT STROKE,** A POTENTIALLY DEADLY FAILURE OF THE NERVOUS SYSTEM CAUSED BY LACK OF ESSENTIAL IONS.

SO IT'S HOMEOSTASIS OR DEATH? NOT QUITE. BETWEEN THE EXTREMES OF PERFECT HEALTH AND CESSATION OF LIFE IS A RANGE OF POSSIBILITIES.

COULD BE AN EPI-SODE OF LOST...

FOR EXAMPLE, WHEN ATTACKED BY SOME DISEASES, WE RUN A **FEVER**. THE HIGHER TEMPERATURE, IT IS THOUGHT, HELPS OUR IMMUNE SYSTEM FIGHT OFF INFECTION. OUR BODY CREATES A NEW "SET POINT" FOR TEMPERATURE.

CHICKEN SOUP, WHY DO YOU ASK?

ONCE THE ILLNESS PASSES, BODY TEMPERATURE RETURNS TO NORMAL. THE CHANGE OF EQUILIBRIUM IS ONLY TEMPORARY.

ON THE OTHER HAND, **ARTHRITIS** VICTIMS SUFFER CHRONIC, INCURABLE, USELESS, BUT SURVIVABLE BOUTS OF INFLAMMATION AND PAIN. THEIR BODIES SETTLE INTO A NEW, UNCOMFORTABLE EQUILIBRIUM AND MAINTAIN IT FOR YEARS.

AT LEAST IT'S BETTER THAN THE ALTERNATIVE!
HOW COULD YOU POSSIBLY KNOW THAT, SONNY-BOY?

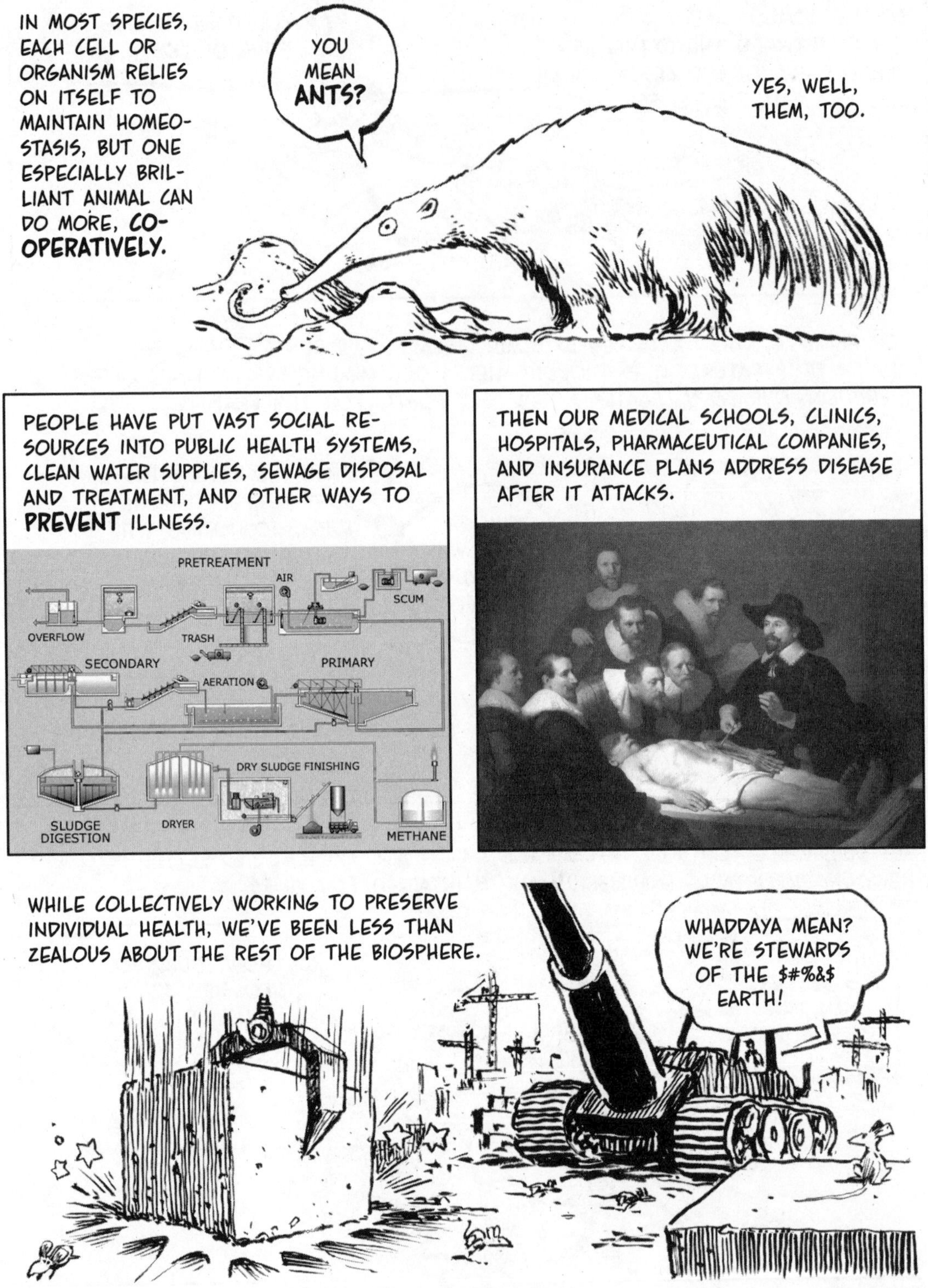
IN MOST SPECIES, EACH CELL OR ORGANISM RELIES ON ITSELF TO MAINTAIN HOMEOSTASIS, BUT ONE ESPECIALLY BRILLIANT ANIMAL CAN DO MORE, **CO-OPERATIVELY**.
YOU MEAN **ANTS?**
YES, WELL, THEM, TOO.
PEOPLE HAVE PUT VAST SOCIAL RESOURCES INTO PUBLIC HEALTH SYSTEMS, CLEAN WATER SUPPLIES, SEWAGE DISPOSAL AND TREATMENT, AND OTHER WAYS TO **PREVENT** ILLNESS.
PRETREATMENT
AIR
SCUM
OVERFLOW
TRASH
SECONDARY
PRIMARY
AERATION
DRY SLUDGE FINISHING
SLUDGE DIGESTION
DRYER
METHANE
THEN OUR MEDICAL SCHOOLS, CLINICS, HOSPITALS, PHARMACEUTICAL COMPANIES, AND INSURANCE PLANS ADDRESS DISEASE AFTER IT ATTACKS.
WHILE COLLECTIVELY WORKING TO PRESERVE INDIVIDUAL HEALTH, WE'VE BEEN LESS THAN ZEALOUS ABOUT THE REST OF THE BIOSPHERE.
WHADDAYA MEAN? WE'RE STEWARDS OF THE $#%&$ EARTH!

A LITTLE HISTORY: AROUND THE YEAR 1500, NORTHEASTERN NORTH AMERICA WAS A FOREST BIOME, HOME TO MILLIONS OF BEAVERS. BEAVERS DAM STREAMS, CREATE PONDS, CHANGE WATER FLOW, MOISTEN THE LANDSCAPE, AND PROFOUNDLY AFFECT THE BIOME'S MIX OF LIFE FORMS. MOSQUITOS MUST HAVE LOVED IT!
MMMMMMMMM
A MODERN ECOLOGIST WOULD CALL THE BEAVER A **KEYSTONE SPECIES,** BUT BACK THEN, THE NEW EUROPEAN ARRIVALS HAD ANOTHER WORD FOR IT:
A PELT.
WITHIN 300 YEARS, HUNTERS NEARLY WIPED OUT THE BEAVER; LOGGERS AND FARMERS FELLED FORESTS AND CLEARED FIELDS OF STUMPS AND STONES.
AND THEY CALL **US** BUSY!
THEN, IN THE 1800S, A MECHANIZED CLOTHING INDUSTRY AROSE; FARMERS MOVED TO TOWN FOR FACTORY JOBS.
FIELDS WENT TO SEED, AND A NEW FOREST, DRIER AND LESS DIVERSE THAN THE OLD ONE (AND STRANGELY FULL OF STONE WALLS), SPROUTED UP.

THE POINT OF THIS STORY IS THAT ECOSYSTEMS, LIKE ORGANISMS, TEND TO MAINTAIN THEIR OWN HOMEOSTASIS.
WE ALL DO OUR PART!
LEFT TO ITSELF, THE OLD FOREST WAS FAIRLY STABLE. INTERACTIONS AMONG ITS SPECIES EVOLVED TO PRESERVE THE HEALTH OF THE WHOLE.

FOR EXAMPLE, PREDATORS AND PREY REGULATE EACH OTHERS' POPULATION. WHEN PREDATORS OVERHUNT, FEWER PREY REMAIN, AND PREDATORS DIE OFF. PREY REBOUND; PREDATORS PROLIFERATE; THEIR POPULATIONS CYCLE IN SYNC. THIS GRAPH COMES FROM THE FUR-TRAPPING RECORDS OF THE HUDSON BAY COMPANY.

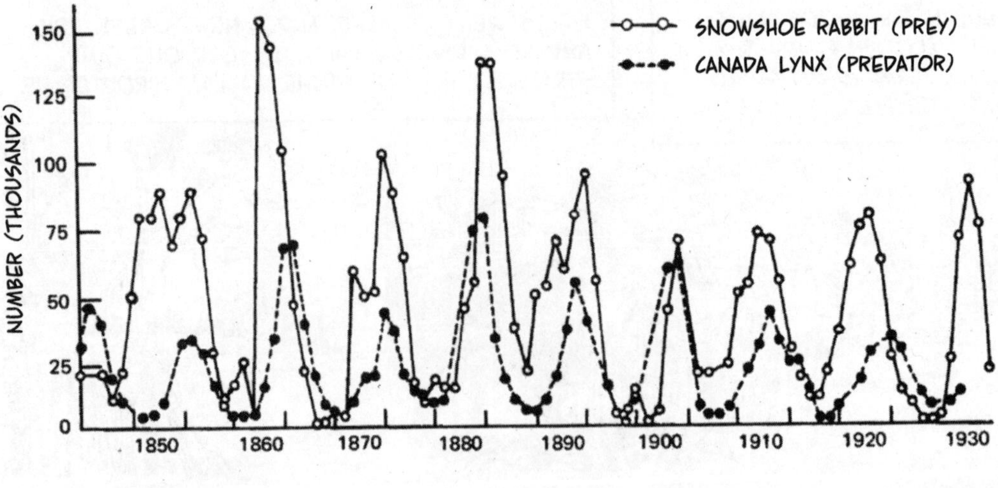

THEN CAME THE DISRUPTIVE BEAVER TRADE. TRAPPERS AND BEAVER DIDN'T CYCLE TOGETHER. COMMERCIAL HUNTERS MAY EVEN STEP UP THEIR ACTIVITY AS PREY BECOMES SCARCE, BECAUSE THEN PRICES GO UP!*

IT'S THE LAW OF SUPPLY, DEMAND, AND DESTROY!

SOME BEAVER SURVIVED, BUT THEY COULD NEVER RECOVER ANYTHING LIKE THEIR ANCESTRAL NUMBERS, ESPECIALLY WHEN FACED WITH OTHER HUMAN CHALLENGES.

HOW THEY GET THAT STRAIGHT CUT I'LL NEVER KNOW.

BUT THE SYSTEM AS A WHOLE SHOWED SOME

RESILIENCE:

A FOREST CAME BACK, BUT A DIFFERENT SORT OF FOREST—ONE THAT COULD DEAL WITH NEARBY TOWNS, ROADS, TOURISTS, SALT ON THE ROADS, MAPLE-SYRUP TAPPERS, AND ALL THAT.

*AT THIS WRITING, A SINGLE BLUEFIN TUNA WAS SOLD FOR $3 MILLION IN TOKYO'S MAIN FISH MARKET. NEEDLESS TO SAY, THIS IS A RARE FISH.

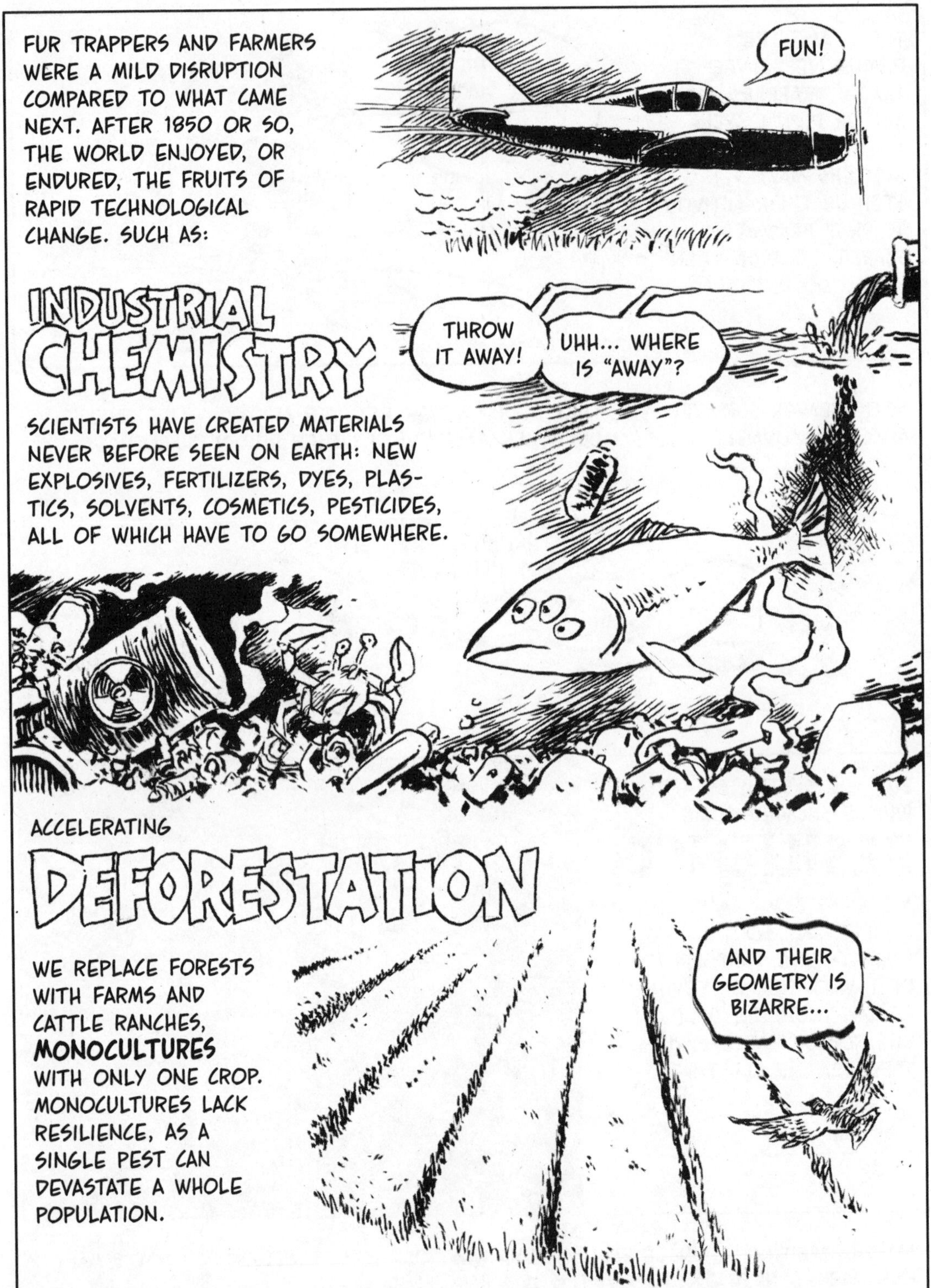
FUR TRAPPERS AND FARMERS WERE A MILD DISRUPTION COMPARED TO WHAT CAME NEXT. AFTER 1850 OR SO, THE WORLD ENJOYED, OR ENDURED, THE FRUITS OF RAPID TECHNOLOGICAL CHANGE. SUCH AS:
FUN!
INDUSTRIAL CHEMISTRY
THROW IT AWAY!
UHH... WHERE IS "AWAY"?
SCIENTISTS HAVE CREATED MATERIALS NEVER BEFORE SEEN ON EARTH: NEW EXPLOSIVES, FERTILIZERS, DYES, PLASTICS, SOLVENTS, COSMETICS, PESTICIDES, ALL OF WHICH HAVE TO GO SOMEWHERE.
ACCELERATING DEFORESTATION
WE REPLACE FORESTS WITH FARMS AND CATTLE RANCHES, MONOCULTURES WITH ONLY ONE CROP. MONOCULTURES LACK RESILIENCE, AS A SINGLE PEST CAN DEVASTATE A WHOLE POPULATION.
AND THEIR GEOMETRY IS BIZARRE...

FOSSIL FUELS

COAL AND PETROLEUM DRIVE OUR BULLDOZERS, CHAINSAWS, TRUCKS, DRILLING AND MINING EQUIPMENT, AGRICULTURAL MACHINERY, AIR-CONDITIONERS, FURNACES, AND ALL THE OTHER GEAR THAT HELPS US EXPLOIT ANYPLACE THAT LOOKS EXPLOITABLE.

THIS FOSSIL CARBON (SEE P. 281) HAS BEEN OUT OF CIRCULATION FOR MILLIONS OF YEARS. NOW HUMANS ARE BRINGING IT BACK INTO THE BIOSPHERE IN A RUSH. IS THIS GOOD OR BAD FOR LIFE?

IN THE LONG RUN, WHO KNOWS? FOR NOW, THE ADDED ENERGY MAINLY FUELS *HOMO SAPIENS* AT OTHERS' EXPENSE. IT DRIVES OUR TRANSPORT, AGRICULTURE, CITIES, AND IN SHORT, OUR ENTIRE SPENDTHRIFT EXISTENCE.

REMEMBER, THERE'S ONLY SO MUCH OF THIS STUFF!

AND THERE'S THE NOT-SO-MINOR MATTER OF GLOBAL
CLIMATE CHANGE
MOST INCOMING SUNLIGHT PASSES THROUGH THE ATMOSPHERE, BUT MUCH REFLECTED RADIATION IS ABSORBED BY CARBON DIOXIDE. THE EXCITED CO_2 MOLECULES WARM THE EARTH. CO_2 TRAPS HEAT. (SO DOES WATER VAPOR, WHICH BUILDS UP WHEN TEMPERATURES RISE.)
AS EXHAUST FROM BURNING CARBON-BASED FUEL STOKES THE AIR WITH MORE HEAT-TRAPPING CO_2, THE WHOLE PLANET GETS WARMER.
AND THE FEVER SHOWS NO SIGNS OF BREAKING...
THE OCEANS ABSORB MUCH OF THIS ENERGY—WATER'S GOOD AT THAT—AND RISING TEMPERATURES AFFECT PREVAILING CURRENTS AND NUTRIENT FLOWS. SEA-DWELLERS HAVE TO FIND NEW NICHES, WHILE PREDATORY NETS PULL WILDLIFE OUT OF THE WATER BY THE SHIPLOAD.
TALK ABOUT THE DEVIL AND THE DEEP BLUE SEA!

THE ATMOSPHERE IS ALSO WARMING, MOST QUICKLY AT THE POLES. THIS EXTRA POLAR WARMING WEAKENS THE WINDS CAUSED BY COLD AIR SINKING AND FLOWING AWAY TO REPLACE RISING WARM AIR IN THE TROPICS.

AS WIND AND RAINFALL PATTERNS SHIFT, MARGINAL BIOMES GET HOTTER AND DRIER. ON LAND, AS IN WATER, CREATURES HAVE TO SQUEEZE INTO NEW HABITATS.
'SCUSE ME.
SORRY.
BEG PARDON.

MELTING GLACIERS AND HEAT-EXPANDED WATER SWELL THE OCEANS. SALT WATER OVERWHELMS LOW-LYING COASTAL REGIONS, INCLUDING CITIES THAT HOUSE HALF OF HUMANITY.

HOW HAS THE BIOSPHERE RESPONDED?
NOT SO WELL. AT WORST, WE SEE
DESERTIFICATION.
BUILD A WALL!
THE SAHARA PUSHES EVER SOUTHWARD, AS THE MARGINAL FARMLAND OF THE SAHEL REGION GROWS EVER DRIER.
SAND DUNES OF THE GOBI, THE WORLD'S FASTEST-GROWING DESERT, HAVE REACHED WITHIN 45 MILES OF BEIJING.
BUILD ANOTHER WALL IN FRONT OF THE FIRST ONE.
IN THE 1960S, THE SOVIET UNION DIVERTED RIVER WATER FEEDING THE ARAL SEA, IN ORDER TO IRRIGATE COTTON PLANTED ON FORMER GRASSLAND. NOW THE LAND IS BARREN, AND THE ARAL SEA NEARLY GONE.
THIS SCHEME WAS TOTALLY CRACKED FROM THE START!

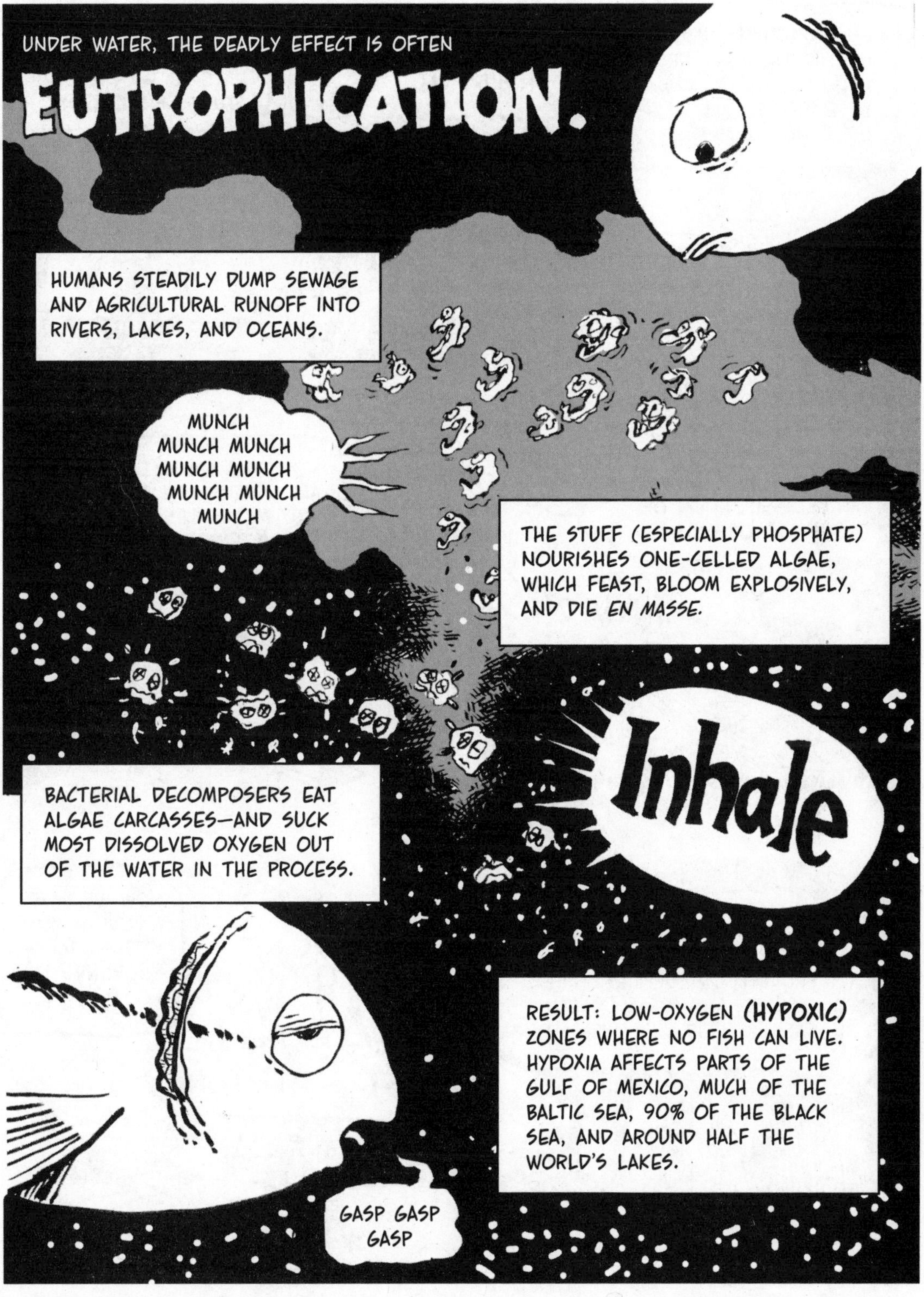
UNDER WATER, THE DEADLY EFFECT IS OFTEN
EUTROPHICATION.
HUMANS STEADILY DUMP SEWAGE AND AGRICULTURAL RUNOFF INTO RIVERS, LAKES, AND OCEANS.
MUNCH MUNCH MUNCH MUNCH MUNCH MUNCH MUNCH MUNCH
THE STUFF (ESPECIALLY PHOSPHATE) NOURISHES ONE-CELLED ALGAE, WHICH FEAST, BLOOM EXPLOSIVELY, AND DIE *EN MASSE*.
BACTERIAL DECOMPOSERS EAT ALGAE CARCASSES—AND SUCK MOST DISSOLVED OXYGEN OUT OF THE WATER IN THE PROCESS.
Inhale
RESULT: LOW-OXYGEN **(HYPOXIC)** ZONES WHERE NO FISH CAN LIVE. HYPOXIA AFFECTS PARTS OF THE GULF OF MEXICO, MUCH OF THE BALTIC SEA, 90% OF THE BLACK SEA, AND AROUND HALF THE WORLD'S LAKES.
GASP GASP GASP

HUMAN ACTIVITY HAS DISRUPTED THE VERY CIRCULATION OF NUTRIENTS ESSENTIAL TO ALL LIFE—INCLUDNG HUMANS.
DUDE, WHERE'S MY PHOSPHORUS?
OUR SPECIES MAY NOW ACTUALLY THREATEN **ITS OWN SURVIVAL.**
IT SOUNDS BAD—IT **IS** BAD—BUT WHAT DOES IT HAVE TO DO WITH THE STUDY OF **BIOLOGY?**
WELL... IN THE PAST, BIOLOGY FOCUSED ON THE STRUCTURE AND CLASSIFICATION OF INDIVIDUAL ORGANISMS.
NOW, THE SUBJECT EMBRACES ALL LEVELS OF LIFE, FROM TINY MOLECULES TO GREAT ECOSYSTEMS.
IN THE PAST, BIOLOGISTS WERE INSPIRED BY THE DESIRE TO IMPROVE FOOD SUPPLIES, CURE DISEASE, AND (OF COURSE!) SATISFY THEIR OWN CURIOSITY.
NOW WE HAVE A FOURTH MOTIVATION: TO **PRESERVE OUR POSTERITY.**
PAST
FUTURE
?

IN ADDITION TO ITS PREVIOUS ROLES, BIOLOGY IS NOW THE SCIENCE OF HEALING ECOSYSTEMS. ECOLOGISTS ARE SOMETHING LIKE PHYSICIANS, WITH LIVING SYSTEMS AS PATIENTS.
AND HUMANITY AS THE INFEC-TIOUS AGENT!
HERE'S WHAT THEY CAN DO.

ANALYZE

DETERMINE ENERGY FLOW AND CHEMICAL CYCLES, INPUTS, OUTPUTS, AND THROUGH-PUTS; ASSESS POTENTIAL EFFECTS OF VARIOUS DISRUPTIVE FACTORS.

PREVENT

WORK WITH GOVERNMENTS AND OTHER ORGANIZATIONS TO PUT SOME ECOSYSTEMS OFF-LIMITS TO HUMAN INFLUENCE (AS MUCH AS POSSIBLE).

MAINTAIN

STRIVE TO MAKE HUMAN ACTIVITY SUS-TAINABLE, NOT DESTRUCTIVE; FIND A REASONABLE, ACCEPTABLE HOMEOSTATIC CONDITION AND PRESERVE IT.

COMPOST
RECYCLE
LANDFILL

CURE

IMPROVE SICK AND DYING ECOSYSTEMS BY CUTTING POLLUTION, RESTORING DEPLETED SOIL, PLANTING TREES, STOCKING FISH, AND THE LIKE.

AND GUESS WHAT? IT WORKS!

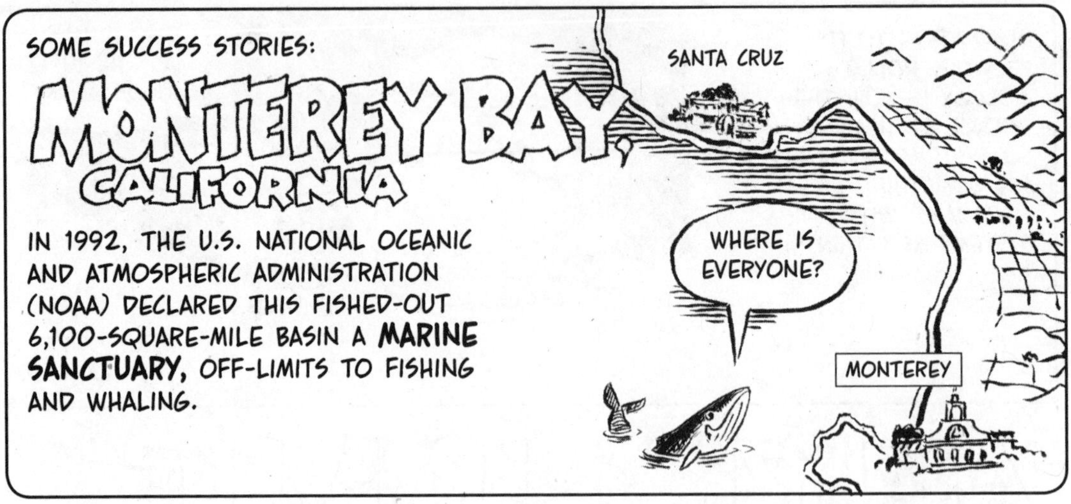
SOME SUCCESS STORIES:
MONTEREY BAY, CALIFORNIA
IN 1992, THE U.S. NATIONAL OCEANIC AND ATMOSPHERIC ADMINISTRATION (NOAA) DECLARED THIS FISHED-OUT 6,100-SQUARE-MILE BASIN A **MARINE SANCTUARY,** OFF-LIMITS TO FISHING AND WHALING.
SANTA CRUZ
WHERE IS EVERYONE?
MONTEREY

TWENTY-FIVE YEARS LATER, THE BAY SUPPORTS FLOURISHING POPULATIONS OF FISH (MORE THAN 300 SPECIES!), MAMMALS, BIRDS, INVERTEBRATES, AND PLANTS.

LET'S SEE... WHAT'S ON TONIGHT'S MENU? OH, RIGHT...

BIOLOGISTS AND GOVERNMENTS, WORKING TOGETHER (OVER THE ENDLESS OBJECTIONS OF ANTI-REGULATORS) HAVE ALSO IMPROVED WATERWAYS WHERE PEOPLE CAN AND DO FISH.

FACELESS BUREAUCRATS!
HEY! WE **HAVE** FACES!
NOAA
SYLVIA EARLE, OCEANOGRAPHER AND DEEP-OCEAN EXPLORER

THE BRITISH GOVERNMENT PUT 25 YEARS AND £100,000,000 INTO SUCCESSFULLY RESTORING THE **MERSEY RIVER** BASIN.

OTHER PROJECTS HAVE TARGETED THE DANUBE RIVER, BOSTON HARBOR IN THE U.S., AND SEVERAL OTHER LOCATIONS. THE EUROPEAN UNION HAS ANNOUNCED PLANS TO ADDRESS THE BALTIC SEA'S PROBLEMS, AND A CLEANER DANUBE IS POURING HEALTHIER WATER INTO THE BLACK SEA.

THESE EFFORTS, WHILE PRAISEWORTHY, ARE, UM, A DROPLET IN THE OCEAN.

CLEARLY, AN ENORMOUS AMOUNT OF WORK NEEDS TO BE DONE.
EQUALLY CLEARLY, IT WILL TAKE POLITICAL EFFORT, THE ENGAGEMENT OF GOVERNMENTS AND OTHER ORGANIZED GROUPS.
THAT, AFTER ALL, IS HOW OUR MEDICAL AND PUBLIC-HEATH SYSTEMS CAME TO BE. CAN WE DO THE SAME FOR OUR SPECIES AS WE DO FOR INDIVIDUALS?
CAN HUMANITY, WITH ITS CURRENT HABITS, CUSTOMS, AND INSTITUTIONS, WORK TOGETHER TO MAKE AND MAINTAIN A VIABLE BIOSPHERE?
WHAT? ARE YOU EXPECTING AN ANSWER FROM A BIOLOGY PROFESSOR AND A CARTOONIST?

THIS MUCH WE CAN SAY FOR CERTAIN: BIOLOGISTS WILL BE AT THE HEART OF ALL ECOSYSTEM MANAGEMENT.
AT THE SAME TIME, MANY BIOLOGISTS WILL PURSUE THE TRADITIONAL SUBJECTS: NUTRITION, MEDICINE, AND PURE RESEARCH.
FUNNY THING ABOUT RESEARCH, THOUGH: IT HAS CROSSOVER EFFECTS. MOLECULAR ACTION CAN AFFECT SYSTEMS—AND VICE VERSA.
AND THERE'S PLENTY LEFT TO DISCOVER. TODAY'S SCIENTISTS FOLLOW THE TRAILS BLAZED BY MENDEL, DARWIN, WATSON, CRICK, FRANKLIN, WOESE, WASHINGTON, MARGULIS, EARLE, AND THE REST—AND WE BLAZE NEW TRAILS OF OUR OWN!
WHO KNOWS WHAT'S STILL OUT THERE, WAITING TO BE FOUND? OR WHERE IT WILL LEAD? OR WHO WILL FIND IT?

IT JUST MIGHT BE YOU!

INDEX

ACKNOWLEDGMENTS

THE CARTOONIST IS GRATEFUL FOR THE GENEROUS SUPPORT OF THE MONTGOMERY FELLOWSHIP PROGRAM AT DARTMOUTH COLLEGE, WHERE SOME OF THIS BOOK WAS CREATED DURING ONE OF THE COLDEST WINTERS OF RECENT YEARS. THANKS TO DAN ROCKMORE FOR WRANGLING THE NOMINATIONS, TO KLAUS MILICH FOR OVERSEEING EVERYTHING WITH VERVE AND HUMOR, AND TO ELLEN HENDERSON FOR FAITHFULLY KEEPING TRACK OF ALL THE ADMINISTRATIVE DETAILS, WITHOUT WHICH NO ACADEMIC PROGRAM COULD EXIST.

Discover the World with Cartoons!

Refreshingly humorous and thorough guides from the author of the bestselling Cartoon Guide series.

THE CARTOON GUIDE TO BIOLOGY

Uses simple, clear, humorous illustrations to make biology's most complex concepts understandable and entertaining.

THE CARTOON GUIDE TO ALGEBRA

Covers algebra essentials, including linear equations, polynomials, quadratic equations, and graphing techniques. Offers a concise overview of algebra's history and its many practical applications in modern life.

THE CARTOON GUIDE TO CALCULUS

Entertainingly teaches all of the course essentials, including functions, limits, and derivatives, as well as more advanced integrals, series, and convergence tests. Concludes with a quick but thorough look at the theoretical basis and history of calculus.

THE CARTOON HISTORY OF THE MODERN WORLD PART 1

Beginning with Columbus's arrival on America's shores, traces the story of humanity through to the United States' emergence as a world superpower, the technological revolution, and the resurgence of militant religion.

THE CARTOON HISTORY OF THE MODERN WORLD PART 2

From the beginning of the Enlightenment to the environmental crisis of today's world. A unique spin on the history, personalities, and big topics that have shaped our world over the past five centuries.

THE CARTOON GUIDE TO CHEMISTRY

A course in college-level chemistry, the book covers electrochemistry, organic chemistry, biochemistry, environmental chemistry, physics as chemistry, and much more.

THE CARTOON GUIDE TO THE ENVIRONMENT

Covers the main topics of environmental science: chemical cycles, life communities, food webs, agriculture, human population growth, sources of energy and raw materials, waste disposal and recycling, cities, pollution, deforestation, ozone depletion, and global warming.

THE CARTOON GUIDE TO STATISTICS

Covers all the central ideas of modern statistics: probability in gambling and medicine, random variables, Bernoulli trials, hypothesis testing, confidence interval estimation—all explained in simple, clear, and, yes, funny illustrations.

THE CARTOON GUIDE TO GENETICS

Ease your way through Mendelian genetics, molecular biology, and the basics of genetic engineering.

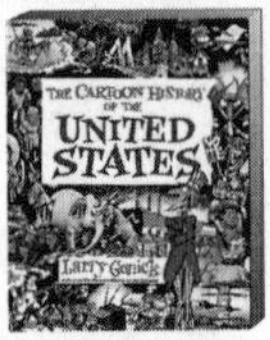

THE CARTOON HISTORY OF THE UNITED STATES

From the first English colonies to the Gulf War and the S&L debacle, Larry Gonick spells it all out from his unique cartoon perspective.

THE CARTOON GUIDE TO SEX

Frank, informative, and written with Larry Gonick's characteristic comic verve and scientific accuracy, this book gives a comprehensive discussion of the spectrum of human sexuality.

THE CARTOON GUIDE TO PHYSICS

Explains many complex ideas through simple, clear, and funny illustrations: velocity, acceleration, explosions, electricity and magnetism, circuits—even a taste of relativity theory—and much more.

AVAILABLE IN PAPERBACK AND EBOOK WHEREVER BOOKS ARE SOLD.
DISCOVER GREAT AUTHORS, EXCLUSIVE OFFERS, AND MORE AT HC.COM.

MORROW